Analog Integrated Circuit Applications

J. Michael Jacob
Purdue University

Prentice Hall
Upper Saddle River, New Jersey   Columbus, Ohio

Library of Congress Cataloging-in-Publication Data
Jacob, J. Michael.
    Analog integrated circuit applications / James Michael Jacob.
      p. cm.
    ISBN 0-13-080909-8
    1.   Linear integrated circuits.   2. Operational amplifiers.   I. Title
  TK7874 . J29 2000
  621.3815 21—dc21   99-042230

Editor: Scott Sambucci
Copyeditor: Maggie Diehl
Production Editor: Stephen C. Robb
Design Coordinator: Robin G. Chukes
Cover Designer: Dean Barrett
Cover illustration: Dean Barrett
Production Manager: Pat Tonneman
Marketing Manager: Ben Leonard

This book was set in Times New Roman and Ariel by J. Michael Jacob and was printed and bound by R. R. Donnelley & Sons Company. The cover was printed by Phoenix Color Corp.

Electronics Workbench® is a registered trademark of Electronics Workbench, Toronto, Ontario, Canada. PSpice® is a registered trademark of OrCAD, Inc., Beaverton, Oregon.

Printed in the United States of America

10 9 8 7 6 5 4 3

ISBN: 0-13-080909-8

Prentice-Hall International (UK) Limited, *London*
Prentice-Hall of Australia Pty. Limited, *Sydney*
Prentice-Hall Canada, Inc., *Toronto*
Prentice-Hall Hispanoamericana, S. A., *Mexico*
Prentice-Hall of India Private Limited, *New Delhi*
Prentice-Hall of Japan, Inc., *Tokyo*
Prentice-Hall (Singapore) Pte. Ltd., *Singapore*
Editora Prentice-Hall do Brasil, Ltda., *Rio de Janeiro*

Contents

3
Digital Control of Analog Functions *96*

4
Power Supplies and Integrated Circuit Regulators   *132*

8
Waveform Generators   *321*

9
Active Filters   *395*

Preface

The integrated circuit, particularly the operational amplifier, is the true building block of analog electronics. Transistors, like vacuum tubes (remember those?), are used in applications where very high power requirements demand their unique abilities. The bulk of analog signal processing is done with integrated circuits.

This book is an extension of *Applications and Design with Analog Integrated Circuits, Second Edition*. It presents a detailed overview of the use of operational amplifiers, analog switches, digital potentiometers, digital to analog converters, both linear and switching integrated circuit regulated power supplies, single supply operation, op amp selection, analog circuit board layout, phase locked loops, direct digital synthesizers, active filters, log amps, and multipliers. Characteristics and limitations of each integrated circuit are discussed. Extensive applications and designs are outlined; the operation of each application is derived or analyzed; and the performance limitations are pointed out. This book enables you to select the appropriate device and circuit configuration for your need, calculate the component values required, build the circuit, and analyze its overall behavior. Throughout, the book is tutorial but not overly brief as with many handbooks.

Three decades of experience teaching analog integrated circuits courses at both the associate degree and the bachelor degree engineering technology levels underlie this book. A reasonable ability in algebra is needed. Passages using calculus have been kept to a minimum and can be skipped without loss in your ability to apply that information. Although this book is intended as a sophomore or upperclass electrical engineering technology text, extensive transistor circuit knowledge is not required. Combined with a basic diode and transistor characteristics supplement, the first five or six chapters work well for the first course in analog electronics in engineering, technology, or industrial electronics, or for personal instruction.

Integrated circuits are treated as functional devices. Analysis of internal transistor circuitry is minimized and used only briefly, when discussing device limitations. This is not a book on the design of analog integrated circuits. It is a text on the application of the integrated circuit devices. As such, circuits using these devices are rigorously analyzed, and all performance or design equations are derived.

To help organize your thoughts, give you specific goals, and help you evaluate what you have learned, each chapter begins with a detailed list of learner outcome objectives. These tell you precisely what you will learn as you work your way through the chapter. The problems at the end of the chapter are based on these objectives. Not only are you required to repeat what was done in the chapter, but many of the problems require that you apply the techniques to new circuits. This gives you the chance to drill and to think and stretch your abilities.

Compared to other linear integrated circuits texts, this book has several noteworthy features. The coverage is as broad as that of many handbooks, but it provides the rigor to help you understand why circuits are built as they are.

Computer simulation is applied as a tool throughout the body of the text. It has not been stuck on as an afterthought at the end of the chapter. Examples use the student version of *Electronics Workbench*® and MicroSim's *PSpice*® to illustrate key points in the performance of the analog IC under study. You can obtain more information about *Electronics Workbench*® at *www.electronicsworkbench.com*. For the latest version of PSpice contact Cadence Design Systems at *www.orcad.com* and select FREE OrCAD Starter Kit. Many end-of-chapter problems are solvable with one of the simulation packages.

There are one or two laboratory exercises at the end of each chapter. These exercises have undergone several iterations of student testing. They provide good hardware verification of the theory in the chapter and are consistent in technology, terminology, and pedagogy with the text. A separate lab manual is no longer needed.

Of major significance is a generic approach to analyzing any negative feedback circuit. Digital control (multiplying D/A, digital pots, switched capacitor cells, etc.) and the applications of nonstandard amplifier configurations (multipliers, several feedback loops, active feedback, composite amplifiers, and cascaded control) are prevalent. It is critical that you be able to determine how any combination of analog ICs function together, even if they do not fit into one of the standard amplifier types. A direct, simple approach is presented in Chapter 2 and then applied throughout the rest of the text.

Microprocessor control is now common in all but the most trivial circuitry. It is important that the uses, and abuses, of digital to analog conversion be clearly integrated with the analog ICs presented. So there is a chapter on digital control of analog functions (Chapter 3) early in the text. These techniques are then included in many subsequent chapters, illustrating how digital circuitry can control the analog ICs being presented.

Chapter 4 gives a unique, practical, simple design approach to linearly regulated power supplies. In addition to the principles and design details of buck and boost switching regulators, continuous operation is compared to discontinuous. Two types of advanced higher power switching regulators that run directly from the line voltage are also explained.

The central element in any op amp circuit is the op amp IC. Its selection is critical to the overall performance of the circuit. Chapter 5 gives a description of how to decide which op amp type fits your circuit application, including an overview of the types of op amps and their performance envelope. CMOS, chopper stabilized, and current feedback op amps are included.

Layout is as critical as any other element in an analog circuit. Poor layout can doom an otherwise fine design. Proper analog layout can be taught; it is not black magic or a rare art form. This section is too often omitted in analog texts. The basics of circuit layout are simple and belong in the hands of the circuit designer, not the graphics artist. Chapter 6 is a short but proven step-by-step recipe for the layout of analog circuits. Since it deals with principles, it applies equally to fabrication on a protoboard with solderless connections, wire-wrap, point-to-point solder fabrication, or printed circuit board design. It is also independent of the mechanics of the software you may choose to use to aid in the board layout.

Active filters are presented rigorously. All transfer functions and tuning parameters are completely derived. However, application of these results is given equal emphasis with many designs and analysis examples. First- through sixth-order filters with varying damping for low pass, high pass, notch, state variable, and voltage controlled circuits are illustrated.

The phase locked loop is a key element in digitally controlled signal generation. Too often, however, either it is treated as magic, with a few diagrams and hand waving, or it is overwhelmed with pages of Laplace transforms. Chapter 8 finds the middle ground. Enough detailed explanation of each block is given to allow you to analyze or design a digitally controlled phase locked loop. However, stability issues are left to more advanced texts, using instead a simple, overdamped response. The phase locked loop is then combined with the XR2206 function generator IC and several op amps. This allows precise control of sinusoidal frequency, amplitude, and offset. Direct digital synthesis allows the creation of sinusoids with purely digital techniques. Each element is explained, with enough detail to allow you to design your own. A complete DDS IC is also presented.

The characteristics and applications of nonlinear circuits are given extended treatment in Chapter 10.

Presented under a new title, this book is effectively a third edition of *Applications and Design with Analog Integrated Circuits*. I would like to express my appreciation to the many students and teachers who have worked through the first two editions. Their comments and encouragement have kept the book practical but rigorous, timely but true to the fundamentals. I hope that you find in reading this text some of the wonder and excitement that I experienced in its creation.

J.M.J.

1

Introduction to
the Operational Amplifier
Integrated Circuit

Integrated circuit technology has allowed the construction of inexpensive, reliable digital systems, such as the pocket calculator and the digital watch. The same basic techniques have allowed the construction of inexpensive, complex analog amplifiers with very stable characteristics. These **integrated circuits** (ICs) are amazingly easy to use. Because of this simplicity, and because of a high performance-to-cost ratio, analog integrated circuits have replaced most discrete transistor amplifier circuits. This chapter introduces the operational amplifier, the most widely used analog IC. It is treated as a functional block. Its characteristics are described, and techniques and considerations for breadboarding are presented.

OBJECTIVES

After studying this chapter, you should be able to do the following:

1. Briefly describe the two IC fabrication techniques.
2. State the differences between an analog and a digital circuit. Give examples of each.
3. List the characteristics of an ideal amplifier.
4. Compare actual operational amplifier characteristics to those of an ideal amplifier.
5. State the requirements and precautions associated with operational amplifier power supplies.
6. Identify IC packages and lead convention.
7. Describe good breadboarding techniques.

1-1 Analog Integrated Circuit

What is an analog integrated circuit? To begin the study of analog integrated circuits, you should first understand what makes them unique. What is the difference between an integrated circuit and a circuit built with discrete transistors? How do analog circuits differ from digital circuits?

An integrated circuit (IC) is a group of transistors, diodes, resistors and sometimes capacitors wired together on a very small substrate (or wafer). Tens of thousands of components can be contained in a single integrated circuit. Functions as simple as single-stage amplifiers to those as complex as complete computers have been built in a single integrated circuit. This drastic decrease in size has also yielded a similar reduction in weight, power consumption, and production cost, while giving a proportional increase in reliability. Integrated circuits have swept every field of electronics and are strongly altering the automotive, medical, entertainment, communications, and business industries.

Integrated circuits are divided into two main groups according to the way they are built (monolithic or hybrid). Two divisions may also be made according to the function of the integrated circuit (analog or digital). The output of an analog IC may be any value whatever and may change continuously from one value to another. The output of a digital IC will be recognized only as either a logical low or a logical high. Only two output values are accepted. Furthermore, in many digital systems, changes from one level to the other can occur only at specific times. An analog IC output and a digital IC output are shown in Figure 1-1.

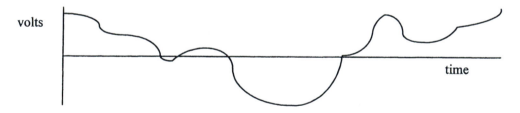

(a) Analog IC output. There is continuous variation between values.

Figure 1-1 Comparison of analog and digital IC outputs

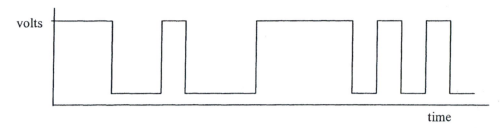

(b) Digital IC output. Only two output levels are allowed.

Figure 1-1 (Cont.) Comparison of analog and digital IC outputs

Computers, calculators, digital watches, communication switching, digital instruments, and some industrial control systems use digital ICs. Analog ICs are used in radios, televisions, stereo amplifiers, voltage regulators, signal generators, filters, test instrumentation, and extensively in industrial measurement and control.

There are two main IC fabrication techniques, monolithic and hybrid. Monolithic integrated circuits are built in a wafer of silicon. Transistors and resistors (and occasionally capacitors) are produced by diffusion processes, just as discrete transistors are. However, instead of being cut apart, the individual transistors are interconnected by very thin metal runs. Isolation between transistors relies on reverse-biased junctions between the transistors and the silicon substrate (wafer).

Design of monolithic integrated circuits requires the tedious preparation of a large number of photographic masks. These are used to control the areas of diffusion for isolation, collectors, bases, emitters (or channels and gates for FETs), resistors, surface insulation, and connection metalization. These masks are reduced in size and are used, one at a time, to produce the desired parts of the monolithic IC. Of course many separate integrated circuits may be made in each wafer, with many wafers processed simultaneously. The wafers are tested, and the good ICs (called chips at this point) are cut apart and placed in one of many different style packages.

For monolithic integrated circuits, the initial design of the masks needed to produce the IC is very time-consuming and expensive. However, hundreds to thousands of monolithic ICs can be produced simultaneously. Consequently, large volume production keeps the price per chip quite low. The major advantage of monolithic integration, then, is low cost per chip if many of the ICs are to be produced.

However, there are several disadvantages. Capacitors are difficult to produce in monolithic ICs. Tuning the IC by adjusting (trimming) internal resistor values to meet custom or very demanding specifications is also difficult. Power dissipation from the small silicon wafer substrate is very limited. Isolation between transistors is not adequate for large voltages. All of these disadvantages place some limits on the use of monolithic ICs.

In applications where these limitations cannot be tolerated, hybrid integrated circuits may be used. In the hybrid IC, discrete resistors, capacitors, diodes, transistors, and even monolithic ICs are placed on a ceramic wafer and then interconnected.

The wafer actually serves as a miniature chassis with discrete components attached and interconnected. This technique provides excellent isolation between components, allows the use of practically any device, and can be easily custom tuned before encasing. Although the hybrid IC can eliminate most of the monolithic IC's disadvantages, the hybrid IC is considerably more expensive because the hybrid ICs must be assembled individually from more expensive components.

1-2 The Ideal Operational Amplifier

The **operational amplifier**, or **op amp** for short, is a high gain, wide band, DC amplifier with high noise rejection ability. The schematic symbol for an op amp is shown in Figure 1-2.

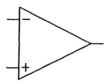

Figure 1-2 Schematic symbol for an operational amplifier

It has two inputs and one output. The output voltage, with respect to circuit common (ground), depends on the difference in potential between the two input voltages (e_{NI} - e_{INV}), as shown in Figure 1-3.

The operational amplifier was first built with vacuum tubes. Originally designed by C.A. Lovell of Bell Laboratories, these high gain, general purpose amplifiers were initially used to control the movement of artillery during World War II. In 1948, George Philbrick produced a single tube version. Soon, these amplifiers were used in analog computers to do addition, subtraction, integration, and scaling. Because the computer's mathematical **operations** were performed by these amplifiers, they became known as **operational amplifiers.**

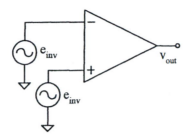

Figure 1-3 Output/input voltage relationship of an op amp

In the early 1960s the tubes were replaced by transistors. A single op amp then required only one printed circuit card full of components. This reduction in size and power consumption broadened its applications. Op amps were used in many signal conditioning areas, test and measuring equipment, and industrial controls.

With the advent of integrated circuit techniques, it became possible to place the entire op amp into a single eight-lead mini-DIP package. The first generation 709 was introduced by Fairchild in 1965, with the still popular second generation 741 op amp available in 1968. Surprisingly, the cost of these op amp ICs is comparable to that of a single discrete transistor. Most small signal analog circuits designed today use operational amplifiers as the basic active device, and most rely on the discrete transistor only when the op amp will not solve the problem.

If you were going to design an ideal operational amplifier, what characteristics would you want? Since the purpose of the circuit is to amplify a signal, you would want an arbitrarily large gain. That is, no matter how small the input signal ($e_{NI} - e_{INV}$ in Figure 1-3), the open loop voltage gain, A_{OL}, would be large enough to provide an output signal (v_{out}) of adequate size.

$$A_{OL} = \frac{v_{out}}{e_{NI} - e_{INV}} = \infty \tag{1-1}$$

This means that the difference between the inputs ($e_{NI} - e_{INV}$) can be negligibly small.

Second, it is important that all of the output signal be applied to the load. However, the output impedance of the op amp (Z_{out}) forms a voltage divider with the load, as shown in Figure 1-4. For all of the op amp's output voltage to be applied to the load (and none dropped across Z_{out}), the output impedance must be zero.

$$V_{load} = \frac{Z_{load}}{Z_{load} + Z_{out}} V_{out}$$

$$V_{load} = V_{out} \quad if \quad Z_{out} = 0$$

So, for an ideal op amp: $Z_{out} = 0$ (1-2)

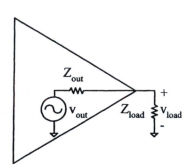

Figure 1-4 Effect of output impedance on the load voltage of an op amp

On the input side of an ideal operational amplifier, it is important to assure that the amplifier does not load down the source. The input impedance of the op amp forms a voltage divider with the output impedance of the source. This is illustrated in Figure 1-5. If all of the source voltage is to be applied to the inputs of the op amp, and none lost across the source output impedance (R_s), then the op amp's input impedance must be arbitrarily large compared to R_s.

$$v_{in} = \frac{Z_{in}}{Z_{in} + R_S} e_{INV}$$

$$v_{in} = e_{INV} \quad if \quad Z_{in} \approx \infty$$

So, for an ideal op amp:

$$Z_{in} = \infty \qquad\qquad\qquad\qquad\qquad (1\text{-}3)$$

This also means that no significant signal current will flow into the input of an ideal op amp.

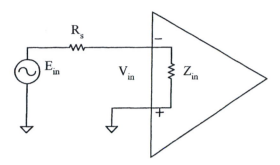

Figure 1-5 Effect of input impedance on input voltage and
source loading for an operational amplifier

At DC you want an input difference in potential of $0V_{DC}$ to produce an output of $0V_{DC}$. Any deviation from this is called **offset.**

$$V_{in} = E_{NI} - E_{INV} = 0 \implies V_{out} = 0$$

or   offset $= 0$ (1-4)

Ideally, it would seem that an AC amplifier would provide the same gain for all frequencies. However, at higher frequencies, the reactance of the parasitic capacitance provides a low impedance between the inputs and the output. This allows the output of the amplifier to be fed back to the input, amplified, fed back to the input, amplified, and so on. The result of too much gain at too high a frequency may be uncontrollable oscillations. Select or set the op amp's bandwidth as follows:

$$\text{bandwidth} \geq \text{just barely enough} \tag{1-5}$$

1-3 Basic Specifications and Requirements

You realize, of course, that the ideal op amp just described cannot actually be built. But, for a remarkably low cost, monolithic op amps can be manufactured which, when properly used, closely approximate the ideal. In this section, key nonideal characteristics will be discussed, along with simple techniques and precautions for minimizing their effects.

1-3.1 Specifications of Initial Importance

The first column of Table 1-1 lists the characteristics of an ideal op amp. The second column lists the typical characteristics the 741C, a popular, general purpose, monolithic op amp IC.

Table 1-1 Characteristics of an op amp

		Ideal	741C (typical)
voltage gain	A_{OL}	∞	2×10^5
output impedance	Z_{out}	0	75Ω
input impedance	Z_{in}	∞	$2M\Omega$
offset current	I_{os}	0	20nA
offset voltage	V_{os}	0	2mV
bandwidth	GBW	just enough	1MHz

Although the open loop gain of the op amp definitely is not infinite, generally only a few percent or less error will be introduced by assuming that it is infinite. This will be true for any circuit using op amps if you

limit the closed loop gain of the circuit to *100* or less.

The typical input and output impedances, though not ideal, do need to be considered only for a few demanding industrial measurement applications.

The output amplitude for most op amps is not limited by the output impedance, typically 75Ω for the 741C. Instead, the amount of current the op amp can source or sink to the load limits the output voltage. In an attempt to make these integrated circuits as foolproof as possible, a short circuit current limit circuit has been added to many op amps. This means that the load impedance can be shorted without overheating the IC. Instead the output current will be limited to a safe level (25mA for the 741C). The output voltage drops appropriately to keep the current below the safe limit as the load impedance falls. This is very handy. However, be careful! Not all op amps are short-circuit protected. You must read the manufacturer's literature. Shorting the output of an unprotected op amp will probably damage it.

Use an op amp which can withstand a short circuit indefinitely.

Interestingly, it is the last three specifications in Table 1-1, along with a few others to be described, which limit an op amp's applications.

Some current must flow into the input of a general purpose op amp to bias its input transistors, as is illustrated in Figure 1-6.

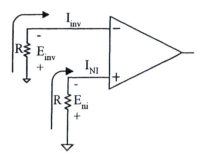

Figure 1-6 Bias currents of an operational amplifier

If, ideally,
$$I_{INV} = I_{NI}$$

then
$$E_{INV} = E_{NI}$$

and
$$V_{out} = A_{OL}(E_{NI} - E_{INV}) = 0$$

The effect of these bias currents on the output voltage is zero. Normally, I_{INV} does not equal I_{NI}. There will be some small difference between the two bias currents.

$$I_{offset} = I_{OS} = |I_{NI} - I_{INV}|$$

You can see in Table 1-1 that for the 741C, the offset current is typically 20nA.

To determine the effect of this offset current, let the resistors in Figure 1-6 equal 1MΩ. The offset current will then cause a difference in potential between the inputs ($e_{NI} - e_{INV}$) of

$$1M\Omega \times 20nA = 20mV$$

Multiplying this 20mV by the typical open loop gain of 200,000 means that the output of the op amp is driven as far up (or down) as it possibly can be. The output is useless.

Even if the op amp is operated with significant negative feedback, 20mV of input error will cause a noticeable offset in the output voltage.

**Reduce the size of the resistance seen from each input terminal
as much as practical.**

Usually, keeping R in the kΩ range is quite adequate.

The bandwidth specification is actually a measurement of the maximum possible gain at a given frequency. As the frequency increases, the gain falls off proportionally. For the 741C, at no point can the product of the voltage gain and the operating frequency exceed 1MHz as shown in Table 1-2. From the table you can see that the open loop gain becomes too small above about 10kHz.

Table 1-2 Possible gains versus frequency for the 741C

Frequency (Hz)	Maximum Possible Gain
10	100,000
100	10,000
1000	1000
10,000	100
100,000	10

For that reason,

**use the 741C only for DC to upper-audio frequencies.
Use a wide band op amp above 10kHz.**

The LM318 has a gain bandwidth product of 15MHz, giving a useful gain up to at least 150kHz.

1-3.2 Power Supply Requirements

In order for an op amp to operate, power supply voltage(s) must be provided. Often, bipolar voltages are required. That is two voltages are needed, one positive and the other equal in magnitude, but negative. Occasionally, these supply voltages are omitted from the schematics. However, Figure 1-7 illustrates the proper schematic and connections.

The point at which these two supplies are connected is called **analog common**, or just **common**. It is given the schematic symbol of a triangle, pointing down. Usually, all sources, loads, and measurements are referenced to this point. If this point is connected to the metal chassis in which the circuits are housed, then the point is at **chassis ground**. If common is tied to the third, green wire of the 115VAC line, or to some other conductor which runs into the earth (such as a cold water pipe), then the common point is called **earth ground**. Connection of analog common or chassis ground to earth ground may be required by safety or noise considerations.

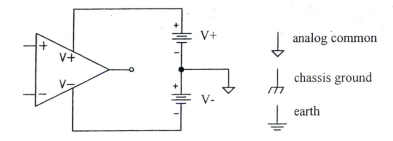

Figure 1-7 Bipolar power supply connections for an operational amplifier

Three op amp specifications that affect the power supply voltages are listed in Table 1-3. The manufacturer specifies that the supply voltages (V^+ and V^-) may not exceed ±18V for the 741C. The output voltage magnitude is limited by these supply voltages. The maximum output voltage will be less than the supply voltage. How much smaller depends on the particular IC you are using. But the output voltage may be as small as two volts below the power supply voltages. This is called V_{SAT}. So select the power supply voltages to assure an adequately large output voltage, about two voltages above the maximum desired output. If there is no major reason to do otherwise, or if achieving rated performance is important, then be sure to use the power supply voltages selected by the manufacturer when the specifications were measured.

Table 1-3 Power supply dependent specifications for the 741C

Absolute Maximum	
power supply voltage	±18V
differential input voltages	±30V
common mode input voltages	±15V

It is specified that the difference in potential between the two inputs (differential input voltages) may be no greater than 30V for the 741C. Also, no more than ±15V may be put on both inputs simultaneously (common mode input voltages). In addition, though not specifically stated, should either input exceed the power supply voltage, damage may occur. If the op amp has a ±9V supply, neither input may exceed ±9V, even though the specifications indicate ±15V. In fact, if you apply an input voltage **before** you turn on the power supplies, during that time the input voltage exceeds the supply voltages, and the IC may be damaged.

Limit the input amplitude to less than the power supply voltages.

One certain way to damage an op amp is to reverse the supply voltages (i.e., +15V connected to the V⁻ terminal, -15V connected to the V⁺ terminal). It is very discouraging to spend several hours wiring a complex circuit using many op amps and then, to accidentally cross the power supply wires, wiping out hours of work and several dollars' worth of ICs.

Use extreme care not to reverse power supply connections.

1-3.3 Packaging

Analog integrated circuits are available in many different package styles. These are illustrated in Figure 1-8.

The dual inline package **(DIP)** of Figure 1-8(a) is popular for commercial applications not requiring tight packaging. It is easy to handle, fits standard mounting hardware, is inexpensive when molded in plastic, and is commonly available. Ceramic DIPs are used for high temperature, high performance (usually military) equipment. The **mini-DIP** [(Figure 1-8(b)] provides a low cost/space efficient package.

The metal can packages [Figure 1-8(d)] allow easy connection to a heat sink and are often chosen when heat dissipation is a chief consideration.

The movement to put more complex electronics into smaller packages has led to the wide use of surface mount technology. Not only are the ICs considerably smaller, but mounting areas on the PC board may be smaller. Also, on surface mount boards is common practice to place components on both sides of the PC board, increasing the circuit density even more. Cost pressures today have forced many electronics manufacturers to automate their production lines. Surface mount packages, as shown in Figure 1-8(e), play a key role in this cost-savings trend because:

1. the mounting of devices on the PC board surface eliminates the expense of drilling holes;

2. the use of pick and place machines to assemble the PC boards greatly reduces the labor costs; and

3. the lighter and more compact assembled products resulting from smaller dimensions of surface mount packages mean lower material costs.

Production processes permit both surface mount and through hole (e.g., DIP) components to be assembled on the same PC board. However, surface mount packaging has become dominant. More and more, surface mount packaging may be the only available package. DIP packages are being phased out.

Pins are identified on the pin diagram included with the manufacturer's specifications. It is standard, however, to view the IC from the **top** and to count pins counterclockwise from the key, mold mark, tab, or square PC board mounting pad. Each package style is identified with a different letter suffix, as shown in Figure 1-8.

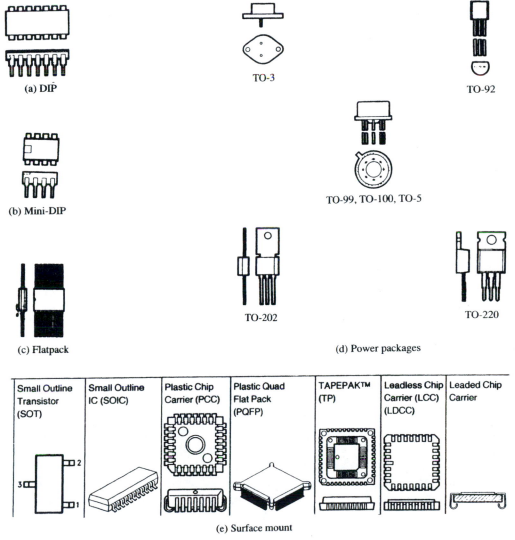

Figure 1-8 Analog IC packaging styles (Courtesy of National Semiconductor)

The op amp terminals connect to different pins in each different package style. For example, the output of the 741C is connected to pin 10 in the DIP, and to pin 6 in the mini-DIP, as shown in Figure 1-9. Since it is critical that the proper connections be made to the op amp, the op amp number, package style suffix, and pin numbers should *always* be recorded on the schematic.

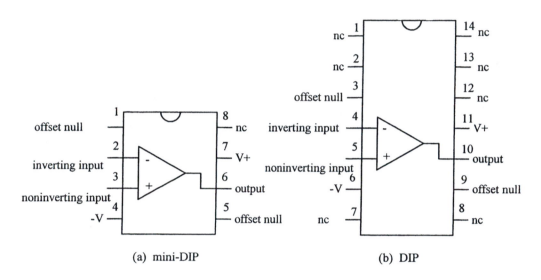

(a) mini-DIP   (b) DIP

Figure 1-9 Pin diagram for the 741 operational amplifier
(Courtesy of National Semiconductor)

1-4 General Breadboarding Hints

Once a circuit has been designed, it must be tested. To do this quickly and reliably, a good breadboarding system is needed. It should allow for the easy interconnection and removal of the analog ICs, discrete components, power supplies, and test equipment. It is **absolutely critical** that connections between the breadboard, the components, the power supplies, and the **test equipment** be mechanically and electrically sound. Most beginners spend more time running down poor or wrong breadboarding connections than they spend actually evaluating the circuit they have built. In this section you will find breadboarding hints which will help you minimize problems and errors in building your circuit for testing.

A universal breadboard socket is illustrated in Figure 1-10.

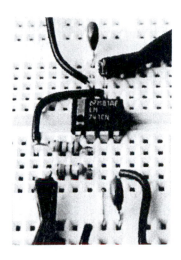

Figure 1-10 Universal breadboarding socket

It provides a popular and convenient technique. It gives two to four busses for power supplies and ground running along the edges. The body provides an array of solderless connections properly spaced and sized for most analog and digital ICs in DIP packages, transistors, diodes, $\frac{1}{4}$ watt resistors, and 22AWG solid wire. Using it, you can construct circuits quickly, compactly, and reliably. These breadboards are available in a variety of styles and qualities from most electronic component suppliers.

The universal breadboarding sockets provide a good interface between components of the circuit, but care must be taken when they are connected to power supplies and test equipment. First, the breadboards are small and quite lightweight. As cables are attached, the board will slide around, turn over, and too often short out the power supply or test leads. To prevent this, all breadboards should be mounted on some larger, sturdier base.

Just as a chain is only as good as its weakest link, test equipment can give results no better than the technique used to connect it to the circuit under test. Generally, power can be brought to the breadboard simply with solid 22AWG wire, bare on both ends. There are many excellent standard leads supplied with banana plugs, or BNC connectors on one end (to go to the instruments), and spring loaded, insulated, self-retracting mini-grabbers on the other end. Use these. Hours of careful design and breadboarding can literally go up in smoke because of a shorted or open wire to a power supply or from two alligator clips which accidentally touch or jump off at just the wrong time. Alligator clips are a major source of trouble. They are often too large for use on a breadboard, short together, or fail to hold adequately.

Use only standard connectors terminated in mini-grabbers to connect test equipment to the breadboard. *Never* use alligator clips.

Probes must also be used carefully. It is far too easy, when you are trying to touch a pin on an IC, for the probe to slip between two pins, shorting them together. This could damage the IC or supporting equipment. Instead of probing IC pins directly, there are two better choices. One is offered by the IC test clip, which easily slips over a DIP and provides well-spaced connections for the probe. A simpler solution is to connect a wire from the point you want to probe to a vacant part of the socket, where it can be secured and safely probed.

Never probe IC pins directly.

Power supply connections to the breadboard and to the breadboard and to the individual ICs, though necessary, can cause some problems. As was stated earlier, one sure way to damage an analog IC is to reverse the power supply connections. This can be easily prevented when you are breadboarding by first labeling each bus in some highly visible way. This should prevent you from connecting the IC to the wrong supply bus. The busses can be protected by two diodes, as shown in Figure 1-11. Should the power supply be incorrectly connected to the busses, the diodes will go off, protecting the components tied to the bus.

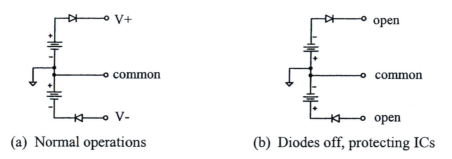

(a) Normal operations   (b) Diodes off, protecting ICs

Figure 1-11 Diode protected power supply busses

The op amp must have good, clean power supplies, providing AC ground at each power pin. But, because of the inductance presented by the cables running from the breadboard to the power supplies, and because of the abundant stray capacitances, high-frequency (often noise) signals run back and forth along the bus. When coupled from one IC to another and amplified, these rf signals can cause the entire circuit to oscillate.

This can be prevented by placing a decoupling capacitor (typically 0.1µF) **immediately** from each power supply pin of each IC to ground. Any rf signals are passed to ground as they leave the IC, before they can contaminate the busses.

Use a decoupling capacitor (0.1µF) *at each* power supply pin of *every* IC.

In addition to the precautions mentioned for connecting test equipment and power supplies to the breadboard system, there are several things to be aware of when you are interconnecting components.

1. Component Selection: Avoid potentiometers.

    a. Potentiometers drift, are usually misadjusted, and are prone to noise.

    b. Wire-wound pots can cause trouble at high frequency.

    c. Many printed circuit size pots will not properly fit into breadboards.

    d. Multiturn pots, although good for making fine adjustments, are often confusing because you may not know where the pot is set.

2. Layout Plan

    a. Simplify the schematic and layout as much as possible for initial testing. Fine tuning, zeroing, and additional stages can easily be added **after** you have the basic circuit working.

    b. Be sure to include IC number, packaging type suffix, and **pin numbers** on each IC on the schematic.

    c. Make the layout look as much as possible like the schematic.

    d. Locate input and feedback resistors as physically close to the IC pins as possible. Long leads, connecting to remotely located resistors, pick up noise. This noise is then amplified by the op amp's **open loop** gain. So it takes very little noise along these leads to create major problems at the output.

    e. Keep the inputs well separated from the output to prevent oscillations.

    f. The supply voltage from the power supply must be fed to a decoupling capacitor **first**, and then to the IC power bus. Each IC must have a decoupling capacitor connected **immediately** to its power pins, sitting between the power pin and the power bus.

3. Circuit Construction

    a. Always completely clear the breadboard of any old circuits before beginning to build a new circuit.

b. Exercise care in inserting and removing ICs. Pins are easily bent and jabbed into your finger.

c. Solder 22AWG solid wire to the leads of components that are too large (power resistors, TO3 packages, ...) or too small (surface mount ICs).

d. Devise, post, and carefully follow a color code scheme for V^+, V^-, common, and signal wires.

e. Avoids jungles. Make all components lie flat. Trim and bend leads and wires to fit the layout. Neat, flat layouts work better and are far easier to troubleshoot than a jumble of components and wires.

f. Do not forget to connect the power supplies to each IC. Although not always shown on a schematic, power *is* required by the IC. This simple oversight is responsible for many lost hours of fruitless troubleshooting.

g. Select one connector as the ground point. Tie the breadboarding socket's common bus, power supply common, and all test instruments' commons to that single point ground. See Figure 1-12.

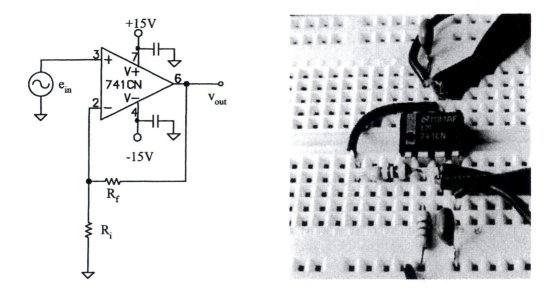

Figure 1-12 Typical amplifier layout with support connections

4. Circuit Testing

    a. Analyze the circuit before applying power to insure that you know what to expect.

    b. Double check all connections, especially power supply connections, before applying power.

    c. Apply power to the ICs before applying the signals.

    d. Check the DC (bias) operation of the circuit before you apply the signal.

    e. Measure voltages with respect to the circuit common. If you need the difference in potential between two points, measure each with respect to ground and then subtract. Test instruments provide much better measurements when referenced to circuit common. Also, digital instruments' commons often inject considerable digital noise through their common lead.

    f. Measure current by determining the voltage across a known resistor. Then calculate the current. Ammeters are rarely sensitive enough, tend to load the circuit, and often inject noise into sensitive nodes.

    g. Remove the signal from the IC before removing power. Change components and connections with the power off.

Summary

This chapter has introduced the operational amplifier integrated circuit. An integrated circuit is defined as a group of transistors, diodes, resistors and occasionally capacitors built on a very small substrate. All of the components in a monolithic IC are produced by diffusion techniques similar to those used in producing discrete transistors. Hybrid ICs consist of discrete components mounted on a ceramic wafer and wired together. The output of an analog IC may be any value and may change at any time. However, a digital IC output is recognized as either of two levels only.

      The ideal operational amplifier has an inverting input (-), a noninverting input (+), and a single output, referenced to circuit common (ground). The value of this output depends on the difference between two inputs.

$$v_{out} = A_{OL}(e_{NI} - e_{INV})$$

The open loop voltage gain, the input impedance, and the gain bandwidth product are extremely large, while the output impedance and offset are negligibly small.

      The fact that no real device can meet these ideal characteristics can be, more or less, ignored if several precautions are observed. These are listed in boldface and are indented in the body of the chapter. Care must be also taken when supplying the op amp with power. These precautions are also identified by indented boldface.

For the beginner, most of the problems encountered in getting a circuit to work are caused by the way it is built, not by the design. Good breadboarding practices can solve many of these problems. Use a universal breadboarding socket, mounted to a base. Connect to test equipment **only** through commercial BNC, banana plugs, and mini-grabbers, **never** alligator clips. Exercise care when probing the circuit, using remote test points rather than the IC pin itself. Ensure that **power supplies are not reversed**, and provide each IC with two decoupling capacitors. Select reliable components, plan the breadboard layout, wire the circuit neatly, and pay close attention to what you are doing (and the order of steps) while testing.

This chapter has introduced the op amp, its characteristics, and basic precautions. In Chapter 2, "Amplifiers and Feedback," you will see how to use the op amp to build comparators, LED drivers, inverting and noninverting amplifiers and summers, and differential amplifiers.

Problems

1-1   Describe the major physical differences between a monolithic IC and a hybrid IC.

1-2   What factors would cause you to select a hybrid IC rather than a monolithic IC?

1-3   Generally, monolithic ICs cost less than hybrid ICs. Explain why.

1-4   List three specific analog circuit application examples not given in this chapter. Why are these considered analog?

1-5   List three specific digital circuit application examples not given in this chapter. Why are these considered digital?

1-6   If an op amp has $e_{INV} = 0V$ (grounded), what must the voltage at e_{NI} be to cause a 5V output, if:
     a. $A_{OL} = 50,000$
     b. A_{OL} is ideal

1-7   In Figure 1-5, find V_{in} if:
     a. $R_s = 10k\Omega$ $Z_{in} = 1M\Omega$ $e_{INV} = 10\mu V$
     b. $R_s = 1M\Omega$ $Z_{in} = 1M\Omega$ $e_{INV} = 10\mu V$
     c. Z_{in} is ideal $e_{INV} = 10\mu V$

1-8   Assume that $I_{INV} = I_{NI}$. What must be the relationship of the two resistors in Figure 1-6 to eliminate the effects of the bias currents on the output voltage?

1-9   You need a gain of 150 at a frequency of 12kHz. Will the 741C op amp work? Explain.

1-10   Explain what internal compensation means. Why is it needed?

1-11   A 741C op amp is connected to a ±12V power supply.
  a. What is the maximum possible output voltage?
  b. What is the maximum allowable input voltage on **both** inputs?
  c. The supplies are raised to ±18V. What is the maximum allowable input voltage with respect to ground?

1-12   Under what conditions would you choose each of the following packages?
  a. metal can
  b. mini-DIP
  c. surface mount

1-13   Design a breadboarding system to meet (or exceed) the following:

> Universal breadboarding socket with at least three power supply busses and a matrix of at least 100 terminals with at least four tie points.
>
> Sturdy mounting base.
>
> Two BNC cable receptacles
>
> Two pair (red and black) 5-way binding posts with $^3/_4$ inch center-to-center spacing between the red and the black post of each pair.
>
> Two 9V transistor batteries to provide $\pm V_{supply}$.

  a. Refer to electronic parts catalogs, or suppliers' World Wide Web pages to obtain part numbers, price and size information.
  b. Draw a mechanical layout diagram (with dimensions) of the system, showing how you would mount all of the parts.
  c. Draw a schematic showing how you would permanently wire the batteries and the connectors to the breadboard socket (with appropriate holders, switches, and fuses).
  d. What is the total cost of your breadboarding system?

2

Amplifiers and Feedback

Practically all circuits using operational amplifiers are based around one of a few fundamental configurations. In this chapter, you will learn about these building blocks. Their schematics, operation, analysis, design, and limitations are discussed. Examples of how these circuits are actually used in laboratory and industrial equipment are given. The chapter ends with a general analysis technique which will work on *any* negative feedback op amp circuit. Every following chapter assumes that you have mastered the basic concepts of comparators and amplifiers, the building blocks of this chapter.

Objectives

After studying this chapter, you should be able to do the following:

1. Describe the open loop operation of an operational amplifier.
2. Discuss the effects of negative feedback on an amplifier.
3. For each of the circuits below,

noninverting comparator	noninverting amplifier
inverting comparator	voltage follower
nonzero reference comparator	inverting amplifier
comparator with hysteresis	inverting summer
simple LED driver	difference amplifier
311 and 339 comparator ICs	general amplifier

do the following:

a. Draw the schematic.
b. Qualitatively describe the operation.
c. Analyze a given circuit to determine quantitative operation.
d. Design a circuit to meet given performance specifications.
e. Discuss specific applications.
f. State limitations.

2-1 Open Loop Operation

The simplest way to use an op amp is shown in Figure 2-1.

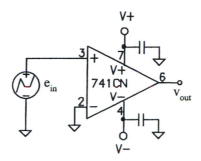

Figure 2-1 Simple op amp comparator

The power supply terminals are properly connected. Decoupling capacitors are connected between each power pin and circuit common. The inverting input (-) is tied to circuit common, and the noninverting input (+) is connected to the source (e_{NI}).

2-1.1   Op Amp Comparators

As with any op amp, the output voltage is

$$v_{out} = A_{OL}\left(e_{NI} - e_{INV}\right) \qquad (2\text{-}1)$$

But e_{INV} is the voltage applied to the inverting terminal, which for the circuit in Figure 2-1 is circuit common, or 0V. So equation (2-1) becomes

$$v_{out} = A_{OL}e_{NI}$$

For an ideal op amp, A_{OL} is arbitrarily large. This means that the circuit in Figure 2-1 has only two possible outputs.

Case 1 $e_{NI} > 0$

$$v_{out} = \infty \times e_{NI}$$

$$v_{out} \rightarrow \infty$$

If the input becomes only **slightly** greater than zero, the arbitrarily large voltage gain forces the output voltage to the maximum possible value, $+V_{SAT}$. So, if

$$e_{NI} > 0$$

then

$$v_{out} = +V_{SAT}$$

Case 2 $e_{NI} < 0$

$$v_{out} = \infty \times (-e_{NI})$$

$$v_{out} \rightarrow -\infty$$

If the input goes even slightly below zero, the large voltage gain drives the output to the maximum negative value, $-V_{SAT}$. So, if

$$e_{NI} < 0$$

then

$$v_{out} = -V_{SAT}$$

To summarize, if the noninverting input voltage is even slightly above the inverting input (ground in this circuit), the output goes to positive saturation. When the noninverting input drops below the inverting input (ground in this circuit), the output goes to negative saturation.

The circuit in Figure 2-1 is called a noninverting comparator. Although it definitely does not provide linear amplification, it can be of major use. Figure 2-2 is a plot of an input voltage applied to the noninverting comparator of Figure 2-1, with the resulting output.

When e_{NI} is greater than 0V, the comparator outputs $+V_{SAT}$. When e_{NI} drops below 0V, the comparator's output switches to $-V_{SAT}$. Consequently, even a change of a few millivolts on the input may cause the output to change as much as 30V. To determine whether e_{NI} is above or below the circuit common, you simply have to look at the output of the op amp.

The noninverting comparator amplifies (to the extreme) any differences between its input, e_{NI}, and the circuit common, giving you a large voltage indication ($\pm V_{SAT}$). Also, the op amp serves as a buffer. Its input impedance is extremely large, so it does not load down e_{NI}. However, it can provide 20mA or more to drive whatever load or indicator you have attached to the output.

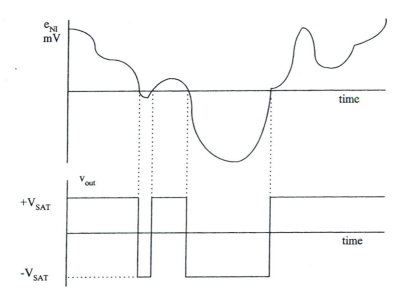

Figure 2-2 Example of noninverting comparator's response to an input

How would the circuit in Figure 2-3 respond?

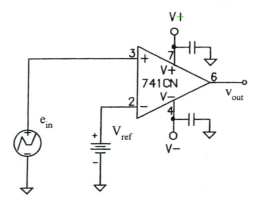

Figure 2-3 Noninverting op amp comparator with a reference voltage

Circuit common at the inverting terminal has been replaced by a reference voltage (V_{ref}). Applying the basic open loop equation, you have

$$V_{out} = A_{OL}\left(e_{NI} - e_{INV}\right)$$

$$e_{INV} = V_{ref}$$

$$V_{out} = A_{OL}\left(e_{NI} - V_{ref}\right)$$

Case 1 $e_{NI} > V_{ref}$

$$V_{out} = \infty \times \text{ some positive number}$$

$$V_{out} \rightarrow \infty$$

If the input becomes **slightly** greater than the reference, the open loop voltage gain drives the output to the maximum positive voltage.

$$e_{NI} > V_{ref}$$

$$V_{out} = +V_{SAT}$$

Case 2 $e_{NI} < V_{ref}$

$$V_{out} = \infty \times \text{ some negative number}$$

$$V_{out} \rightarrow -\infty$$

If the input becomes **slightly** less than the reference, the open loop voltage gain drives the output to the maximum negative voltage.

$$e_{NI} < V_{ref}$$

$$V_{out} = -V_{SAT}$$

Figure 2-4 shows the input/output response of a noninverting comparator with a reference voltage.

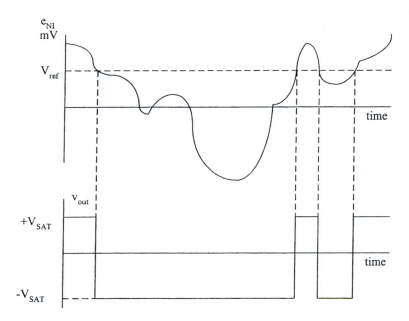

Figure 2-4 Input/output response of a noninverting comparator with a reference voltage

Figure 2-5 gives the schematics of several practical references.

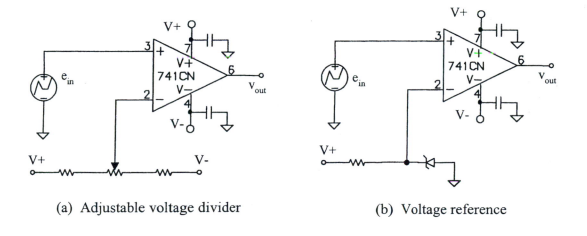

(a) Adjustable voltage divider   (b) Voltage reference

Figure 2-5 Comparators with practical reference voltage

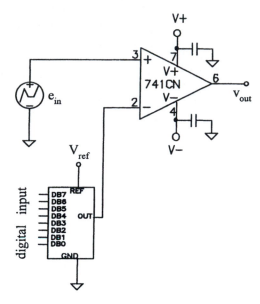

(c) Digitally adjusted reference

Figure 2-5 (Cont.) Comparators with practical reference voltage

Replacing the circuit common at the inverting terminal with a reference voltage changes the comparison level. For the simple comparator of Figure 2-1 and Figure 2-2, switching occurs when the input crosses **common**. Switching for the referenced comparator of Figure 2-3 through Figure 2-5 occurs when the input crosses the **reference voltage**.

The circuit in Figure 2-6 is an inverting comparator. The input signal is applied to the inverting input (-) while the noninverting input (+) is tied to circuit common. An input slightly above zero forces the output to $-V_{SAT}$, while an input just below zero causes the output to go to $+V_{SAT}$. Figure 2-7 illustrates the input to output response of the inverting comparator.

$$v_{out} = A_{OL}\left(e_{NI} - e_{INV}\right) \qquad\qquad e_{NI} = 0V$$
$$v_{out} = -A_{OL}e_{INV}$$

Case 1   **Case 2**

$$e_{INV} > 0V \qquad\qquad\qquad\qquad e_{INV} < 0V$$
$$v_{out} = -A_{OL} \times e_{INV} \qquad\qquad v_{out} = -A_{OL} \times e_{INV}$$
$$v_{out} = -V_{SAT} \qquad\qquad\qquad v_{out} = +V_{SAT}$$

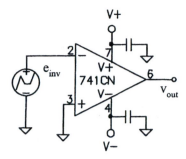

Figure 2-6 Inverting comparator

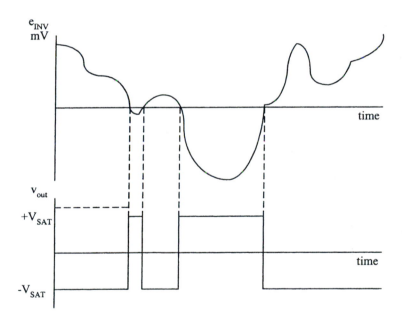

Figure 2-7 Example of an inverting comparator's response to an input

A reference voltage can be applied to the inverting comparator, as it was to the noninverting comparator. The schematic and the input/output response are shown in Figure 2-8.

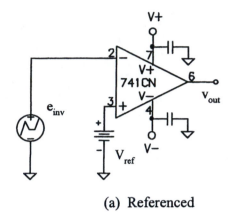

(a) Referenced

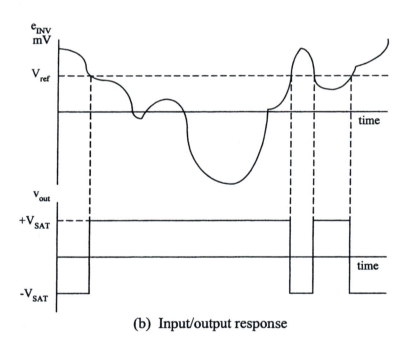

(b) Input/output response

Figure 2-8 Inverting comparator with a nonzero voltage reference

An op amp comparator outputs $+V_{SAT}$ when the voltage at the noninverting input exceeds the voltage at the inverting input. It outputs $-V_{SAT}$ when the voltage at the inverting input exceeds the voltage at the noninverting input.

2-1.2 Hysteresis

The simple op amp comparator produces **false** output transitions under two conditions. When the input is only a few tens of millivolts peak or less, or when the input changes very slowly compared to the output, noise is coupled from the output of the comparator back to the input. This is illustrated in Figure 2-9.

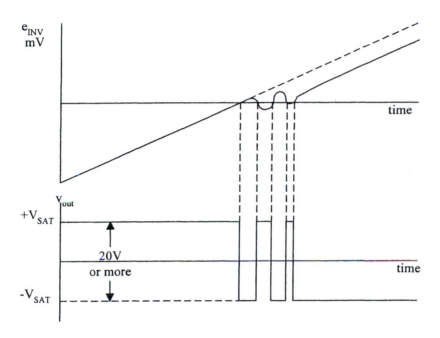

Figure 2-9 Noise produces false triggering in an inverting comparator

When the input crosses above circuit common, the output is driven down to $-V_{SAT}$. This 20V rapid fall couples a small negative-going voltage through the interwiring capacitance from the output back to the inverting input. The input is driven below circuit common by the noise voltage. This, in turn, sends the output up 20V to $+V_{SAT}$.

Now the inverting input is back above circuit common. So the output of the comparator falls 20V back down to $-V_{SAT}$. This 20V rapid fall couples a small negative-going voltage through the interwiring capacitance from the output back to the inverting input. The input is driven below circuit common by this noise voltage. This, in turn, sends the output up 20V to $+V_{SAT}$.

This cycle continues to repeat itself, sending the output up and down, until the input has moved far enough above circuit common that the noise no longer is large enough to drive the input through circuit common. This is certainly a problem, producing many large transitions at the output of the comparator, when only one should occur.

The solution is to add **hysteresis** to the comparator. Hysteresis is the technique of forcing the comparator to switch at two slightly different input levels, rather than always switching when the input passes back and forth through circuit common. These two switching levels are automatically produced by positive feedback added around the circuit, as shown in Figure 2-10.

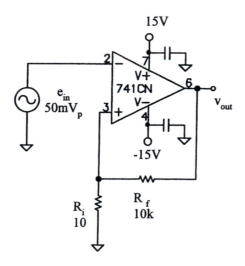

Figure 2-10 Hysteresis added to an inverting comparator

Instead of being compared to circuit common, the input is now compared to a voltage-divided version of the output. When the input is well below this reference voltage, the output is $+V_{SAT}$. The voltage divider provides a positive reference of

$$V_{ref} = \frac{R_i}{R_i + R_f} V_{SAT}$$

For the component values shown in Figure 2-10 the voltage on the comparator's noninverting input is held at about $+13\text{mV}$.

When the input crosses circuit common, nothing happens. Only when the input rises above $V_{ref} = 13mV$ does the output switch to $-V_{SAT}$. As soon as the output is driven down to $-V_{SAT}$, the reference is forced to

$$V_{ref} = -\frac{R_i}{R_i + R_f} V_{SAT}$$

because the output has gone negative.

Now, for the comparator's output to switch back positive, noise must be large enough to pull the inverting input down to -13mV (a 26mV swing). Any smaller swing, even through circuit common, will not cause the comparator to switch. The false triggering has been eliminated. Of course, the comparator no longer switches at precisely 0V. However, this small error is usually acceptable in order to eliminate the false transitions.

Example 2-1

Analyze the performance of the inverting comparator with hysteresis shown in Figure 2-10. Set the input as a 100Hz sine wave. Plot the input, the output, and the feedback on the op amp's noninverting input pin.

Solution

The evaluation version of *MicroSim's Schematic*, *PSpice*, and *Probe* provide the results shown in Figure 2-11. For more information about PSpice see page ix.

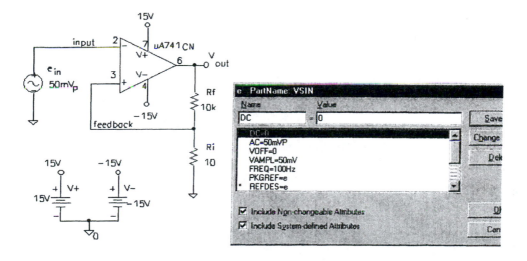

Figure 2-11 (a) Schematic with nodes labeled and source dialog box

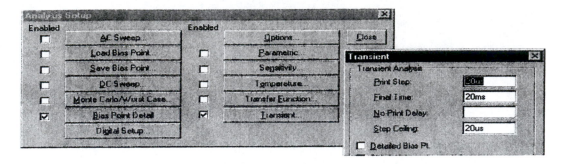

Figure 2-11 (b) Transient analysis dialog boxes

The input and feedback node are labeled in the schematic of Figure 2-11(a). Double clicking on the source brings up its dialog box. Both the AC and the transient values are set, as well as the frequency.

The Analysis dialog boxes of Figure 2-11(b) call for a transient analysis, with data points taken every 20µs for 20ms. This gives 1000 points, over two cycles of the input. That is enough to give a detailed picture.

The resulting plots are shown in Figure 2-11(c). The output is plotted in the top graph. The inverting input and the positive feedback (noninverting input) are plotted together in the lower graph.

When the input is above the feedback ($e_{INV} > e_{NI}$), the op amp's output is in negative saturation. This happens before about 5.47ms. Unfortunately, PSpice sets the uA741's saturation only a part of a volt below the ±15V supplies. A real 741 loses at least ±2V, giving for this example ±V_{SAT} = ±13V or less.

An output of -V_{SAT} (shown as -14.5V by PSpice) feeds back a voltage of

$$v_{feedback} = \frac{10\Omega}{10k\Omega + 10\Omega}(-14.5V) = -14.5mV$$

to the op amp's noninverting input.

When the inverting input falls below -14.5mV at about 5.5 ms, the op amp's output is driven to positive saturation, +14.5V in Figure 2-11(c). In practice, this positive saturation is more like +13V. This, in turn, feeds +14.5mV back to the noninverting input.

The op amp will not switch again until the voltage on the inverting input rises 29mV, up to +14.5mV.

What do you expect to happen if the input amplitude is reduced to 20mV peak?

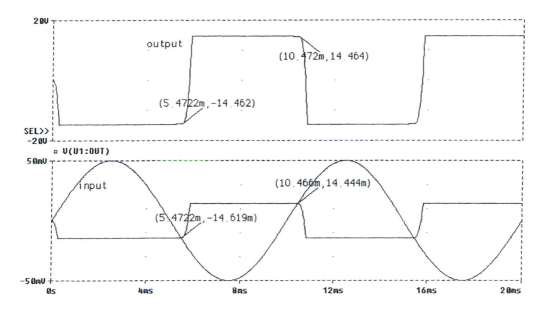

Figure 2-11   (c) Probe plots of input, feedback, and output

2-1.3 The LM339 Comparator Integrated Circuit

While you most certainly may use an op amp to perform comparisons, as described above, it is rarely a good idea. Operational amplifiers are designed to vary their output voltage **linearly** in response to changes in the input voltages. To pull its transistors out of saturation, drive the output voltage from $-V_{SAT}$ up to $+V_{SAT}$, and then force the transistors into saturation again takes a 741 op amp 40µs or longer. For many comparison operations, this is far too slow. At higher frequencies, the difference between the noninverting input and the inverting input pins may have to be 40mV or more to force the 741 to swing its output from $-V_{SAT}$ to $+V_{SAT}$. This inaccuracy is unacceptable in most precision applications. The output voltage levels from the 741 op amp are $V^{-}_{supply} + 2V$ a logic low to $V^{+}_{supply} - 2V$ for a logic high. This is not compatible with many families of logic ICs. The 741 also limits its output current to 20mA into a short circuit. Loads such as relays, lamps, buzzers, small motors, and laser diodes require far more current than the 741 can provide.

    The LM339 is a quad comparator IC. Within a single package are four circuits specifically designed to operate open loop, driving their outputs from one logic level to the other quickly, accurately, and efficiently. With one input drive 5mV above the other, the output will switch in 1.3µs. Output levels can easily be made compatible with many logic families. The IC may be powered from +5V and ground or may use a split supply ranging from ±1V to ±18V.

The schematic and connection diagrams for the LM339 are given in Figure 2-12. Look at the output configuration of the 741 in Figure 2-13.

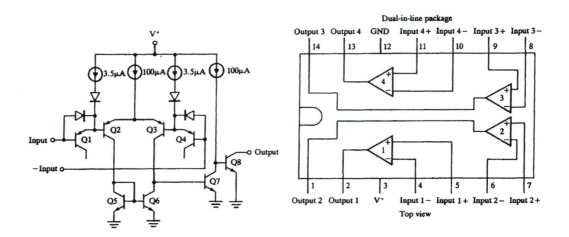

Figure 2-12 LM339 schematic and connection diagram
(Courtesy of National Semiconductor Corp.)

The op amp has a totem pole output, actively driving the output voltage to whatever level is appropriate. The LM339 has a single, open collector transistor at its output. This provides both opportunity and obligation. When the voltage at the inverting input of the LM339 is greater than the voltage at the noninverting input, the output transistor is saturated, providing a logic low. When the noninverting input voltage exceeds the inverting input's voltage, the output transistor is turned off. The output is **open**. You must provide a properly sized resistor between the output pin and the positive supply voltage. The resistor must be small enough to allow parasitic capacitance to charge rapidly, but large enough to limit the current through the output transistor, when it is on, to 6mA or less. The pull-up voltage, to which the resistor is connected, is the logic high output level, provided by the comparator. It need not be the positive supply voltage of the IC.

Example 2-2

For the circuit in Figure 2-13, select the size of the pull-up resistor which will provide the shortest rise time. Calculate that rise time.

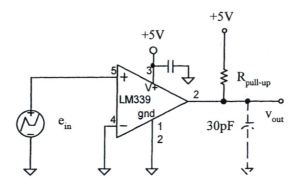

Figure 2-13 Schematic for Example 2-2

Solution

The 30pF capacitance at the output is not intentionally wired to the IC. It is parasitic, formed by such things as sockets, leads, and printed circuit board runs. When the LM339 opens its output for the output voltage to rise, that parasitic capacitance must charge through the pull-up resistor. For a short rise time, the pull-up resistor must be small.

However, when the output transistor is turned on, the pull-up resistor must limit the output transistor's collector current. The manufacturer's specifications guarantee that the output can sink at least 6mA. So, the pull-up resistor must be selected to limit the current from the pull-up voltage supply to 6mA.

$$R_{pull-up} = \frac{V_{pull-up} - V_{Q\,on}}{6mA}$$

$$R_{pull-up} = \frac{5V - 0.2V}{6mA} = 800\Omega$$

Select $R_{pull-up}$ as the next available resistor above 800Ω.

It requires 5RC for a charging capacitor to fully (99%) charge. However, the rise time is measured from the 10% to the 90% points on the rising wave. A little manipulation of

$$v_c(t) = V_{max}\left(1 - e^{-t/RC}\right)$$

reveals that it takes 0.1RC for the output to rise to 10%, and 2.3RC for the output ro rise to 90%. This gives

$$\text{rise time} = 2.2\text{RC} \qquad (2\text{-}2)$$

$$= 2.2 \times 800\Omega \times 30\text{pF}$$

$$= 52.8\text{ns}$$

There are several other application considerations. Since there are four comparators in the IC package, unused comparators may cause interference. Ground the **inputs** of all unused comparators. Feedback from the output to the input through parasitic capacitance may cause oscillations (false triggering), particularly if you place the IC in a socket. To eliminate this effect, lower the input impedance to 10kΩ by placing a 10kΩ resistor from each input to ground. This reduces the effect of noise currents. Also, provide 10mV of hysteresis. Input voltages in excess of the power supply may damage the IC. This is a real problem when the comparator is powered from a single 5V supply. Any input voltage greater than 5V is a threat. **All** negative input voltages are below the ground pin of the IC and could also damage the part. Inputs must not go more than -0.3V below the negative supply. Diode limiting with a Schottky diode may be necessary.

Traditional op amps have a totem pole output, actively driving a voltage onto the output pin at all times. As with all voltage sources, you may **not** wire op amp outputs together. Trying to source different voltages, they would fight each other, currents being limited only by their low output impedance (and any special internal circuitry). The op amps may be damaged.

However, the output of the LM339 is an open collector transistor. You may treat it as a single direction switch to ground (or wherever you have tied pin 12). Wiring multiple LM339 outputs together provides a logical operation. The common node will be low if **any** of the comparators pull it low. Only when **all** comparators output a high (open) does the common output go high.

Called a **wired OR**, this connection is used in the window comparator of Figure 2-14. Its function is to provide an alarm when the input is within (or outside) a band or window, set by E_{HI} and E_{LOW}. This type of monitoring is often used when it is important to maintain a key parameter (e.g., power supply voltage, temperature, level, pressure, speed, etc.) within a prescribed window, or to initiate some corrective action should that parameter become too high or too low.

Comparator U1a checks the upper limit. When the input voltage is too high, its inverting pin is above E_{HI}. This sends its output to ground. The window comparator's output is shorted to ground. This occurs before time **a** in Figure 2-14(b).

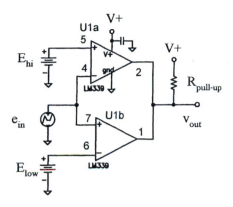

(a) Schematic

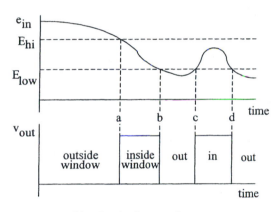

(b) Typical waveforms

Figure 2-14 Window comparator

When the input voltage falls below the high limit, the output of U1a opens. This input is also applied to the noninverting input pin of U1b. If this voltage is above E_{Low}, then U1b's output is also open. With both outputs open, the window comparator's output voltage is pulled to V_{logic} through the shared pull-up resistor. This is illustrated between times **a** and **b** and also between times **c** and **d** in Figure 2-14(b).

Should the input fall below E_{Low}, comparator U1b's noninverting input is below its inverting input. Its output is shorted to ground. Although U1a's output is still open, U1b pulls the window comparator's output to ground. This occurs between times **b** and **c** in Figure 2-14(b).

2-1.4 The LM311 Comparator Integrated Circuit

The LM339 is considerably faster than many op amps. However, it may still be too slow when compared to typical TTL switching times.

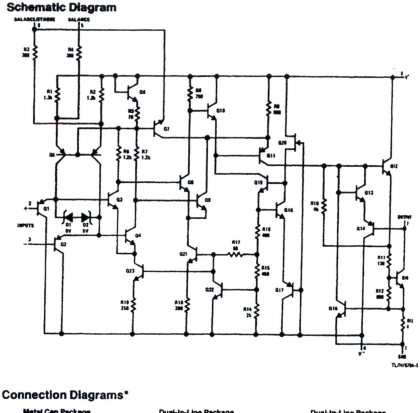

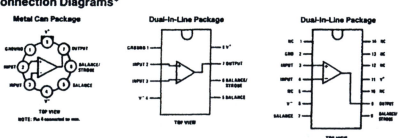

Figure 2-15 LM311 schematic and connection diagram
(*Courtesy of National Semiconductor Corp.*)

Its output current may be no more than a few milliamperes, far too small to drive any significant load. The output must be taken from an open collector transistor, whose emitter is tied to whatever negative supply voltage you have provided to the IC.

These limitations are corrected by the LM311. It is a single comparator, requiring only 200ns to respond to a 5mV overdrive. To obtain this speed, you must rigorously decouple the power supply pin and use hysteresis for inputs whose frequencies are in or below the audio range.

The schematic and connection diagram for the LM311 are shown in Figure 2-15. Look carefully at the output transistor, Q15. Its collector is on pin 7, called the output. But, its emitter is connected through a 4Ω resistor to pin 1, called gnd. This pin is **not** tied to the IC's negative power supply. You may (must) use it.

When the noninverting input is above the inverting input, the output of a traditional comparator is high. The LM311's output transistor is **open**. Conversely, when the inverting input voltage exceeds the noninverting input voltage, a traditional comparator provides a low. The LM311 saturates its output transistor, providing a short between pins 1 and 7.

These two output pins allow you to drive either a load which is tied to the positive supply, or a ground referenced load. Look at Figure 2-16. When the noninverting input voltage (e_{in}) is above the inverting input voltage (e_{ref}), the output transistor is open. The pull-up resistor sets the output voltage to V_{logic}. When the noninverting voltage falls below the inverting input voltage, the output transistor shorts, connecting the pull-up resistor to ground. This looks like a traditional comparator's output. A grounded load is shown in Figure 2-17.

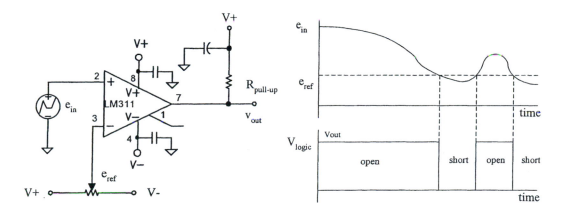

Figure 2-16 LM311 driving a load tied to V_{logic}

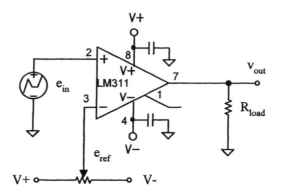

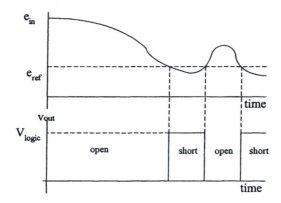

Figure 2-17 LM311 driving a grounded load

For a noninverting input voltage, the open output transistor means that no current flows through the grounded load. The output voltage is

$$V_{load} = I \times R_{load} = 0 \times R_{load} = 0$$

When the input voltage (noninverting input) falls below the reference (inverting input), the output transistor shorts. The positive supply is connected to the load.

Voltage supplies for the LM311 are quite flexible. The main power connections (pins 8 and 4) may be +5V and ground or may be ±V. The pull-up resistor (Figure 2-16) or the output itself (Figure 2-17) may be tied to the positive supply pin (pin 8) or may be pulled to some other logic level. Similarly, pin 1 may be tied to ground, or to some other voltage, as long as it is below the pull-up supply voltage. However, the **input** voltages must always be within the voltages tied to the supply pins. So, if you expect that the input signal may go negative, then you cannot power the LM311 from +5V and ground. The negative power pin must be more negative than the most negative input voltage. So, if you expect a bipolar input, be sure to select ±V for the power connections. However, if the input will **always** be above ground, then logic power should work.

The output transistor can pass as much as 50mA. This allows you to drive more realistic loads (LEDs, relays, lamps, alarms, etc.) than you could with the LM339 or with an op amp comparator. This load current must be limited by the size of the load (or pull-up) resistor. At 50mA the output transistor may require as much as 1.5V, rather than the 0.2V you usually allow for a saturated transistor.

There are three other pins, to provide offset balancing and to strobe the comparator on and off. You may leave them open for prototyping, but since any open pin is an invitation to noise and unreliable operation, be sure to review the manufacturer's recommendations for these pins when you implement a production version circuit using the LM311.

When it comes time to build a comparator circuit, do **not** just grab an op amp, an LM339 or an LM311. In most situations, one will probably work tolerably well. However, there is such a wide selection of comparator ICs that you should spend as much effort selecting the IC as you do the other components in the circuit. Consider each of the following: input voltage level, available supply voltage(s) and supply current, speed requirements, packaging, support components (e.g., a reference voltage) needed, and availability. Then go shopping.

For example, the MAX921 provides a comparator, 1% reference, and built-in hysteresis in its surface mount package. It will run from a single low voltage power supply, drawing a stingy 3µA from the supply battery. But it has very poor load drive ability and requires 12µs to pass a signal. This would be fine for a slow, battery-operated biomedical or music application. On the other extreme is the MAX905. It can respond in 1.8ns, over 6000 times faster than the MAX921. However, it requires 24mA of supply current and is configured for ECL compatibility. So, when building a comparator circuit, look around. There is bound to be an IC just right for your needs.

2-1.5 LED Drivers

The light emitting diode (LED) had found wide application as an indicator, communications interface, and isolation device. To properly drive the LED, its anode must be made 1.8V more positive than its cathode, **and** 10 to 25mA (or more) of current must be provided. However, often the signal being sensed cannot provide this voltage at the needed current. A driver circuit is needed. A transistor could be used to provide the current for the LED. However, bias resistors, reference offsets, and an LED current limit resistor must all be provided.

Many of the comparators in the previous sections can drive the LED directly. An op amp version is illustrated in Figure 2-18. When e_{in} rises above V_{ref}, the output of the op amp climbs toward $+V_{SAT}$. However, when the output reaches $+1.8V$, D1, the red LED, begins to go on. It demands more and more current from the op amp as it illuminates. The op amp enters its short circuit current limiting mode, maintaining enough voltage to keep the output current at about 20mA. This is enough to turn the red LED on brightly, but should not damage it. The output voltage, instead of being at $+V_{SAT}$, is held at $+1.8V$ by the diode and the short circuit current limit circuitry within the IC. When the input voltage exceeds the reference voltage, the red LED is turned on. This red LED could be a panel indicator warning of a dangerously high voltage (or other condition).

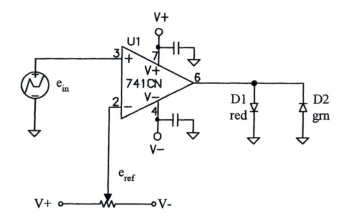

Figure 2-18 Op amp comparator driving two LEDs

When e_{in} falls below V_{ref}, the output of the op amp drops toward $-V_{SAT}$. The red LED goes off. When the output reaches $-1.8V$, the green LED's cathode is now 1.8V below its anode (which is tied to circuit common), and D2 begins to go on. It demands more and more current, which flows from ground, through D2, and into the op amp (sink current). The current limit is reached, and the op amp maintains the output at $-1.8V$, 20mA. The green LED glows. This could be taken as a panel indication that the parameter represented by e_{in} is below the dangerous level (set by V_{ref}).

The LM741 op amp limits current to 20mA or less. For some communications or power interface applications it is often necessary to provide 50mA or more to the LED. In addition, communications signal frequencies usually exceed the speed available from the LM741. Single $+5V$ supply operation is also desirable.

The LM311 comparator can provide 50mA through the LED at frequencies well into the 100's of kHz, from a single +5V supply. Look at Figure 2-19. The IC is powered from +5V and ground, with its V- pin tied to ground along with its ground pin (pin 1). Do not forget the decoupling capacitor at pin 8 (V+). When the input e_{in} goes high, the comparator's output transistor saturates. At 50mA there is typically a 0.75V drop between pin 7 and pin 1. Current flows from the +5V supply, through R_{limit}, through

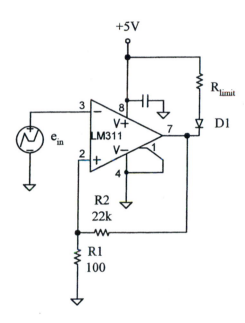

Figure 2-19 Comparator IC driving an LED

the LED, into the comparator at pin 7, and out of the comparator at pin 1 to ground. R_{limit} sets the magnitude of this current.

$$I = \frac{V^+_{supply} - V_{LED} - V_{pin7}}{R_{limit}}$$

$$R_{limit} = \frac{5V - 1.8V - 0.75V}{50mA}$$

$$R_{limit} = 49\Omega$$

The 22kΩ and the 100Ω resistors add about 20mV of hysteresis to minimize the effects of noise, preventing false triggering.

The circuits of Figures 2-18 and 2-19 give you an indication that the input voltage is above or below a single, specific voltage. For more resolution, more comparators are needed. For most reasonable displays, this soon multiplies into far too many components.

The LM3914, LM3915, and LM3916 are single IC dot/bar display drivers. Analog voltage is applied to the input. Ten levels are detected and ten external LEDs are driven accordingly. The LM3914 provides a **linear** relationship between the input analog voltage and the level of the LED(s) which is illuminated. The LM3915 is similar in function, but provides a 3dB difference between LEDs. This is particularly handy for audio power displays. VU meters, popular on the front panel of consumer audio electronics, have a slightly different arrangement. The LM3916 is configured specifically for this application.

The LM3914's block diagram is given in Figure 2-20. Its most conspicuous feature is the string of 10 comparators and the 1kΩ reference divider resistors. The other input to all of the comparators is fed by the buffer and its 20kΩ resistor and diode limiter. The voltage reference source provides a fixed 1.25V between the reference out and the reference adjust pins. By properly cascading low level and high level references, as many as 10 LM3914s can be stacked, to provide a 100 LED display.

Power requirements are flexible. The V+ supply may range from 3V to 25V. However, it must be at least 1.5V above the top end of the analog display range. Internal protection is provided at the analog input. This allows the analog voltage to vary between ±35V without damaging the IC. The LEDs need not be powered from the same supply as the IC, as long as a common ground is provided. The LED power may be fluctuating DC, largely unfiltered. This eliminates the need for heavy filtering or regulation.

When the mode pin is at V+, the 10 LEDs are driven in bar mode. Pulling the mode pin at least 0.2V below V+, or leaving that pin open, causes a single dot to be driven up and down in response to the analog voltage.

Each LED is driven with a constant current which is adjustable by a single external resistor. This allows brightness adjustment and removes the need for current limiting resistors. The current to each of the 10 LEDs is 10 times the current out of the reference out pin ($V_{REF\ OUT}$ = 1.25V). So a resistor or resistor and rheostat combination between the reference out pin and ground allows you to set the current to the LEDs, and therefore their brightness.

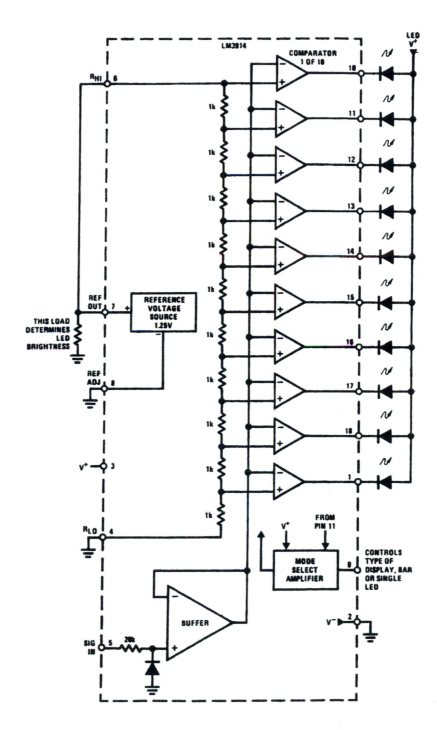

Figure 2-20 LM3914 block diagram (*Courtesy of National Semiconductor Corp.*)

2-2 Effects of Negative Feedback

The comparator will not provide linear amplification because of the extremely large open loop gain of the IC. Also, this gain varies drastically. For the 741C it is typically 200,000 but may be as small as 20,000 with no maximum specified. Although this factor of 10 or more variation in gain will not affect a comparator (since the output is being driven into saturation), variation of this magnitude could cause very serious problems for linear amplifiers.

       The same negative feedback techniques used in transistor circuits can be used with op amps to reduce the overall circuit gain to a usable level, to allow the user to set this circuit's gain precisely, and to make the circuit gain independent of changes in the op amp's open loop gain.

       Figure 2-21 is a block diagram of a negative feedback amplifier. The output signal is v_{out}. It is fed back through an attenuator, which reduces it by a factor of β.

$$e_f = \beta \times v_{out}$$

where $\qquad\qquad\qquad\beta < 1 \qquad\qquad$ Remember, it's an attenuator.

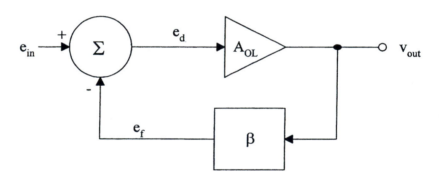

Figure 2-21 Block diagram of a negative feedback amplifier

       This feedback signal, e_f (a reduced version of the output), is subtracted from the input signal, e_{in}, by the summer (Σ).

$$e_d = e_{in} - e_f$$

The difference between the circuit input and the feedback signal is applied to the op amp. Here it is increased by the factor A_{OL} (a very large number).

$$v_{out} = A_{OL}e_d$$

The output is, in turn, sampled and fed back, beginning the process again.

Properly designed, this circuit can become virtually independent of the open loop gain, A_{OL} (if A_{OL} is large). For example, should A_{OL} increase, the output signal (v_{out}) would increase. An increase in v_{out} will increase e_f, the negative feedback. Increasing e_f increases the amount **subtracted** from the input, e_{in}. This causes a smaller difference signal, e_d. A decrease in e_d will cause v_{out} to go down. Since the increased gain, A_{OL}, originally increased the output signal, this drop in output should bring v_{out} back to approximately where it was before A_{OL} increased. This process is outlined in Figure 2-22.

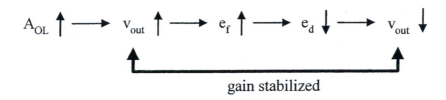

gain stabilized

Figure 2-22 Gain stabilization by negative feedback

This same effect can be proven more rigorously by deriving the overall closed loop gain (A_v). From definitions in Figure 2-21,

$$v_{out} = A_{OL}e_d \tag{2-3}$$

$$e_f = \beta \times v_{out} \tag{2-4}$$

$$e_d = e_{in} - e_f \tag{2-5}$$

The closed loop gain is defined as

$$A_v = \frac{v_{out}}{e_{in}} \tag{2-6}$$

Substituting equation (2-3) into equation (2-6) yields

$$A_v = \frac{A_{OL}e_d}{e_{in}} \tag{2-7}$$

Equation (2-5) can be rearranged to give

$$e_{in} = e_d + e_f \tag{2-8}$$

Substituting equation (2-8) into equation (2-7), you obtain

$$A_v = \frac{A_{OL}e_d}{e_d + e_f} \tag{2-9}$$

Substituting equation (2-4) into equation (2-9) gives

$$A_v = \frac{A_{OL}e_d}{e_d + \beta v_{out}} \tag{2-10}$$

Substituting equation (2-3) into equation (2-10), you get

$$A_v = \frac{A_{OL}e_d}{e_d + \beta A_{OL}e_d} \tag{2-11}$$

Grouping and rearranging terms in the denominator gives

$$A_v = \frac{A_{OL}e_d}{e_d\left(1 + \beta A_{OL}\right)}$$

And, finally,

$$A_v = \frac{A_{OL}}{1 + \beta A_{OL}} \tag{2-12}$$

Since A_{OL} is very large for an op amp, it is reasonable to expect that

$$\beta A_{OL} \gg 1$$

So the 1 in the denominator of equation (2-12) can usually be ignored. Then equation (2-12) becomes

$$A_v = \frac{A_{OL}}{\beta A_{OL}}$$

This simplifies to

$$A_v = \frac{1}{\beta} \tag{2-13}$$

The parameter of the op amp (open loop gain, A_{OL}) fell out of the gain equation, and the closed loop (circuit) gain is **solely** dependent on the feedback attenuation factor (β). Setting β with stable, fixed resistors external to the op amp will allow you to establish a fixed, linear gain, independent of the op amp.

Use of negative feedback will give stable, linear circuit gains which are independent of the op amp. The gain is determined solely by feedback resistors. This assumes that appropriate negative feedback (βA_{OL}) is provided.

2-3 Voltage Follower

One of the most common **linear** op amp circuits (one that uses negative feedback) is the voltage follower. Before beginning the analysis of this circuit, however, you must recall two facts about the op amp itself. These are illustrated in Figure 2-23.

First, since the input impedance is very high, the **signal** current that flows into either input terminal (- or +) is negligibly small.

$$i_{INV} \approx i_{NI} \approx 0$$

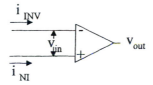

Figure 2-23 Fundamental op amp characteristics

Second, the open loop gain, v_{out}/v_{in}, is extremely large. This means that for any reasonably sized output (any output in the linear region, not $\pm V_{SAT}$), v_{in} must be negligibly small.

$$v_{in} \approx 0$$

The two inputs leads are at virtually the same potential. You certainly will never be able to measure any significant potential difference between the two inputs during linear operation.

The schematic for a voltage follower is given in Figure 2-24.

The input is applied directly to the noninverting input. The output is fed back directly to the inverting input, so there is no attenuation of the negative feedback signal.

$$\beta = 1$$

$$A_v = \frac{1}{\beta}$$

$$A_v = 1$$

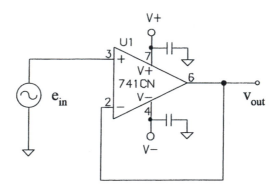

Figure 2-24 Voltage follower

From negative feedback theory, a voltage gain of 1 is obtained. This means that the input and the output are identical.

Another way to analyze the voltage follower is to recall that there is no difference between the two inputs. This means that since the noninverting input is at the source potential, so is the inverting input. However, the inverting input is tied **directly** to the output. The output voltage equals the voltage at the noninverting input, which is the input voltage. So the output voltage equals the input voltage, giving a gain of 1.

Example 2-3

Analyze a voltage follower circuit to determine input and output voltages, gain, and input impedance with $Z_{in} = 1M\Omega$, Z_{out}, and

a. $A_{OL} = 20,000$
b. $A_{OL} = 1000$

Solution

The schematic needed for the computer simulation is shown in Figure 2-25, along with the Analysis Setup. The op amp is modeled as a resistor of $1M\Omega$ between its noninverting and inverting inputs, and a dependent voltage source with a gain of 20E3. The value of the dependent voltage source's output voltage is 20E3 times larger than the voltage between its inputs.

The key parts of the output file (*.out) indicate that

```
E_E1  $N_0001 0 in out 20e3
****         SMALL-SIGNAL CHARACTERISTICS
      V(out)/V_Ein =  1.000E+00
      INPUT RESISTANCE AT V_Ein =   2.000E+10
      OUTPUT RESISTANCE AT V(out) =   5.000E-03
```

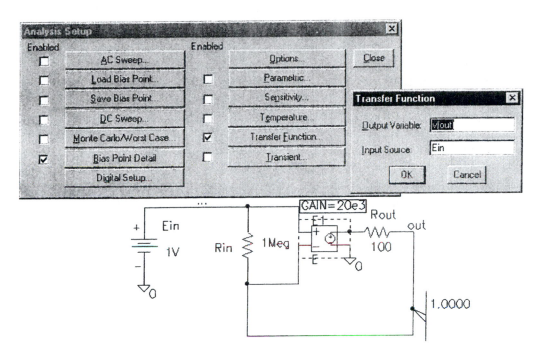

Figure 2-25 *PSpice for Windows* schematic and setup for a voltage follower using a simple op amp model

The voltage at the input is 1V and the output voltage is 1.000V as indicated on the schematic. This is precisely what the voltage follower should do. Small signal characteristics, as determined by the transfer function analysis, show a closed loop gain of 1.000, an input impedance of 20GΩ, and an output impedance of 0.005Ω. Changing the gain of the dependent voltage source to 1000 gives

```
E_E1   $N_0001 0 in out 1e3
****          SMALL-SIGNAL CHARACTERISTICS
       V(out)/V_Ein =   9.990E-01
       INPUT RESISTANCE AT V_Ein =   1.001E+09
       OUTPUT RESISTANCE AT V(out) =   9.990E-02
```

In this second run, the op amp has an open loop gain of 1e3, a factor of only ¹⁄₂₀ of its value in the first problem. However, the **closed loop** gain has fallen to 0.990, or only a 1% reduction. The input impedance is still outrageously high, and the output impedance is still negligible.

A recurring problem in electronics is loading down high impedance voltage sources. If the sources' internal impedance is 10kΩ and the load impedance is 10kΩ, only 50% of the source voltage is applied to the load. If the load is raised to 100kΩ, 91% of the source voltage reaches the load. Though this is an improvement, either of these cases causes significant error in transferring the source voltage to the load. Placing a voltage follower between the source and the load places a GΩ load on the source, preventing any loading of the source. Also, the output impedance of a voltage follower is negligibly small. (Look back at the output impedances in Example 2-3.) So even a small load impedance will not load down the op amp.

Although the voltage follower does not increase the **voltage** amplitude of the signal, it gives good current drive and low output impedance to the load while presenting a very high impedance to the source. The voltage follower makes an excellent impedance transformer, or buffer.

2-4 Noninverting Amplifier

The voltage follower you saw in the previous section used 100% negative feedback. But, as a result, it produced an output voltage no larger than its input. To allow a larger output, you must reduce the amount of negative feedback. This is done by the voltage divider made of R_f and R_i in Figure 2-26.

Also, look back at the block diagram of the general negative feedback amplifier in Figure 2-21. The voltage divider R_f and R_i in the noninverting amplifier of Figure 2-26 provides the attenuation, β (from the block diagram in Figure 2-21). The feedback voltage, e_f, is at the inverting input of the op amp. Both the summer and the high gain amp, A_{OL}, are inside the op amp. So, the general derivation of the closed loop gain for the block diagram also applies to the op amp based, noninverting amplifier. The final result of that derivation was

$$A_v = \frac{1}{\beta}$$

where β is the attenuation ratio of the feedback.

For the noninverting amplifier of Figure 2-26, the attenuation is provided by R_f and R_i.

$$\beta = \frac{e_f}{v_{out}} = \frac{R_i}{R_f + R_i}$$

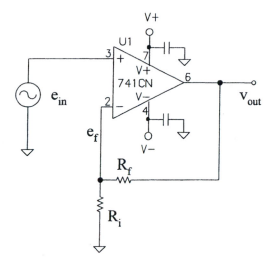

Figure 2-26 Noninverting amplifier

The closed loop voltage gain is the inverse of this attenuation,

$$A_v = \frac{R_f + R_i}{R_i}$$

This is often written

$$A_v = 1 + \frac{R_f}{R_i}$$

You may also derive the voltage gain by recalling that there is virtually no difference between the two inputs of the op amp (during linear operation with negative feedback). Look at Figure 2-27.

$$v_{Ri} = e_{in}$$

$$i_{Ri} = \frac{v_{Ri}}{R_i} = \frac{e_{in}}{R_i}$$

Since no significant current flows into the inputs of an op amp, the current through R_i is also the current through R_f.

$$i_{Ri} = i_{Rf}$$

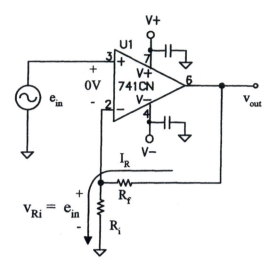

Figure 2-27 Alternate derivation of noninverting amplifier voltage gain

The voltage dropped by this current across R_f is

$$v_{R_f} = i_{R_f} \times R_f$$

$$= \frac{e_{in}}{R_i} \times R_f$$

The output voltage is the voltage across R_i plus the voltage across R_f.

$$v_{out} = v_{R_i} + v_{R_f}$$

$$= e_{in} + \frac{e_{in}}{R_i} R_f$$

$$= e_{in}\left(1 + \frac{R_f}{R_i}\right)$$

Since the gain is the output voltage divided by the input voltage,

$$A_v = \frac{v_{out}}{e_{in}} = 1 + \frac{R_f}{R_i}$$

The input impedance of the noninverting amplifier is

$$Z_{\text{in noninverting amp}} = \beta A_{\text{OL}} Z_{\text{in op amp IC}}$$

Remember, β is a fraction, but rarely smaller than 0.01, since it is the reciprocal of the closed loop gain. A_{OL} is the open loop gain of the op amp, usually several thousand or more before the amplifier's gain begins to rolloff. This means that even though you may not know precisely what the amplifier's input impedance may be, and that the impedance will vary, it will be **very large**, usually in the GΩ range.

Example 2-4

Analyze the circuit in Figure 2-27, given that

$$e_{\text{in}} = 1V_{\text{DC}}$$
$$R_f = 13.3k\Omega$$
$$R_i = 2.2k\Omega$$

Determine the voltage at the op amp's noninverting input, inverting input, and output.

Solution

The results of a simulation run with *Electronics Workbench* is shown in Figure 2-28. As expected, the inverting input is at the same value as the noninverting input. According to the voltage divider law, to drive the inverting input to 1V, the output must be at

$$1V = \frac{2.2k\Omega}{2.2k\Omega + 13.3k\Omega} \times V_{\text{out}}$$

$$V_{\text{out}} = \frac{2.2k\Omega + 13.3k\Omega}{2.2k\Omega} \times 1V$$

$$= 7.05V$$

The simulation shows $7.07V_{\text{DC}}$ at the output.

You could also calculate the output voltage by using the gain formula.

$$V_{\text{out}} = E_{\text{in}} A_v$$

$$= 1V\left(1 + \frac{13.3k\Omega}{2.2k\Omega}\right)$$

$$= 7.05V$$

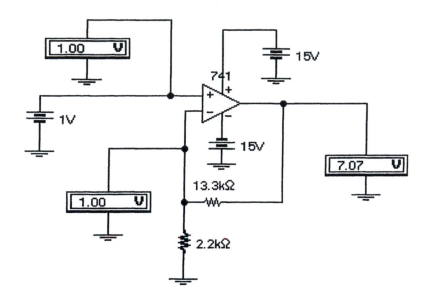

Figure 2-28 *Electronics Workbench* simulation of Example 2-4

2-5 Inverting Amplifier

The schematic of an op amp inverting amplifier is shown in Figure 2-29. The op amp provides the gain. Resistors R_i and R_f attenuate the feedback signal (from v_{out} to the negative input), establishing both β and therefore the circuit's closed loop gain. The output signal is 180° out of phase with the input, e_{in}, since the inverting terminal of the op amp is being driven. Consequently, when the feedback combines with the input signal at the inverting terminal, subtraction takes place (as required for a **negative** feedback amplifier). All of the requirements for a negative feedback amplifier outlined in Section 2-2 are met.

The output signal, v_{out}, is 180° out of phase with the input, e_{in}. The circuit's voltage gain is determined by the feedback components, R_i and R_f.

$$A_v = -\frac{R_f}{R_i}$$

The negative gain indicates the 180° phase shift. The input impedance is set by R_i.

$$Z_{in} = R_i$$

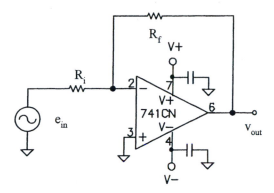

Figure 2-29 Op amp inverting amplifier

You can derive these characteristics from the following analysis. Also, please refer to Figure 2-30.

With an open loop gain of several thousand or more, if the output is not in saturation, then

$$v_{out} = A_{OL} \times v_{in}$$

$$v_{in} = \frac{v_{out}}{A_{OL}} \approx 0$$

The voltage at the junction of R_i, R_f, and the inverting input of the op amp must be at practically 0V. The inverting input is said to be at **virtual ground**.

With the inverting input at virtual ground, the input impedance is the resistance between the circuit input (e_{in}) and ground. This is just R_i.

$$Z_{in} = R_i$$

The current flowing through R_i is

$$i_{R_i} = \frac{e_{in}}{R_i} 1$$

Since no significant signal current flows into the op amp's inputs,

$$i_{in} = 0$$

So all of i_{Ri} must flow through R_f.

$$i_{R_f} = i_{R_i}$$

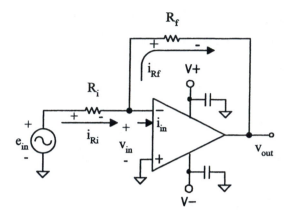

Figure 2-30 Analysis of an op amp inverting amplifier

This means that the voltage dropped across R_f is

$$v_{R_f} = i_{R_f} \times R_f = i_{R_i} \times R_f$$

$$v_{R_f} = \frac{e_{in}}{R_i} R_f$$

But look at the polarity of v_{Rf}. The left end of R_f is at virtual ground, and the current flows from left to right. That makes the right end of R_f negative. That is,

$$v_{out} = -v_{R_f}$$

$$= -\frac{e_{in}}{R_i} R_f$$

The voltage gain of the amplifier, A_v, is

$$A_v = \frac{v_{out}}{e_{in}} = -\frac{R_f}{R_i}$$

With the inverting amplifier you can set the input impedance by picking R_i and the voltage gain with R_f. However, there are several practical considerations.

1. Setting R_i too high will give problems with the effect of the bias current. It is usually restricted to less than 10kΩ.

2. Remember, there is an upper limit to the gain bandwidth product ($A_v \times f$). Do not try for a gain which is so high that at upper frequencies this gain bandwidth product is violated. Usually A_v is held below 100.

3. The peak output is limited by the power supplies. At about 2V less than the supply, the op amp goes into saturation.

4. The output current may or may not be short-circuit limited. If it is not, heavy loads (low load resistance) may damage the op amp. If there is short circuit protection, a heavy load may drastically distort the output voltage wave shape.

Example 2-5

Select the components necessary to configure an amplifier with a gain of -7 and an input impedance of 4.7kΩ.

Solution

The input impedance is set by R_i.

$$R_i = Z_i = 4.7k\Omega$$

The gain is set by

$$A_v = -\frac{R_f}{R_i}1$$

So

$$R_f = -A_v \times R_i$$

$$= -(-7) \times 4.7k\Omega$$

$$= 33k\Omega$$

2-6 Inverting Summer

A simple addition to the inverting amplifier produces a circuit with much versatility. This inverting summer is shown in Figure 2-31. An additional input (e_{in2}) and input resistor (R_{in2}) have been added to the inverting amplifier.

The input impedances and the output voltage equation are derived very much as the inverting amplifier equations were. Since the circuit uses negative feedback to obtain linear operation, there is no significant difference in potential between the input pins of the op amp; the inverting input pin is at virtual ground. The input impedance seen from e_{in1} is the resistance from that point to ground. This is just R_{in1}. Similarly, the input impedance seen from e_{in2} is R_{in2}.

$$Z_{in1} = R_{in1}$$

$$Z_{in2} = R_{in2}$$

The current from e_{in1} through R_{in1} is

$$i_1 = \frac{e_{in1}}{R_{in1}}$$

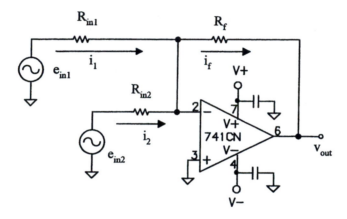

Figure 2-31 The inverting summer

Similarly, for i_2,

$$i_2 = \frac{e_{in2}}{R_{in2}}$$

As you saw with the inverting amplifier, no significant signal current flows into the op amp itself. This means that i_1 and i_2 must combine at the junction of R_{in1}, R_{in2}, and R_f to produce i_f. Applying Kirchhoff's current law at that node, you have

$$i_1 + i_2 = i_f$$

The voltage drop produced by i_f flowing through R_f must equal the output voltage. But do not forget the current direction and therefore the polarity inversion.

$$v_{out} = -i_f R_f$$

$$= -(i_1 + i_2)R_f$$

$$= -\left(\frac{e_{in1}}{R_{in1}} + \frac{e_{in2}}{R_{in2}}\right)R_f$$

This can be rearranged to look a bit more familiar.

$$v_{out} = -\left(\frac{R_f}{R_{in1}}e_{in1} + \frac{R_f}{R_{in2}}e_{in2}\right)$$

The ratio $\dfrac{R_f}{R_{in1}}$ is the gain for e_{in1}, and $\dfrac{R_f}{R_{in2}}$ is the gain for e_{in2}.

The inverting summer outputs a voltage (v_{out}) which is the inverted, weighted sum of the two inputs. The weighting or scaling factors are the gains.

Since this is only a slight modification of the inverting amplifier, you should observe all of the precautions listed in the preceding section. Also, although three, four, or more inputs could be summed just as two were, the amplifier's noise goes up with each input added. So you should limit the number of summing inputs to as few as possible.

The inverting summer allows you to alter the zero and the full scale (span) of a signal. This is particularly important in adjusting the output of industrial transducers, allowing them to be displayed in convenient engineering units, or to match the full range of an A-D converter.

Example 2-6

The LM35 temperature transducer outputs 10mV/°C. Alter this output so that 100mV/°F can be displayed.

Solution

	input (x)	output (y)
freezing	0V	3.2V
boiling	1V	21.2V

We will never actually subject a silicon sensor to boiling water.

$$y = mx + b$$

$$m = gain = slope = \frac{\Delta y}{\Delta x}$$

$$= \frac{21.2V - 3.2V}{1V - 0V} = 18$$

So, the circuit must have a gain of 18.

$$b = \text{offset} = \text{output when the input is } 0V$$

$$b = 3.2V$$

Both the zero and the span must be considered. Begin with the inverting summer in Figure 2-31. However, a second inverting amplifier must be added to keep the overall relationship noninverting.

The output equation is

$$V_{out} = -\left(\frac{R_f}{R_{in1}} e_{in1} + \frac{R_f}{R_{in2}} e_{in2} \right)$$

Let e_{in1} be the input from the sensor. Compare the summer's equation to the equation of the line that describes the input to output relationship.

$$y = mx + b$$

$$y = 18 e_{transducer} + 3.2V$$

$$y = v_{out}$$

$$m = \text{gain} = 18 = \frac{R_f}{R_{in1}}$$

$$x = e_{in1}$$

$$b = \text{offset} = 3.2V = \frac{R_f}{R_{in2}} e_{in2}$$

Pick $R_f = 18k\Omega$. $R_{in1} = \frac{R_f}{18} = 1k\Omega$

This assures that changes on the input will cause the correct **change** at the output. The offset is set by applying a DC voltage for e_{in2} and adjusting R_{in2} so that

$$3.2V = \frac{R_f}{R_{in2}} e_{in2}$$

The positive supply is a convenient DC voltage. Pick $e_{in2} = 15V$. Now, solve the offset equation for R_{i2}.

$$R_{in2} = \frac{R_f}{3.2V} e_{in2} = \frac{18k\Omega}{3.2V} \times 15V$$

$$= 84k\Omega$$

The accuracy of this design can be checked using a DC Analysis in *MicroSim's PSpice for Windows*. The schematic is shown in Figure 2-32(a).

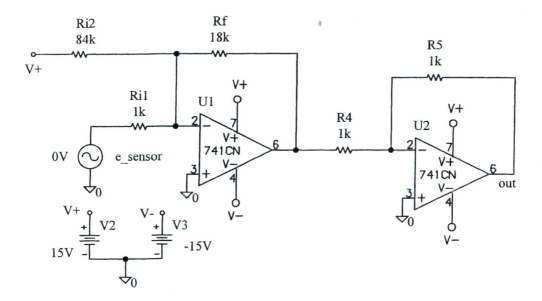

Figure 2-32 (a) Schematic

The Analysis setup is shown in Figure 2-32(b). Notice that under Analysis Setup a DC Sweep has been selected. This selection brings up the DC Sweep dialog box. Sweep the Voltage Source, e_sensor, from 0V to 0.6V in 1mV steps.

The resulting Probe plot is shown in Figure 2-32(c). From the two points identified on the plot, the zero and span design works correctly.

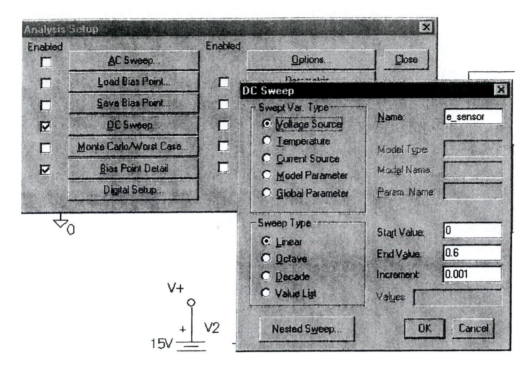

Figure 2-32 (b) DC Analysis setup

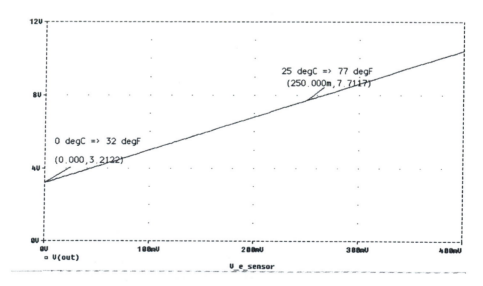

Figure 2-32 (c) Signal conditioner's transfer function

2-7 Difference Amplifier

To this point you have seen how to amplify or amplify and add signals. Often, it is also important to be able to obtain the difference between two signals. This could be used to tell you how far apart the two signals are, or to eliminate (subtract out) any portion of the two signals which is the same (or common), giving an output only when the two signals are different.

The schematic for a difference amplifier is shown in Figure 2-33. Input signals are connected through **identical** resistors (R_i) to both the inverting and the noninverting op amp terminals. A sample of the output is fed back through the upper R_f to the inverting terminal to assure negative feedback. The lower resistor R_f attached between the noninverting terminal and ground balances operation, insures 0V out when e_{INV} equals e_{NI}, and significantly simplifies circuit analysis and design.

With the two input resistors (R_i) identical and the other two resistors (R_f) selected to be equal, the output voltage equation is

$$v_{out} = \frac{R_f}{R_i}\left(e_{NI} - e_{INV}\right)$$

The output is the difference between the two inputs times a gain set by R_f and R_i.

You can derive this relationship by first calculating the voltage at the noninverting op amp input. Since there is no significant current flowing into the op amp, the voltage at the noninverting op amp input is set by the voltage divider formed by R_f and R_i.

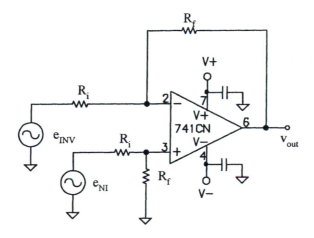

Figure 2-33 Difference amplifier

$$v_{NI} = \frac{R_f}{R_f + R_i} e_{NI}$$

There is no significant difference between the inverting and noninverting inputs. So

$$v_{INV} = v_{NI} = \frac{R_f}{R_f + R_i} e_{NI}$$

The current from e_{INV} through R_i to the op amp's inverting input node is

$$i_{R_i upper} = \frac{(e_{INV} - v_{INV})}{R_i}$$

$$= \frac{\left(e_{INV} - \dfrac{R_f}{R_f + R_i} e_{NI} \right)}{R_i}$$

All of this current flows through the upper feedback resistor, R_f, because no significant signal current flows into the op amp's inputs.

$$i_{R_f upper} = i_{R_i upper} = \frac{\left(e_{INV} - \dfrac{R_f}{R_f + R_i} e_{NI} \right)}{R_i}$$

This current produces a voltage drop across the upper R_f with the positive on the left and the negative on the right.

$$v_{R_f upper} = i_{R_f upper} R_f$$

$$= \frac{\left(e_{INV} - \dfrac{R_f}{R_f + R_i} e_{NI} \right)}{R_i} R_f$$

The voltage at the output is the voltage at the inverting input **minus** the voltage dropped across the upper R_f.

$$V_{out} = V_{INV} - V_{R_f upper}$$

$$= \frac{R_f}{R_f + R_i} e_{NI} - \frac{\left(e_{INV} - \dfrac{R_f}{R_f + R_i} e_{NI} \right)}{R_i} R_f$$

$$= \frac{R_f}{R_f + R_i}e_{NI} - \frac{R_f}{R_i}e_{INV} + \frac{R_f^2}{R_i(R_f + R_i)}e_{NI}$$

Now, it's just a matter of grouping like terms and simplifying the algebra.

$$V_{out} = \left[\frac{R_f}{R_f + R_i} + \frac{R_f^2}{R_i(R_f + R_i)}\right]e_{NI} - \frac{R_f}{R_i}e_{INV}$$

$$= \frac{R_iR_f + R_f^2}{R_i(R_f + R_i)}e_{NI} - \frac{R_f}{R_i}e_{INV}$$

$$V_{out} = \frac{R_f(R_f + R_i)}{R_i(R_f + R_i)}e_{NI} - \frac{R_f}{R_i}e_{INV}$$

$$V_{out} = \frac{R_f}{R_i}(e_{NI} - e_{INV})$$

Example 2-7

Determine the output of a difference amplifier with $R_i = 1k\Omega$ and $R_f = 10k\Omega$, with

a. $e_{INV} = e_{NI} = 5V_{DC}$

b. $e_{INV} = 4.95V_{DC}$ $e_{NI} = 5.05V_{DC}$

Solution

a.   Theoretically, with identical inputs, the difference amplifier output should be

$$V_{out} = \frac{10k\Omega}{1k\Omega}(5V - 5V) = 0V$$

This, however, does not consider the nonideal characteristics of the devices. *PSpice for Windows* gives a more realistic result. The schematic with DC viewpoints are shown in Figure 2-34. The simulation indicates only a -1.3mV error at the output, a pretty good result for such an inexpensive op amp.

b.   With two different inputs,

$$V_{out} = \frac{10k\Omega}{1k\Omega}(5.05V - 4.95V) = 1.0V$$

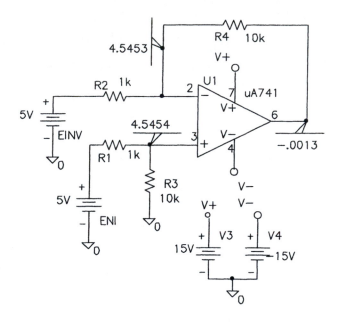

Figure 2-34 Simulation results for Example 2-7a

The common mode rejection ratio (CMRR) is a measure of how well the differential amplifier increases the size of the signals that are different, while rejecting any portion of the inputs which are identical. First, two gains must be defined. The differential gain is the output voltage divided by the difference in input voltages.

$$A_{diff} = \frac{V_{out}}{e_{NI} - e_{INV}}$$

The common mode gain is obtained when a single, common voltage is applied to each input.

$$A_{cm} = \frac{V_{out}}{e_{in\ common}}$$

Now, the common mode rejection ratio can be defined.

$$CMRR = 20 \times \log\left(\frac{A_{diff}}{A_{cm}}\right)$$

Ideally, the output for identical inputs is zero. This means that the theoretical CMRR is infinite. Practically, the CMRR of the amplifier circuit is limited by the CMRR of the op amp itself, how well the R_i's match each other, and how well the R_f's match each other. You may expect CMRRs in the 40dB to 80dB range.

Example 2-8

Industrials sensors often have to transmit small signals very accurately over long distances. Usually, the information is differential. That is, two signals are sent, one positive and the other equally negative. Along the way, these signals often encounter very large quantities of 60Hz interference. This noise infects each signal exactly the same. So the noise is **common** to both signals and can be rejected by the CMRR of a good difference amplifier. A typical example is shown in Figure 2-35.

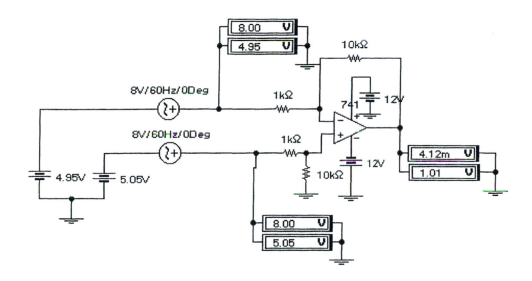

Figure 2-35 Simulation for Example 2-8

The sensor signals are represented by the 5.05V and the 4.95V DC sources. That is, the sensor outputs ±50mV riding on a $5V_{DC}$ level. The noise injected into the signals' lines is simulated by the two 8V/60Hz sine wave sources. These are identical. The voltmeters at the amplifier's inputs show $8V_{RMS}$ at each input, and $5.05V_{DC}$ at the amplifier's noninverting input and $4.95V_{DC}$ at its inverting input. The output DC should be

$$V_{out\ DC} = \frac{10k\Omega}{1k\Omega}(5.05V - 4.95V) = 1.00V$$

The *Electronics Workbench* simulation shows $1.01V_{DC}$ at the output. This is only a 1% error.

The $8V_{AC}$ at each input completely buries the 50mV signal (as does the 5V offset). But the difference amplifier, ideally, rejects all of these common mode signals, giving no AC output. According to the simulation, there is a $4.12mV_{AC}$ output. All but 4mV of the 8V input common mode signal has been rejected! The ±50mV signal transmitted from the sensor, offset by 5V and buried in 8V of AC noise, has been recovered with 1% error in its amplitude and 0.4% noise.

To calculate the CMRR for this example,

$$A_{diff} = \frac{1.01V}{5.05V - 4.95V} = 10.1$$

$$A_{cm} = \frac{4.12mV}{8V} = 515 \times 10^{-6}$$

$$CMRR = 20 \times \log\left(\frac{10.1}{515 \times 10^{-6}}\right) = 86dB$$

2-8 A General Approach

In the previous sections of this chapter you saw several different types of amplifiers. Many signal processing requirements can be satisfied by some combination of these circuits. However, as the mix of analog and digital functions continues to grow, as more exotic parts become available, and as system requirements expand, you will need a general approach. Below is a six-step analysis technique which you can use in any **linear negative feedback** system. Following the explanation of the six steps are several examples illustrating how to apply the technique to a variety of nontraditional, but fairly common circuits.

1. **Noninverting input voltage**
    Begin by determining the voltage on the noninverting input of the op amp. This may simply be the ground to which that pin is tied. Or you may need to apply Kirchhoff's voltage law if several sources and resistors are involved.

2. Inverting input voltage

Given that there is negative feedback and that the op amp is not saturated, then the voltage on the inverting input pin is virtually the same as the voltage you just calculated for the noninverting input pin. (This ignores the input offset voltage. You will see how to handle V_{ios} later.)

3. Input resistor currents

Next you must determine the current through each **input** resistor. Input resistors are all resistors tied to the inverting input **except** the resistor to the output. Be sure to keep track of each current's direction.

4. Feedback resistor current

These currents sum at the inverting input node and flow through the feedback resistor to the output of the op amp. Apply Kirchhoff's current law to calculate this current. Watch their directions carefully, to decide whether you add or subtract the currents and what the direction of the final, resultant current is. The op amp then sinks or sources this current, to keep the difference in potential at its input pins negligible.

5. Feedback resistor voltage

The voltage across the feedback element, then, is just this current times the impedance of the feedback. Again, watch the current direction to be sure to get the polarity of the voltage correct.

6. Output voltage

Finally, the output voltage is the voltage at the inverting input pin plus (or minus, depending on the direction of the current) the voltage across the feedback resistor.

Example 2-9

Verify all indicated currents and voltages in the circuit in Figure 2-36. This schematic shows another way to add a DC offset to an inverting amplifier.

Solution

1. Noninverting input voltage

$$V_{NI\,DC} = 0.5V_{DC}$$

$$v_{NI\,AC} = 0V_{AC}$$

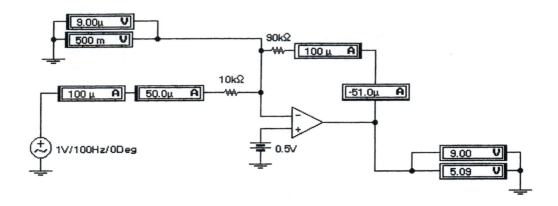

Figure 2-36 Inverting amplifier with offset simulated with *Electronics Workbench*

2. Inverting input voltage

$$V_{INV} = V_{NI\,DC} = 0.5V_{DC}$$

$$v_{INV\,AC} = v_{NI\,AC} = 0V_{AC} \approx 9\mu V_{AC}$$

3. Input resistor currents

$$I_{R_i\,DC} = \frac{500mV_{DC}}{10k\Omega} = 50\mu A_{DC} \quad \leftarrow$$

Since the DC voltage at the inverting input node is positive, the DC current flows from right to left through the $10k\Omega$ R_i and through the AC source.

$$i_{R_i\,AC} = \frac{1V_{AC}}{10k\Omega} = 100\mu A_{AC} \quad \rightarrow$$

4. Feedback resistor current
No significant signal current flows into the input of the op amp. So

$$I_{R_f\,DC} = \frac{500mV}{10k\Omega} = 50\mu A_{DC} \leftarrow$$

Watch the direction of current flow. It is from the output of the op amp into the ammeter, through R_f and to the node at the op amp's inverting input. That is why the simulation indicates that the current is -50µA.

$$i_{R_f\,AC} = \frac{1V_{AC}}{10k\Omega} = 100\mu A_{AC} \quad \rightarrow$$

5. Feedback resistor voltage

$$V_{R_f\,DC} = 50\mu A_{DC} \times 90k\Omega = 4.5V_{DC}$$

Since this current is flowing from right to left, the right end of R_f is more positive than its left end.

$$v_{R_f\,AC} = 100\mu A_{AC} \times 90k\Omega = 9.0V_{AC}$$

6. Output voltage

$$V_{out\,DC} = V_{INV\,DC} + V_{R_f\,DC}$$

$$= 0.5V + 4.5V = 5.0V_{DC}$$

$$v_{out\,AC} = v_{INV\,AC} + v_{R_f}$$

$$= 0V + 9.0V = 9.0V_{AC}$$

Example 2-10

Verify all indicated currents and voltages in the circuit in Figure 2-37. This is the schematic of a more advanced difference amplifier, called a two op amp **instrumentation** amplifier. Each input signal has a very high input impedance. The output is the amplified difference between the two inputs. The gain can be adjusted by changing the single 200Ω resistor. Though a little unusual looking, this very handy circuit can be analyzed using the six general steps.

Solution

1. Noninverting input voltages

$$V_{NI\,U1} = 4.95V$$

$$V_{NI\,U2} = 5.05V$$

2. Inverting input voltage

$$V_{INV\,U1} = V_{NI\,U1} = 4.95V$$

$$V_{INV\,U2} = V_{NI\,U2} = 5.05V$$

3. Input resistor currents, U1

$$I_{R2} = \frac{4.95V}{9k\Omega} = 550\mu A \qquad \downarrow$$

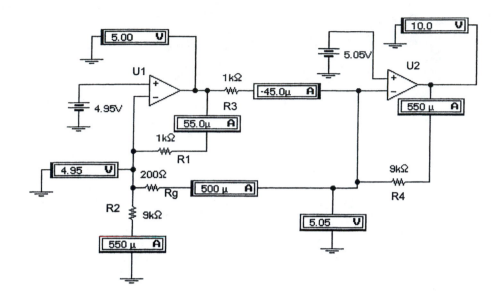

Figure 2-37 Two op amp instrumentation amplifier simulation with *Electronics Workbench*

$$I_{R_g} = \frac{5.05V - 4.95V}{200\Omega} = 500\mu A \qquad \leftarrow$$

4. Feedback resistor current, U1
    No significant signal current flows into the input of the op amp. So

    $$I_{R1} = 550\mu A - 500\mu A = 50\mu A \qquad \leftarrow$$

5. Feedback resistor voltage, U1

    $$V_{R1} = 50\mu A \times 1k\Omega = 50mV \qquad - \text{ to } +$$

6. Output voltage, U1

    $$V_{out\,U1} = V_{INV\,U1} + V_{R_1} = 4.95V + 0.05V = 5.0V$$

Now, it's time to repeat the last three steps for U2.

3. Input resistor currents, U2

    $$I_{R3} = \frac{5.05V - 5.0V}{1k\Omega} = 50\mu A \qquad \leftarrow$$

4. Feedback resistor current, U2

$$I_{R4} = I_{R3} + I_{R_g} = 50\mu A + 500\mu A = 550\mu A \quad \leftarrow$$

5. Feedback resistor voltage, U2

$$V_{R4} = 550\mu A \times 9k\Omega = 4.95V \quad\quad - \text{ to } +$$

6. Output voltage, U2

$$V_{out\ U2} = V_{INV\ U2} + V_{R_4}$$

$$= 5.05V + 4.95V = 10.0V$$

Example 2-11

The schematic shown in Figure 2-38 illustrates how an op amp can be used to allow a digital to analog converter to set the output voltage from a voltage regulator.

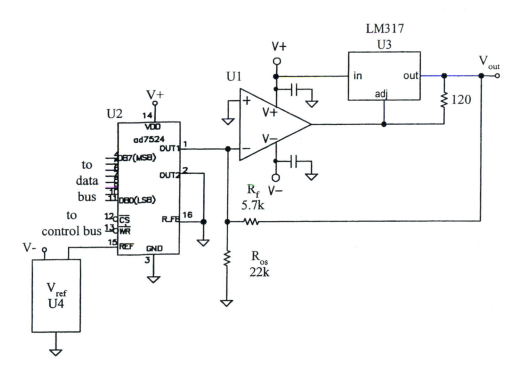

Figure 2-38 Digital to analog converter control of voltage regulator

Solution

The evaluation version of *MicroSim's PSpice for Windows* does not support the AD7524 DAC or the LM317. However, to the rest of the circuit, the DAC looks like a current source. Changing the digital code or the reference into the DAC changes its sink current. For this example, that current is picked at 210µA Also, between the output and the adjust pins of the LM317, is a constant 1.2V. These two substitutions have been made, and the resulting simulation is shown in Figure 2-39. Finally, to investigate the effect of a 60mV input offset voltage of the op amp, the V_{ios} source is added. The resulting simulation with DC voltages (viewpoint) and currents (iprobe) is shown in Figure 2-39.

1. Noninverting input voltages

$$V_{NI\,U1} = E_{ios} = -60mV$$

2. Inverting input voltage

$$V_{INV\,U1} = V_{NI\,U1} = -60mV$$

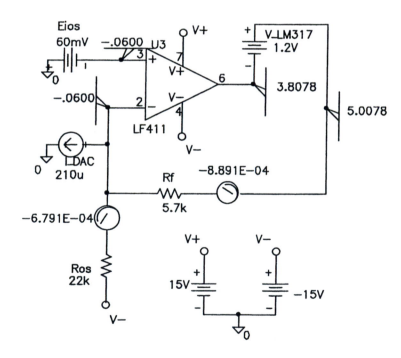

Figure 2-39 *PSpice for Windows* simulation of DAC controlled voltage regulator

3. Input resistor currents, U1

$$I_{R_{os}} = \frac{-15V - (-60mV)}{22k\Omega} = 679\mu A \qquad \downarrow$$

4. Feedback resistor current, U1
   No significant signal current flows into the input of the op amp. So

$$I_{R_f} = 679\mu A + 210\mu A = 889\mu A \qquad \leftarrow$$

5. Feedback resistor voltage, U1

$$V_{R_f} = 889\mu A \times 5.7k\Omega = 5.07V \qquad - \text{ to } +$$

6. Output voltage

$$V_{out} = V_{INV\ U1} + V_{R_f} = -60mV + 05.07V$$

$$= 5.01V$$

This is the output of the **circuit**. However, the output of the op amp is shifted by the voltage between the output and the adjust pins of the LM317. This is a fixed 1.2V offset. To keep the difference in its input pins at 0V, the op amp drives its output voltage down another 1.2V.

$$V_{op\ amp\ output} = 5.01V - 1.2V = 3.81V$$

So, without knowing the internal operation of the LM317, or even the relationship between the input and output, you have been able to calculate all of the currents and voltages by "making an end run around" through the negative feedback elements.

Summary

Operating open loop (no negative feedback), an op amp's output will normally be at either $+V_{SAT}$ or $-V_{SAT}$. When the voltage at the noninverting input is larger than the voltage at the inverting input, the op amp's output voltage is forced to positive saturation. On the other hand, when the voltage at the inverting input is larger, the output goes to negative saturation. No other outputs will normally occur.

This characteristic allows the op amp to be used open loop as a voltage comparator. Inverting and noninverting comparators can be built which compare the input with either ground or a reference voltage. For high speed ICs, or for slow, small input voltages, noise coupled from the rapid, large output transition may cause false triggering. This is solved by adding a small amount of positive feedback, called hysteresis.   But even the best op amps make only fair comparators.   There are comparator ICs which are much faster, more accurate, capable of higher load currents, and operation from a single supply.   The LM339 IC provides four comparators in a single package.   The LM311 is a single comparator which is faster than the LM339 and has a higher current drive and more flexible output connections.

Short-circuit protected op amps usually limit the output current to about 20mA.   This is just right to drive a light emitting diode.   For stacks of 10 or more LEDs, the LM3914/15/16 are a wise choice.

Negative feedback is necessary to allow the op amp to operate linearly.   This also, to a large degree, removes a circuit's dependence on the op amp's open loop gain. The closed loop gain of the circuit can be strictly determined by the characteristics of the negative feedback network (usually fixed, stable, linear resistors).

Voltage followers, noninverting amplifiers, inverting amplifiers, summers, and difference amplifiers can be built with op amps and negative feedback networks. Analysis of each shows that the relationship between the input and output voltage is linear.   It is set solely by the circuit providing the negative feedback.

There is a wide variety of applications for these linear op amp circuits.   They allow you to build circuits which directly implement mathematical equations. Multiplication by a constant is amplification.   Addition gives you mixing and DC level shifting.   Subtraction allows for common mode rejection.   All of these can be done for AC signals, DC signals, or combinations.   Also, the circuits can present a high impedance to the source (to prevent loading it down), while giving a low output impedance to the load.

As the complexity of analog circuits increases, the mix of analog, digital, and power components becomes more common.   You will see more and more circuits using op amps in novel ways.   **Any** circuit using op amps with negative feedback can be analyzed with six steps: (1) calculate $V_{NI\ input}$, (2) calculate $V_{INV\ input}$, (3) calculate the currents through the input elements (often resistor), (4) calculate the current through the feedback resistor, (5) calculate the voltage across the feedback resistor, and (6) calculate the output voltage.

Analog electronics provide the senses and the muscle of most electronics systems, but the microprocessor forms the brains for all but the simplest systems. Combining the intelligence and flexibility of the microprocessor with the accuracy, precision, speed, and power of analog ICs allows you to build some rather amazing electronics with just a handful of parts, **if** you understand how to glue the two types of ICs together.   The next chapter will feature the uses and limitations of analog switches and multiplexers, digitally controlled potentiometers, and voltage out and multiplying digital to analog converters.   With these, your circuits can be sensitive, powerful, **and** intelligent!

Problems

2-1   Draw the input and the output of the op amp shown in Figure 2-40 for
  a. $e_{ref} = 0V$.
  b. $e_{ref} = +3V$.
  c. $e_{ref} = -4V$.
  d. $e_{ref} = 6V$.

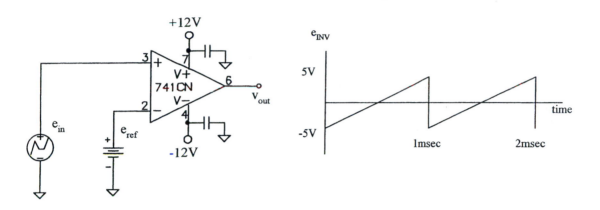

Figure 2-40 Schematic and input for Problem 2-1

2-2   Draw the input and the output of the op amp shown in Figure 2-40 for the set of inputs shown in Figure 2-41.

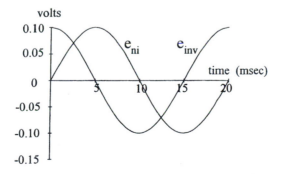

Figure 2-41 Inputs for Problem 2-2

2-3   Draw the input and the output of the op amp shown in Figure 2-40 for the set of inputs shown in Figure 2-42.

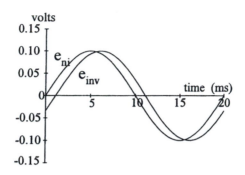

Figure 2-42 Inputs for Problem 2-3

2-4   Design an inverting op amp comparator using a 741 operating from ±15V power supplies which has a ±3V hysteresis. Select the smaller resistor to be 100Ω.

2-5   For the circuit in Figure 2-43, calculate the
   a. voltage at the op amp's noninverting input pin.
   b. input signal level necessary to cause the output to switch from +V$_{SAT}$ to -V$_{SAT}$.

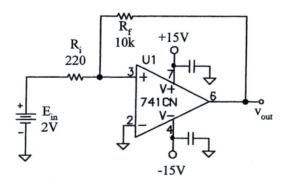

Figure 2-43 Schematic for Problem 2-5

2-6   For the circuit in Figure 2-44, calculate the
    a. pull-down resistor's value for the shortest rise time.
    b. rise time.

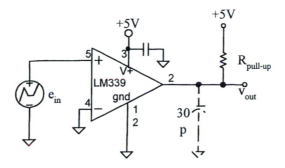

Figure 2-44 Schematic for Problem 2-6

2-7   Design a window comparator using an LM339 to meet the following specifications:
        Operates from a single +5V supply.
        Has window voltages of 2.5V and 3.2V.
        Provides a logic low when the input voltage is within the window.
        Also lights an LED when the input is within the window.
        Has a 10mV hysteresis.

2-8   Repeat Problem 2-7 using LM311s. Can you take advantage of using pin 1 for an output?

2-9   For the circuit in Figure 2-45, with a pull-up voltage of 9V and a parasitic capacitance of 20pF, select the following:
    a. the pull-up resistor's value for the shortest rise time.
    b. that rise time.

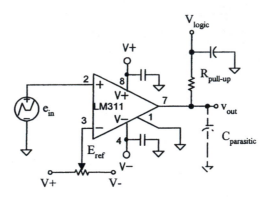

Figure 2-45 Schematic for Problem 2-9

2-10   For the circuit in Figure 2-46, what is the condition of each LED for
   a. $e_{in} = 1V$?
   b. $e_{in} = 2V$?
   c. $e_{in} = 3V$?

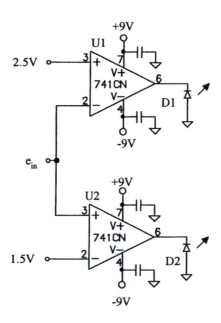

Figure 2-46 Schematic for Problem 2-10

2-11   Describe how negative feedback compensates for a decrease in open loop gain of an op amp.

2-12   What range of gains and what input impedance would you expect from the circuit in Figure 2-47 ?

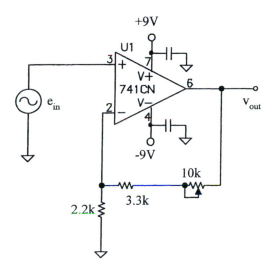

Figure 2-47 Schematic for Problem 2-12

2-13   Design an amplifier that will give a gain which can be adjusted from 1.0 to 1.2.

2-14   Design an inverting amplifier with an input impedance of precisely 10kΩ and a gain that can be adjusted between -10 and -20 without affecting the input impedance.

2-15   Draw the output of the op amp in Figure 2-48 for
    a. $e_{in\,1} = 0.5V_{PP}$ sine wave $E_{in\,2} = +2V_{DC}$
    b. $e_{in\,1} = 1V_{PP}$ sine wave $E_{in\,2} = -4V_{DC}$

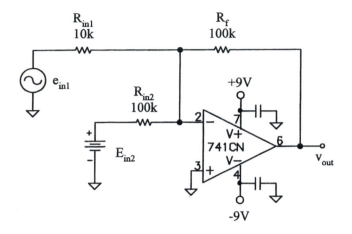

Figure 2-48 Schematic for Problem 2-15

2-16   Design a signal conditioner using an inverting summer followed by an inverting amplifier that will convert a 10mV/°F signal (from a temperature transducer) into 0V when the temperature is freezing and 5.00V when the temperature is 150°F.

a.   Tabulate at freezing and at 150°F the temperature in °F, the input voltage, and the output voltage.

b.   Calculate the change in input voltage as the temperature goes from freezing to 150°F.

c.   Calculate the gain necessary to produce the desired output change from the given input change (step b).

d.   Given that the overall circuit's equation is

$$V_{out} = -\left(\frac{R_f}{R_{i1}} e_{in1} + \frac{R_f}{R_{i2}} e_{in2} \right)$$

select $R_{i1} = 2.2k\Omega$. Calculate the required value for R_f.

e.   Select e_{in2} = -15V. At freezing, calculate the value of R_{f2} needed to produce the correct output.

f.   Verify that the circuit produces the correct results at 150°F.

2-17   What is the output voltage from the circuit in Figure 2-49 if the input signal is a 7.5mA current?

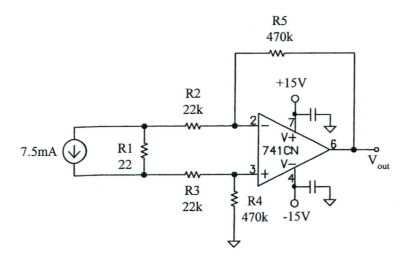

Figure 2-49 Schematic for Problem 2-17

2-18   For the circuit in Figure 2-50, the **span** potentiometer is set to 23kΩ. The
        zero potentiometer is set to give a voltage at point B of 1.26V. Calculate the
        voltages at the three other nodes (A, C, and D) given an input current of
        a.   4mA.
        b.   20mA.

2-19   A differential amplifier has R_i = 22kΩ, R_f = 330kΩ, and a CMRR of 58dB.
        Calculate the nonideal output in response to a 5V signal, common to both
        inputs.

2-20   Calculate all node voltages for the circuit in Figure 2-51.

2-21   Derive the equation for the output voltage for the circuit in Figure 2-52.

2-22   Design a circuit that will give the following output:

$$v_{out} = A_3 e_{in3} - A_2 e_{in2} - A_1 e_{in1}$$

        where A_1, A_2, and A_3 are fixed gains and
                 e_{in1}, e_{in2}, and e_{in3} are input voltages

2-23   Calculate all of the node voltages in the circuit in Figure 2-53.

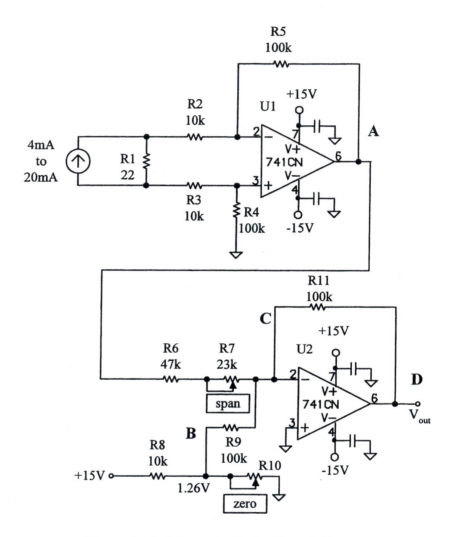

Figure 2-50 Schematic for Problem 2-18

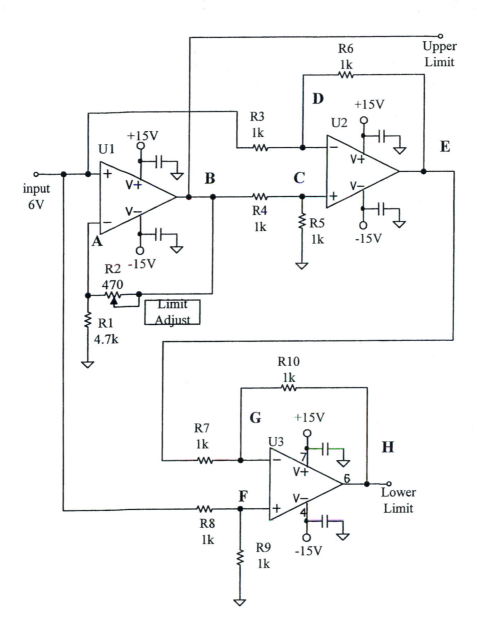

Figure 2-51 Schematic for Problem 2-20

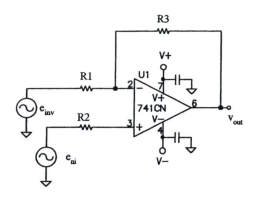

Figure 2-52 Schematic for Problem 2-21

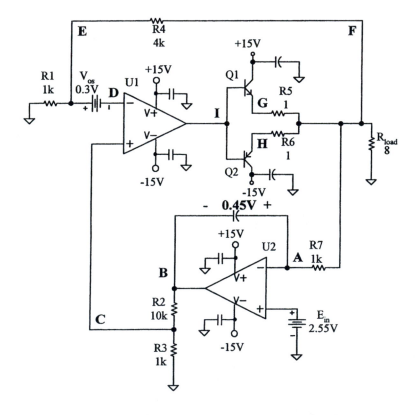

Figure 2-53 Schematic for Problem 2-23

Comparator Analysis Lab Exercise

Purpose

During this lab exercise you will examine the performance of the noninverting, the inverting and the referenced comparator circuits built with an LM741 op amp. The input to output relationships, delay time, rise time, overdrive (minimum input) voltage, and high frequency responses will be measured.

A. Noninverting Comparator

    1. Build the circuit in Figure 2-54.

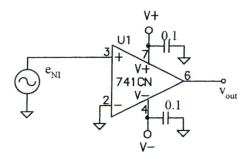

Figure 2-54 Noninverting comparator

    2. Set $\pm V$ to $\pm 15V$.

    3. Apply $e_{NI} = 1V_P$, 100Hz, sine wave with a $0V_{DC}$ level (offset).

    4. Display both the input and the output voltages on the oscilloscope.

    5. Accurately record the display, labeling the DC level, peak positive and peak negative voltages, and the phase relationship between the input and the output voltages.

    6. Measure and record the delay time and the rise time. The **delay time** is the time from when the input signal crosses ground until when the output signal rises 10%. The **rise time** is the time it takes for the output voltage to rise from the 10% point to the 90% point.

    7. Discuss the results obtained in steps A5 and A6 to those predicted by theory.

    8. Decrease the input amplitude until the output no longer enters **both** positive and negative saturation.

    9. Record this peak input amplitude. This is the **minimum** overdrive voltage.

10. Return the amplitude to $1V_P$. Increase the input's frequency to 100kHz and accurately record the output's wave shape. Be sure that the oscilloscope's channel displaying the output is DC coupled. Explain this shape.

B. Noninverting Comparator

    1. Repeat steps A1-A6 for the inverting comparator of Figure 2-55.

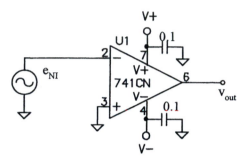

Figure 2-55 Inverting comparator

    2. Discuss the differences and similarities between the results you obtained for the noninverting and the inverting comparator.

C. Comparator with a Reference Voltage

    1. Build the circuit in Figure 2-56. Use a multiturn potentiometer with a resistance $\geq 1k\Omega$.

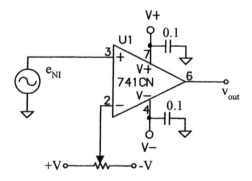

Figure 2-56 Comparator with adjustable reference

2. Apply $e_{NI} = 1V_P$, 1000Hz sine wave with a $0V_{DC}$ level (offset).

3. Adjust the reference voltage (pin 2 to ground) to $0V_{DC}$.

4. Display both the input and output voltages on the oscilloscope.

5. Accurately record the display, labeling the DC level, peak positive and peak negative voltages, and the phase relationship between the input and the output voltages.

6. Discuss the comparison of the data in step C5 to that from the simple noninverting comparator (step A5).

7. Adjust the reference voltage (pin 2 to ground) to 0.7V.

8. Repeat steps C4 and C5.

9. Adjust the reference voltage (pin 2 to ground) to -0.7V.

10. Repeat steps C4 and C5.

11. Adjust the reference voltage (pin 2 to ground) to 1.3V.

12. Repeat steps C4 and C5.

13. Discuss the comparison of output level and wave shape of the results obtained with
    a. $E_{ref} = 0V$.
    b. $E_{ref} = 0.7V$.
    c. $E_{ref} = -0.7V$.
    d. $E_{ref} = 1.2V$.

D. Comparator IC with Hysteresis
   1. Design an inverting comparator using an LM311 to meet the following:
      a. 0V or 5V output levels.
      b. 0.5V hysteresis.
      c. minimum output rise time.
      d. input level $\leq \pm7V$.

   2. Build the circuit that you designed.

   3. Apply $e_{NI} = 1V_P$, 100kHz sine wave with a $0V_{DC}$ level (offset).

   4. Display both the input and output voltages on the oscilloscope.

   5. Accurately record the display, labeling the DC level, peak positive and peak negative voltages, the phase relationship between the input and the output voltages, and the hysteresis.

6. Discuss the comparison of the data in step D5 to that from the simple noninverting comparator (step A5).

7. Measure and record the delay time and the rise time. The **delay time** is the time from when the input signal crosses ground until when the output signal rises 10%. The **rise time** is the time it takes for the output voltage to rise from the 10% point to the 90% point.

8. Discuss the results obtained in steps C7 to those of the 741 from section A.

Noninverting Amplifiers Lab Exercise

Purpose

During this lab exercise you will examine the performance of the voltage follower and the noninverting amplifiers. Input to output amplitude and phase relationships will be verified. Actual gain will be compared to the theoretical values.

A. Voltage Follower
1. Build the circuit in Figure 2-57.

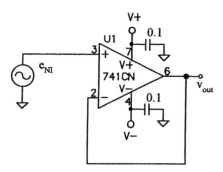

Figure 2-57 Voltage follower

2. Set ±V to ±15V. Set e_{in} to a sine wave of $0.1V_{RMS}$ 1kHz, $0V_{DC}$ level (offset).

3. Display both the input and the output on the oscilloscope (DC coupling). Record their wave shapes.

4. Measure the input and output's DC and RMS values with the digital multimeter. Record all stable digits. Its important to get **at least** three significant digits. More is better.

5. Calculate the gain (don't forget the phase).

6. Discuss the comparison between the actual gain (measured in steps A4 and A5) and the theoretical gain.

B. Noninverting Amplifier
    1. Build the circuit in Figure 2-58.

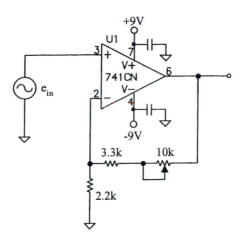

Figure 2-58 Noninverting amplifier

2. Apply an input sine wave of $0.1V_{RMS}$, 1kHz, $0V_{DC}$ level (offset).

3. Set the 10kΩ potentiometer to 10kΩ.

4. Display both the input and the output signals on the oscilloscope (DC coupled).

5. Measure the input and output's DC and RMS values with the digital multimeter. Record all stable digits. It's important to get **at least** three significant digits. More is better.

6. Calculate the gain (don't forget the phase).

7. Measure $V_{pin2\ RMS}$ and compare the results with what you should obtain.

8. Set the potentiometer to 0Ω while watching the signal. Describe the effect of step B8.

9. Repeat steps B4-B7. Discuss the comparison of the results you obtained in steps B1-B10 to the theoretical results from Problem 2-12.

3

Digital Control of Analog Functions

During the previous chapters you have seen how to adjust the amplitude (gain) and offset (DC level) of an analog signal. With careful selection of the type of op amp to be used, this can be done **very** accurately and inexpensively. By properly combining these circuits with power transistors, you have also seen how to apply considerable power to a load with equal precision.

In the electronic "body," the sensing and the muscles are analog, composed of op amps, resistors, capacitors, diodes, and power transistors. To a large degree, where electronics touches the outside world, analog is critical. However, to an ever-increasing degree, the intelligence that uses the information gathered by the analog circuits, deciding how to use the power transistor muscles, is digital.

Microprocessor control is now common in all but the simplest systems. This chapter covers the circuitry needed to allow the microprocessor to accurately and efficiently provide that control. You will see three techniques: the analog switch, a digital potentiometer, and the digital to analog converter. General considerations for each will be presented, followed by details of a specific part, and several applications.

Objectives

After studying this chapter, you should be able to do the following for the analog switch, digital potentiometer, and digital to analog converter:

1. Discuss their applications and limitations.
2. Select an appropriate model for a given application.
3. Analyze a given circuit.
4. Apply the details of both the analog and the digital sides of a specific analog switch, digital potentiometer, and digital to analog converter to give digital control of an analog circuit.

3-1   Analog Switch

The analog switch is the simplest of digital to analog converters. Like its mechanical counterpart, the analog switch opens or closes a signal path, perhaps moving that signal among a variety of connections. But instead of physically moving an arm, you alter the logic level of one or more digital control bits. In order to provide this control, the analog switch must be made of silicon. It contains a variety of digital electronics to interpret the digital input(s), level-shifting analog circuits to boost the logic level signals, and metal oxide field effect transistors (MOSFETs) to be driven hard ON or hard OFF. Remember, the switch itself is just a MOSFET being saturated or cut off. In the final analysis, the characteristics and limitations of the MOSFET set the performance of the analog switch.

Analog switches are specified as **normally open**, **N.O.**, or **normally closed**, **N.C.** That defines the condition of the switch when a logic low is applied to the digital input. When a logic high is applied, the analog switch changes state. These two types of switches are shown in Figure 3-1.

Figure 3-1 Normally open and normally closed analog switches

The second distinction that is used when describing an analog switch is the number of **poles** and the number of **throws**. **Poles** refers to the number of circuits that are being controlled, typically single or double. **Throw** describes the number of places that each of those circuits can be put, also usually either single or double. So there are four common configurations: single-pole single-throw (**SPST**), single-pole double-throw (**SPDT**), double-pole single-throw (**DPST**), and double-pole double-throw (**DPDT**). Finally the single pole switches may be normally open (SPST-NO, DPST-NO), or they may be normally closed (SPST-NC, DPST-NC). Each of these types is shown in Figure 3-2.

The silicon version of the rotary switch is called an analog multiplexer. It passes an input signal to one of eight or sixteen possible outputs. Of course, there must be several **channel select** digital control lines as well. A **one-of-eight** multiplexer is shown in Figure 3-3.

Unlike their mechanical counterparts, analog switches have several key parameters which you must understand and seriously consider. The DG211, a quad SPST-NC switch, is shown in Figure 3-4.

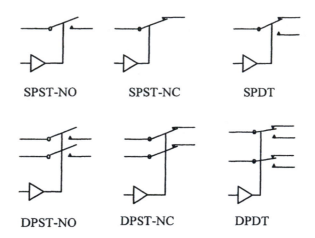

SPST-NO   SPST-NC   SPDT

DPST-NO   DPST-NC   DPDT

Figure 3-2 Types of analog switches

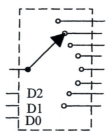

Figure 3-3 One-of-eight analog multiplexers

The supply voltages of the DG211 may range from ±4.5V to ±18V. The V_L pin should be set to the digital system's supply level, typically +5V. As with most integrated circuits, all input voltages must be less than the supply voltages. So if you know that your analog signal will be $8V_P$, then you cannot use ±5V supplies. In fact, careful review of the specifications reveals that most of the switch's tests were run with ±V = ±15V and V_{analog} = 10V. Given no other criteria, it is reasonable to select these as your normal operating parameters. At these conditions the **ON** resistance of one of the switches is less than 175Ω at 1mA. These are central to the correct use of an analog switch. When closed, it is **not** a dead short. It **cannot** pass any significant amount of current. It is a signal device, not a power IC. Similarly, in the **OFF** state, as much as

5nA of current may leak through it. In the building of precision circuits, these two parameters will set the limits on performance.

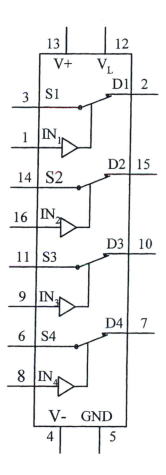

Figure 3-4 DG211 quad SPST-NC analog switch

If you expect to use the analog switch in high frequency applications, be sure to look carefully at the switch's capacitance. For the DG211, that is 5pF to ground. When *ON*, the switch looks like a 175Ω resistor between the input and the output, and a 15pF capacitor to ground. The time that it takes this switch to open or close is 1000ns or less, not exactly fast in the world of 200MHz personal computers. So, if your logic system is running at a high rate, do not forget to allow the switch 1μs to settle.

Example 3-1

Simulate the DG211 and verify that your simulation is correct for both the open and the closed case, taking into account the on resistance, the leakage current, and its logic levels.

Solution

Using *Electronics Workbench*, the voltage controlled switch is selected and set up as shown in Figure 3-5.

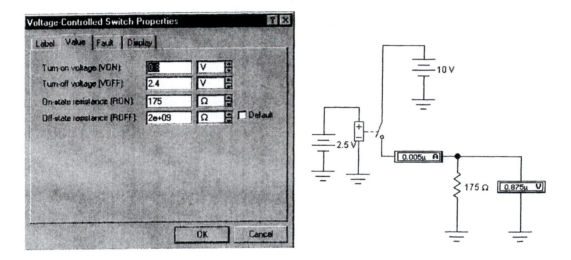

Figure 3-5 *Electronics Workbench* setup of a DG211

The **turn on** and **turn off** voltages assure that the switch is normally closed and will open when a TTL high is applied. The R_{ON} is set to 175Ω. To account for the 5nA of off state leakage current, R_{OFF} is set to

$$R_{off} = \frac{10V}{5nA} = 2G\Omega$$

Also in Figure 3-5, the control voltage is 2.5V. This opens the switch. The ammeter indicates 5nA of leakage current. This current flowing through the 175Ω load resistor produces $0.875\mu V$.

In Figure 3-6, the logic voltage has been changed to 0.7V, a logic low. The switch closes. This is shown on the schematic.

There are two major problems with this circuit. First, with an internal resistance of 175Ω, **half** of the 10V is dropped across the switch, and only half is passed to the load. Secondly, the current through the switch is

$$I = \frac{5V}{175\Omega} = 28.7mA$$

Although current this high does not present a problem for the simulated switch, actual currents greater than 1mA may produce overheating. This will force the switch's on resistance to increase, eventually causing the switch to fail.

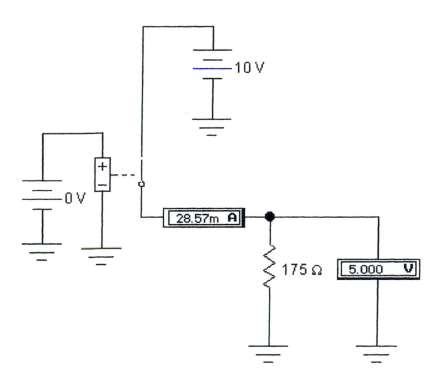

Figure 3-6 Analog switch in the ON state

Example 3-2

Design a switchable attentuator with an input impedance $\geq 1M\Omega$. When a logic low is applied to the control input, the analog output will not be attenuated ($V_{out} = E_{in}$). When the logic high is applied to the control input, the analog output will be attenuated by a factor of 10 ($V_{out} = 0.1E_{in}$).

Solution

The schematic is shown in Figure 3-7, configured for simulation with *PSpice for Windows*. When the switch is closed, R1 is shorted, $R_{analog\ in} = 1M\Omega$, and $V_{out} = E_{in}$.

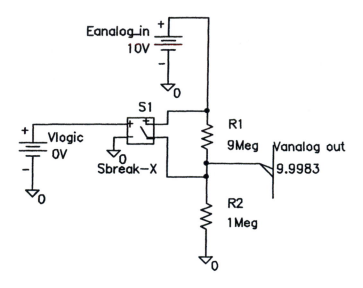

Figure 3-7 Schematic for Example 3-2 with logic low in

To properly simulate the DG211, you must change the model of the Sbreak-X. This is done by selecting the switch, then **Edit, Model, Edit Instance Model (text)**. Alter the dialog box as shown in Figure 3-8.

When the logic input is changed to 2.5V, a logic high, the switch opens. There is now a 9MΩ resistor paralleled by a 2GΩ resistor, effectively 9MΩ. So 90% of the analog voltage is dropped across that resistor, leaving only 10% across the 1MΩ, at the output. This is shown in Figure 3-9.

Model Editor

Model Name: Sbreak-X

.model Sbreak-X VSWITCH
Roff = 2e9
Ron = 175
Voff = 2.4
Von = 0
*$

Figure 3-8 Sbreak-X model setup

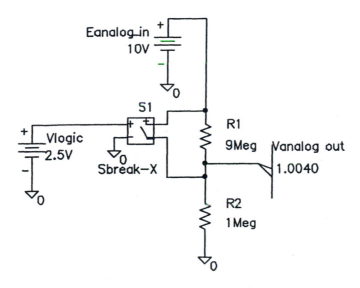

Figure 3-9 *PSpice for Windows* simulation with a logic high command

3-2 Digital Potentiometers

A digital potentiometer consists of 255 matched resistors and 256 normally open analog switches. The resistors are connected in series. There is one analog switch at each end of this series chain, and one at each of the resistors' junctions. A digital control section decides which **one** of the switches is closed. Closing the switch connects that point in the series circuit to the output terminal labeled **wiper**. Entering a different digital code connects a different point in the chain to the wiper, effectively sliding the wiper up or down on the body of the potentiometer.

Currently, there are three major manufacturers of digital potentiometers: Analog Devices, Dallas Semiconductor, and Xicor. Each offers a variety of digital potentiometers.

Several items are common among the digital potentiometers. They are all eight-bit, 256-position devices. **Log taper** is available, just as it is in mechanical potentiometers used for audio dB amplitude adjustment. For the **linear taper** potentiometers, the match between resistors is good, typically 1% variation from resistor to resistor. However, the total resistance may vary as much as ±20% from that specified. So, if you move the potentiometer from position 26 to position 52, the voltage will accurately change from 10% of the applied voltage to 20% of that applied to the potentiometer. However, when it is set to position 26, one 100kΩ IC may have a resistance between its wiper and its bottom of 8125Ω, while a different IC may have a resistance of 12,188Ω at position 26. Used as a potentiometer, these devices work well. If you need to use them as a rheostat, care must be exercised.

Each switch has a finite ON resistance. Between models this may vary widely. Some may be as low as 50Ω, while others may be as large as 1kΩ. This resistance is in series with the wiper. In a mechanical potentiometer, we usually assume that that the wiper's resistance is insignificant. That is **not** a valid assumption with digital potentiometers.

The amount of current that can flow from the wiper also varies widely among models. Some may allow as much as 20mA through the wiper, while others require that you limit that current to less than 1mA.

Most digital potentiometers are powered for ground and +5V. This coincides with the power requirements of the logic circuit. There are also some models designed to work from 3.3V supplies to match this lower voltage logic family. Since all voltages applied to an IC must be within its supply voltage range, using a single +5V (or +3.3V) supply means that the analog input to the potentiometer section must also be between 0V and +5V. The signal may **not** go negative. For many applications this may be a handicap. There are a few digital potentiometers that allow supplies of ±5V. Only these may be used for signals that go both positive and negative.

There are a wide variety of digital control schemes. Some models allow a full parallel data bus with typical microprocessor chip select and data available control signals. Others require three-wire, serial communications. This drastically reduces the traces needed to the IC but slows the communications to it.

What happens when power is removed and restored also varies widely. Some models set the wiper to a defined location when power is applied. Others have nonvolatile memory built in. When so commanded, the IC stores its wiper location. When power is cycled, the wiper is sent to that position. There are many choices of how the digital potentiometer will perform, both on the analog side and on the digital. Carefully review all of your system requirements before you begin looking at the digital potentiometer.

The DS1267 is a dual digital potentiometer IC from Dallas Semiconductor. It comes in three models. The difference among models is the total resistance, -10 (10kΩ), -50 (50kΩ), and -100(100kΩ). There are also three package styles available. The traditional 14-pin DIP requires through-hole mounting. There are also two surface mount packages. The pinout for each package is different, so be sure you know which you are dealing with. There are three power pins: V_{CC}, V_B, and GND. V_{CC} must be between +4.5V and +5.5V (with respect to GND). You may apply any voltage between $-5.5V$ and 0V to V_B. So, if you are working in a single polarity system, connect the V_B pin to ground. However, if the analog signal to the potentiometers might go negative, you can power the DS1267 from ±5V. Assure that no signals are applied to the resistors that are more positive than V_{CC} or more negative than the supply tied to V_B. With ±5V supplies, the analog signal must be constrained to ±5V as well.

The accuracy of the total resistance is ±20%. So for the DS1267-100, each individual resistor is

$$\frac{100k\Omega \pm 20\%}{256} = 312\Omega \text{ to } 469\Omega, \text{ nominally } 391\Omega$$

Within a particular IC, the match between resistors is better than 1%. But in one IC, the resistors may all be between 312Ω to 315Ω, while in a different DS1267-100, the resistors may all be between 464Ω and 469Ω. Figure 3-10 gives an overview of the DS1267.

The wiper's resistance is between 400Ω and 1000Ω. This value is dependent on the instantaneous voltage applied to the active switch. So the wiper's resistance changes with the voltage applied to the potentiometer, and with the wiper setting. It may even change with time if the voltage is varying (i.e., AC). The manufacturer requires that you limit the current through the wiper to less than 1mA. This must be done by careful consideration of the voltages and other resistances applied to the IC. Actually, since the wiper's resistance is considerable and varies quite a bit, prudent design dictates that you should severely limit wiper current (into the microampere range) to minimize the impact of this wiper resistance. Also, to minimize the impact of this wiper's resistance, choose the 100kΩ model, DS1267-100.

The analog switch inside the DS1267 controls which wiper is connected to S_{out}. This allows you to connect the two potentiometers in series to produce one 200kΩ, 512-step pot. Connecting S_{out} to the upper potentiometer allows you to run this composite wiper from the top, down to the 50% point. Then switch S_{out} to the lower pot, and you can run the composite wiper from 50% down to the bottom.

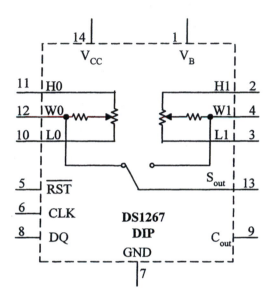

Figure 3-10 DS1267 diagram

For high frequency applications, be sure to consider the IC's input capacitance (<5pF) and output capacitance (<7pF). Because of this C-R-C configuration, choosing a DS1267-10 (R=10kΩ) gives much better high frequency performance than the DS1267-100 (R=100kΩ). The f_{-3dB} for the DS1267-10 is 1MHz. For the DS1267-100, f_{-3dB} drops to 100kHz.

When power is first applied, the wipers are sent to the 50% point. To change the position of these wipers, you must send 17 data bits to the IC. This data is clocked in using the three serial control bits, $\overline{RST}$, DQ, and CLK. The digital section is inactive while $\overline{RST}$ is low. Begin communications with both $\overline{RST}$ and CLK low. Taking $\overline{RST}$ high selects the IC. Data, setting each of the two potentiometers' wipers, is clocked in, one bit at a time, along with the S_{out} control bit. There must always be 17 data bits. Once $\overline{RST}$ has gone high and is stable (≥30ns), the first data bit is applied to DQ and allowed to stabilize (≥20ns). Then CLK is taken high for at least 50ns. The CLK signal is taken low and the next data bit may be entered. Again the CLK is taken high, to enter the second data bit, and then returned low. Continue clocking the data into the DS1267, on the rising edge of the CLK pulse, until all 17 bits have been entered. When you take the $\overline{\text{RST}}$ low, to end communications, the two potentiometer wipers and the S_{out} analog switch are all simultaneously updated to the new values just serially shifted in.

Transmit the S_{out} bit first. Follow this by the eight bits which define the position of potentiometer 1, with the most significant bit first. Finally, send the eight bits which set potentiometer 0, again, with the most significant bit first.

Example 3-3

Determine the data string needed to set the following:
a. Potentiometer 1 to 72.5kΩ between the wiper and its LO pin.
b. Potentiometer 0 to 30%.
c. S_{out} to the wiper of potentiometer 0.

Solution

$S_{out} = 0$

Potentiometer 1

$$\frac{256 \text{ steps}}{100\text{k}\Omega} = 2.56 \times 10^{-3} \frac{\text{steps}}{\Omega}$$

$$72.5\text{k}\Omega \times 2.56 \times 10^{-3} \frac{\text{steps}}{\Omega} = 185.6 \text{ steps}$$

Select $n_{pot1} = 186_{10} = 10111010_2$

Potentiometer 0

$$\frac{256 \text{ steps}}{100\%} = 2.56 \frac{\text{steps}}{\%}$$

$$30\% \times 2.56 \frac{\text{steps}}{\%} = 76.8 \text{ steps}$$

Select $n_{pot0} = 77_{10} = 01001101_2$

The final 17-bit data word is   0   10111010   01001101. Enter the data left to right, with the 0 (S_{out}) entered first.

The C_{out} bit allows you to communicate with several DS1267s using only three lines **total**. Apply the data to the DQ pin of the first IC. Then daisy chain to the next, by connecting its DQ pin to the C_{out} pin of the previous. CLK and $\overline{\text{RST}}$ are tied to all ICs. When the $\overline{\text{RST}}$ bit goes high, all DS1267s are ready to accept data. Seventeen clock pulses enter the data into the first IC's shift register. As the clock continues to pulse, these bits are shifted out of the first IC (as more bits are shifted in), appear on the C_{out} pin, and are clocked into the second DS1267. Always send a multiple of 17 bits, one set of 17 for each IC in the chain. When the $\overline{RST}$ bit finally goes low, all potentiometers in the chain are updated simultaneously.

To properly apply a digital potentiometer, you must choose a circuit configuration that will minimize both the effect of the wiper's resistance and the impact of the variation in the potentiometer's bulk resistance.

Example 3-4

Design an amplifier whose gain is digitally adjustable to a maximum of 100. What is the effect on gain of the $\pm 20\%$ potentiometer resistance variation?

Solution

One solution is shown in Figure 3-11. The op amp must be selected to work well on $\pm5V$. A 741 is **not** acceptable. (There is discussion about op amp selection in Chapter 5.) The digital potentiometer is being used as R_f in a noninverting amplifier configuration. For a gain of 100, with R_f at its maximum of 100kΩ, calculate R_i.

$$A = 1 + \frac{R_f}{R_i}$$

$$R_i = \frac{R_f}{A-1} = \frac{100k\Omega}{100-1} = 1.01k\Omega$$

Pick $R_i = 1k\Omega$. Also, remember to decouple the DS1267. The pins of the unused potentiometer are tied to ground to minimize noise pick-up. That completes the design.

But how well will this configuration perform? When the digital potentiometer is set to its maximum, the gain should be 100. At this maximum setting, R_f is 100k$\Omega \pm 20\%$ = 80kΩ to 120kΩ. So, depending on the particular IC,

$$A_{upper-smallest} = 1 + \frac{80k\Omega}{1k\Omega} = 81$$

$$A_{upper-largest} = 1 + \frac{120k\Omega}{1k\Omega} = 121$$

This is not exactly a reliable result. Secondly, when the potentiometer is set to its minimum,

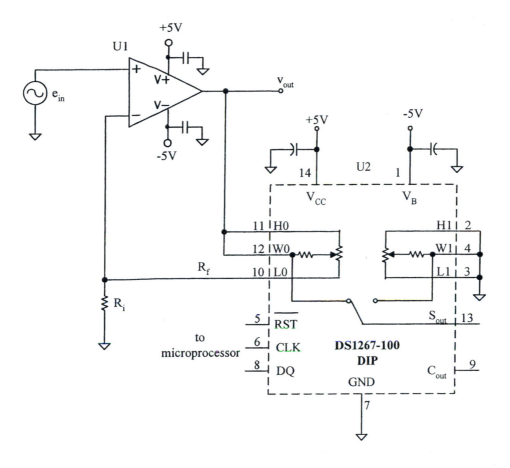

Figure 3-11 Initial solution to Example 3-4

$$R_f = \frac{100k\Omega}{256} = 390\Omega$$

But don't forget the wiper's resistance of up to 1000Ω. So at the bottom end, the gain may vary.

$$A_{lower-smallest} = 1 + \frac{390\Omega}{1k\Omega} = 1.39$$

$$A_{lower-largest} = 1 + \frac{390\Omega + 1k\Omega}{1k\Omega} = 2.39$$

This design will produce widely varying results, depending on the particular ICs, the input voltage, and the position of the wiper.

Now look at the alternate design in Figure 3-12. The potentiometer has been moved to the input. Its wiper is tied to the noninverting input of the op amp. No current flows through the wiper, so its variation in resistance will have no effect. Also, this arrangement eliminates the effect of the ±20% bulk resistance uncertainty. If the potentiometer is set at 50%, its output will be half of its input, regardless of the resistance of the potentiometer. This design provides the required gain and is independent of the variation in the DS1267's parameters.

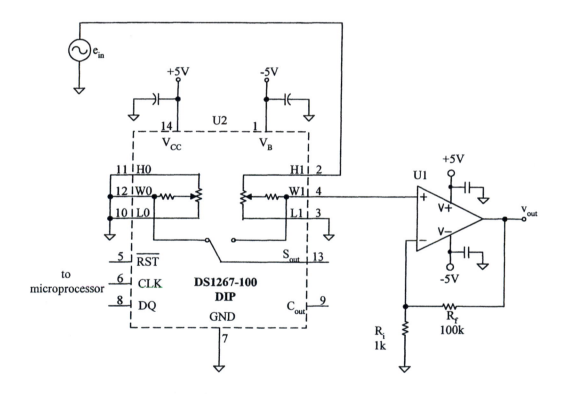

Figure 3-12 Improved solution to Example 3-4

But what if you cannot find a configuration that is independent of changes in the bulk resistance, or what if the wiper's resistance cannot be ignored? There are two ways to use the intelligence of the microprocessor controlling the potentiometer to compensate for potentiometer uncertainties. The easier is to recognize that there are many circumstances under which you really do not care what the resistance is. The microprocessor is observing some quantity that is altered by the potentiometer. If that quantity is wrong, the microprocessor moves the wiper and then looks at the effect; and moves the wiper and looks at the effect; and moves the wiper and looks at the effect; and so on.

For example, the potentiometers could be used to set the timing of two monostable multivibrators, one triggering the other. The result is a variable duty cycle, 50kHz, TTL signal. (See Problem 3-7.) In turn, this signal would drive a power MOSFET switch. The switch sets the power delivered to a motor. The motor's speed is detected by a shaft encoder, that is fed back to the microprocessor. The objective here is to control the speed of the motor. If the speed is too low, the microprocessor increases the resistance of one potentiometer and decreases the resistance of the other. This raises the time that the TTL signal is high and reduces the time that the signal is low. As a result, the power switch is on longer and more voltage is applied to the motor. The motor speeds up.

The second way to allow the microprocessor to compensate for potentiometer uncertainty is to incorporate a calibration cycle into the system's operation. In fact, this technique allows you to correct for 20% tolerance components in an entire chain of circuits and to still obtain results no worse than 1%. Remember, in a production run, the bulk resistance of the potentiometer may vary from one IC to the next by as much as $\pm20\%$. However, the step-to-step linearity of any particular IC is better than 1%. So, when the circuit is first powered up, and periodically thereafter, use an analog switch to move the input of the circuit from the actual input and tie it to ground. The output from the circuit is its **offset**. Next move the input of the circuit to a known reference. Set the potentiometer to a defined position. Again read the output of the circuit with the microprocessor. You now have two points. The circuit is linear within 1% (regardless of the initial accuracy of the components). So the microprocessor can use the $y = mx + b$ calculation to compensate for all of the uncertainties in the chain of circuits. Even with "popcorn" priced parts, you now have a circuit that is tuned to better than 1% accuracy!

3-3   Digital to Analog Converters

Digital potentiometers allow you to control resistance and, through that resistance, to set a wide variety of analog parameters: time delay, cut-off frequency, oscillation frequency, gain, offset, voltage, and current. Each of these requires circuitry in addition to the digital potentiometer. And, currently, there are only 8-bit (256-step) potentiometers available. Digital to analog converters output either current or voltage directly, with little or no additional circuitry needed. They also come in a wide variety of resolutions, from 8 bits (one part in 256) all the way up to 24 bits (one part in 16,777,216). You will even find 8 converters packaged into a single surface mount IC.

3-3.1   Overview of Digital to Analog Converters

The simplified block diagram for a generic digital to analog converter is shown in Figure 3-13. In order to output 1.000V when the digital code commands it, the digital to analog converter must know exactly how big a volt is. The reference provides that information. The more accurate the reference, the better the output can be. Some converters have the reference built inside the IC. This is handy, but allowing an external reference lets the user establish the level of accuracy needed, and that can be afforded. Also, if the converter can tolerate large swings in its external reference, or even AC, then you have a much more flexible analog part. More on that topic follows shortly.

The dividers are arranged in an R-2R ladder. Remember, the digital potentiometer needed 256 identical resistors to respond to an 8-bit digital command. With this ladder configuration, only 2n resistors are needed for n bits (16 resistors for an 8-bit command, not 256). This **greatly** simplifies the fabrication. The characteristics of this R-2R ladder are illustrated in Example 3-5.

Example 3-5
Calculate the input impedance and the output current from the circuit in Figure 3-13, given that R = 10kΩ, the voltage from the reference is 2.56V, and the inputs to the buffer are at either ground or virtual ground.

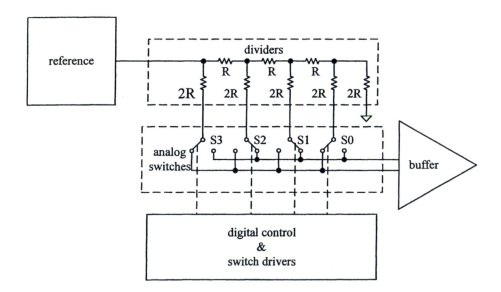

Figure 3-13 Basic digital to analog converter block diagram

Solution

Starting at S0, 2R in parallel with 2R gives R, that is in series with R, making 2R. At S1 this 2R is in parallel with 2R, giving R, which is in series with R, giving 2R going into S2. This arrangement is repeated at every switch. Regardless of the switch positions, the impedance looking into an R-2R divider network is R, 10kΩ in this example. This resistance is also independent of the number of switches and R-2R pairs.

So the current going into the divider from the reference is

$$I_{from\ reference} = \frac{2.56V}{10k\Omega} = 256\mu A$$

At the first node, above S3, the current sees 20kΩ looking down through S3 to ground or virtual ground and 20kΩ looking to the right. So the current divides at that node, sending 128μA to the lower bus and 128μA on toward switch S2.

At the node above S2, this 128μA sees 20kΩ looking down through S2 to ground or virtual ground and 20kΩ looking to the right. So the current divides again, sending 64μA to the upper bus and 64μA on toward switch S1.

At the node above S1, this 64μA sees 20kΩ looking down through S1 to ground or virtual ground and 20kΩ looking to the right. So the current divides again, sending 32μA to the upper bus and 32μA on toward switch S0.

At the node above S0, this 32μA sees 20kΩ looking down through S0 to ground or virtual ground and 20kΩ looking to the right. So the current divides again, sending 16μA to the lower bus and 16μA to ground.

At each node, the current is divided in half. The switches decide where the current goes. For this example,

$$I_{upper\ bus} = 64\mu A + 32\mu A = 96\mu A$$

$$I_{lower\ bus} = 128\mu A + 16\mu A = 144\mu A$$

In earlier models, the analog switches were made with bipolar transistors. The major consequence of this technology is that current can flow in only one direction through the switch. (For an NPN transistor, current must flow from collector to emitter; it cannot reverse.) This unidirectional limitation means that digital to analog converters made from bipolar transistors must have a reference that is unipolar, usually always positive. These ICs are also referred to as **one-quadrant,** as is shown in Figure 3-14.

Although the reference may vary, it is allowed to be only positive, keeping the current always flowing in the same direction. Changing the digital code changes the relationship between the input voltage and the output current. The larger the code, the

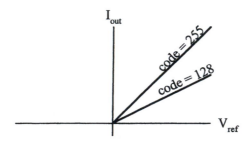

Figure 3-14 One-quadrant digital to analog converter performance

more current a particular input voltage produces. Changing the code changes the device's slope and its resistance. Keep in mind that the resistance seen by the source is always R, regardless of where the switches are set (what the input code is). But the relationship between input voltage and output current changes as the code changes.

To allow an AC input (reference) into bipolar switches, the offset, two-quadrant converter was developed. Its characteristic curve is shown in Figure 3-15. A zero and span amplifier was included within the IC. Negative input voltages were shifted up, to allow current to continue to always flow in the same direction through the bipolar switches. Although this **does** allow an AC input, it creates a major problem. Zero input (reference) voltage does not produce zero current. Worse than that, the amount of current out at 0V input changes if you change the code. This does not well duplicate the performance of a digitally controlled resistor.

For a MOS transistor, as long as the gate to source bias voltage is kept high enough, the channel is driven wide open and appears as a very low resistance. Current

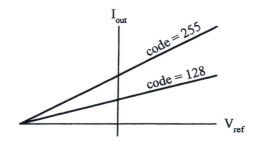

Figure 3-15 Offset two-quadrant converter characteristic curve

can flow in either direction through the channel (the switch). This technology allows true bidirectional, two-quadrant operation. Look at Figure 3-16.

     With a true two-quadrant multiplying DAC, as the input voltage passes through zero, the output current also passes through zero, regardless of the code. This is a proper emulation of a resistor. Four-quadrant multiplying converters are also available. For those, the response can be sent into the second and fourth quadrants by changing a polarity bit sent along with the code. This allows the IC to provide an inverting as well as a noninverting response.

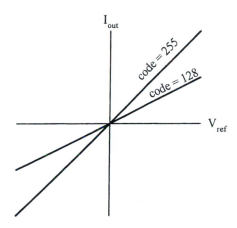

Figure 3-16 Two-quadrant multiplying converter operation

3-3.2 Current Out Digital to Analog Converter

The AD7524 is an example of an 8-bit current out, multiplying digital to analog converter. Its schematic symbol is shown in Figure 3-17. Power, V_{DD}, must be between 5V and 15V. Unlike with most ICs, you may apply an input to V_{ref} which is beyond this supply voltage. The maximum rating for V_{ref} is ±25V without damaging the IC, but it is normal to limit this voltage to ±10V.

     The R-2R dividers are 10kΩ and 20kΩ each. This means that the input impedance looking into V_{ref} is 10kΩ. Specifications indicate that this value may vary from 5kΩ to 20kΩ from one IC to the next. That is, within one IC all of the resistors may be 5kΩ and 10kΩ, while in the next IC out of the tube, all of the resistors may be 20kΩ and 40kΩ. This is much worse than the digital potentiometer of the previous section. However, the match among resistors within the same package is good enough to give 1% error or less. This bulk resistance variation **could** be a problem. Just as with

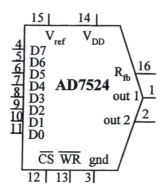

Figure 3-17 AD7524 current output multiplying DAC

the digital potentiometer, you will have to choose the circuit configuration carefully to minimize its impact; or use the converter in applications where the microprocessor can monitor the output, adjusting the converter until it gets what it wants; or implement some form of calibration cycle.

There is no output buffer in the AD7524. The two output busses from the switches go the **out1** and **out2**. For the resistors to properly divide the current, both must be at ground. Typically that means you will tie **out2** to ground and **out1** to virtual ground at the inverting input of an op amp. With this done, the current going into V_{ref} is

$$I_{in} = \frac{V_{ref}}{R}$$

The current coming out of **out1** is

$$I_{out1} = \frac{V_{ref}}{R} \times \frac{code}{256}$$

The rest of the current comes out of **out2**.

Since V_{ref} is actually an input, and the output of the convert is current, you can picture the converter as a digitally controlled resistor (that must be tied to a ground potential).

$$\frac{I_{out1}}{V_{ref}} = \frac{1}{R} \frac{code}{256}$$

$$R_{in-out} = \frac{V_{ref}}{I_{out1}} = R \frac{256}{code}$$

So depending on the code (and that particular IC's bulk resistance R), the effective resistance may be set between $\sim$1kΩ (code = 255) to 2.56MΩ (code = 1). Since **code** is in the denominator, this is a nonlinear relationship.

The code bits are loaded in parallel. When both $\overline{\text{CS}}$ and $\overline{\text{WR}}$ are low, the output currents change in response to changes on the data lines. Taking either high freezes the switches. Changes in V_{ref} will still be passed to the output current, but **code** is set.

Example 3-6

     a.   Design an amplifier, using an AD7524, with a digitally adjustable gain up to -100 and with a digitally adjustable offset that varies from -5V to +5V.

     b.   What is the gain at midrange (code = 128)?

     c.   What is the variation in this gain when accounting for the IC to IC variation in R?

Solution

     a.   Look at the schematic in Figure 3-18. The upper converter, U1, provides the gain, by playing the role of R_i.

$$A = -\frac{R_f}{R_i}$$

Maximum gain is set when the converter provides a minimum resistance. Assuming a nominal R of 10kΩ,

$$R_i = 10k\Omega \frac{256}{code}$$

$$= 10k\Omega \frac{256}{255} \approx 10k\Omega$$

So pick $$R_f = -A \times R_{i\,max} = -(-100) \times 10k\Omega = 1M\Omega$$

The lower converter, along with R_{os}, creates the offset. With +15V into U2, as its code increases, more and more current is dumped into the inverting input node of the op amp, through R_f, positive to negative. This forces the output voltage negative, so U1 drives the output down. When the code into U2 is 0, no current flows from it, and the output is at one of its DC extremes. Since increasing the code makes the output go negative, at 0A from U2, the DC output must be at +5V. This is produced by current flowing from the output, through R_f, positive to negative, through R_{os} and to -15V.

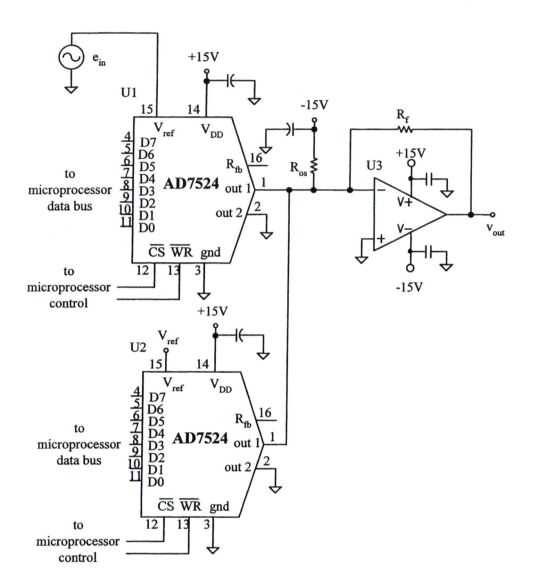

Figure 3-18 Solution to Example 3-6

To get +5V out, this offset current must be

$$I_{OS} = \frac{5V}{1M\Omega} = 5\mu A$$

This current must be sunk through R_{os}, to -15V. Remember, the op amp's inverting input is at virtual ground.

$$R_{os} = \frac{-V}{I_{os}} = \frac{-15V}{5\mu A} = 3M\Omega$$

When the code to U2 goes to 255, its output current reaches a maximum, sending current through R_f, overcoming $I_{Rf\ offset}$, and driving the output DC to -5V.

$$I_{DC\ max} = \frac{5V}{1M\Omega} = 5\mu A$$

$$I_{U2} = I_{Rf} + I_{os} = 5\mu A + 5\mu A = 10\mu A$$

To get 10μA from U2 at maximum code, its reference must be set to

$$I_{U2} = \frac{V_{ref}}{10k\Omega} \times \frac{code}{256} = \frac{V_{ref}}{10k\Omega} \times \frac{255}{256} = 99.6\frac{\mu}{V} \times V_{ref}$$

$$V_{ref} = \frac{10\mu A}{99.6\frac{\mu}{V}} = 100mV$$

b. At midrange, code = 128.

$$R_{U1} = 10k\Omega\frac{256}{code} = 10k\Omega\frac{256}{128} = 20k\Omega$$

$$gain = -\frac{R_f}{R_i} = -\frac{1M\Omega}{20k\Omega} = -50$$

It appears that there is a linear relationship between the code to U1 and the gain. Check several other points to verify this.

c. R may be as low as 5kΩ. For ICs with that internal resistance, the gain would be:

$$R_{U1} = 5k\Omega\frac{256}{code} = 5k\Omega\frac{256}{255} = 5k\Omega$$

$$gain = -\frac{R_f}{R_i} = -\frac{1M\Omega}{5k\Omega} = -200$$

Following similar calculations, for those ICs with internal resistance of 20kΩ, the maximum gain falls to −50.

This 200% variation in gain is characteristic of current out converters when used to emulate resistors. Manufacturing techniques are able to assure a close match between resistors within an IC. But there is wide variation from one production run to the next. This means that you will have to employ the same techniques you did for the digital potentiometer to compensate. Ignore the resistance and measure the **effect**, adjusting to whatever resistance produces the effect you want. Or use an initial calibration cycle to determine the internal resistance of the IC, and then employ a calculation to compensate.

3-3.3 Voltage Out Digital to Analog Converter

If your circuit design can be controlled with voltage rather than current, the effects of the bulk resistance variation within the R-2R divider can be removed. This gives a converter that is **accurate** to better than 1 LSB (least significant bit). For an 8-bit converter, that is

$$\frac{1}{2^8} \times 100\% = \frac{1}{256} \times 100\% = 0.39\%$$

For a 12-bit converter, the accuracy jumps to

$$\frac{1}{2^{12}} \times 100\% = \frac{1}{4096} \times 100\% = 0.024\%$$

This is more like it! A wide variety of voltage output, four-quadrant, multiplying, digital to analog converters are available. Look carefully at the number of converters within the package. There may be as many as eight. Also be sure to get the digital control scheme that best suits your system. Parallel and serial are both available. There are also converters with a variety of power supply requirements. These requirements need to match the logic family you are using (5V, 3.3V, 3V, …). But they also set the limits on the analog output, since you cannot get more out than you put in.

The MAX512 is a triple, 8-bit, four-quadrant, multiplying, voltage output digital to analog converter. There are three voltage output converters. The top two, A and B, share the same reference, and each has an output voltage follower buffer. Converter C has its own reference but has no buffer. The R-2R dividers within each converter have been inverted. This allows them to output a voltage to their buffers, rather than a current to virtual ground. You may look at these in more detail in the problems at the end of the chapter.

The schematic symbol for a MAX512 is given in Figure 3-19. V_{DD} is the positive supply and should be +5V. V_{SS} can be set anywhere between ground and -5V. Be sure to connect a 0.22μF capacitor **directly** from each power pin to digital ground, for supply decoupling. Also, remember that all inputs and the output must be within these supplies. So if you want AC in and out, choose ± 5V power.

RefAB, the inputs for DAC A and DAC B, may be bipolar, extending from +3.5V to -5V (**not** +5V). It must be driven from a high-quality source, with an output impedance of 5Ω or less. So you cannot drive this input directly from a voltage divider. You may have to add an op amp in front of the converter. It is important that this pin be **clean**. Add a 0.1μF to 4.7μF capacitor directly from pin 12 to analog ground. But, be aware that some op amps may not drive this capacitor well. So you will have to select the input driver carefully. The input voltage to RefC may extend from ground to +5V. This allows OutC to be driven to the + rail. But RefC cannot go negative. It is unipolar. The input impedance to RefAB and RefC varies with code, from infinity down to 8kΩ. This should not be a problem since you are going to drive these pins from a source with less than 5Ω of resistance anyway.

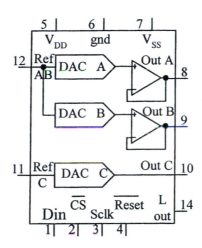

Figure 3-19 MAX512 voltage out digital to analog converter

OutA and OutB each have an output impedance of about 50Ω. Although the specifications indicate that these can output 5mA, a careful review of the graphs reveals that at currents above 10μA, the output voltage begins to sag significantly. So restrict the current load applied to these outputs if you expect to reach rated accuracy. Also, to assure stable operation, you must add a 0.1μF capacitor between OutA and analog ground. OutB needs at least a 0.01μF capacitor. OutC is not buffered. Its output impedance is about 24kΩ. So it too loads down if you try to draw much current. It needs no output capacitance.

The value at any one of the three outputs is

$$V_{out} = V_{Ref} \frac{code}{256}$$

Remember, OutA and OutB share the same input reference, and that reference may be AC (if ±5V power is supplied). OutC, like its reference, can be only positive. Pulling $\overline{Reset}$ low forces OutA and OutB to their full scale value, and OutC to zero. This also clears all other registers.

Data is clocked into the MAX512 serially. A 16-bit word is needed to set **each** register. Communication begins when $\overline{CS}$ is taken low. On every rising edge of the serial clock, SCLK, the logic level on the Data In line, Din, is shifted into the input shift register. Send the most significant bit first. Taking the $\overline{CS}$ high loads the data and command signals and updates the appropriate converter. Each converter must receive its own 16-bit word, so you cannot update all three converters simultaneously. However, with a maximum clock rate of 5MHz, it may take less than 10µs to set each one in its turn. Since the settling time for each output is 70µs, you will never notice that the converters were changed sequentially, rather than all at the same instant.

The 16 bit control and data word is structured as follows:

Q2	don't care	
Q1	1 = set L_{out}	0 = reset L_{out}
SC	1 = shut down DAC C	0 = do not shut down DAC C
SB	1 = shut down DAC B	0 = do not shut down DAC B
SA	1 = shut down DAC A	0 = do not shut down DAC A
LC	1 = DAC C is addressed	0 = DAC C is not addressed
LB	1 = DAC B is addressed	0 = DAC B is not addressed
LA	1 = DAC A is addressed	0 = DAC A is not addressed
B7-B0	data	most significant bit first

It is a good idea to shut down any of the converters that you are not currently using. It saves power and minimizes interference generated at the outputs. Since you can address only one converter at a time, only one of the **LC**, **LB**, and **LA** bits should be high. The other two should be low. **Q2** is the first bit to be presented to the IC. **B0** is the last.

When testing the converter, as with any serial logic, you must be sure that you provide a clean edge for the SCLK and the $\overline{CS}$ signals. So use debounced buttons or switches if you are not driving these pins from a microprocessor. Otherwise, the inevitable bounce that occurs when you press the button will cause several transitions, producing random results.

Example 3-7

   a. Design an amplifier, using a MAX512, with a digitally adjustable gain up to
   -100 and with a digitally adjustable offset that varies from -5V to +5V.

   b. What is the gain at midrange (code = 128)?

Solution

   This is a repeat of Example 3-6. Look at Figure 3-20. There are several
   considerations for the general layout of the circuit. Both an AC input and a DC
   offset are needed. So, DAC A is used to provide the AC control, and DAC C
   adjusts the offset. Notice all of the capacitors around the converter IC. It is
   assumed that the +5V input into DAC C is accurate to better than 1%.
   Otherwise, you will need to provide the DC input into DAC C, and into R_{os} from
   a voltage reference, not the ±5V supply. Also, it is assumed that the output is to
   be less than $5V_P$. If a larger output is required, then the op amp will need higher
   supply voltages, though the converter must still be powered from ±5V. Finally,
   it was assumed that 1% accuracy is acceptable. This is well within the abilities
   of an 8-bit converter (1 part in 256 = 0.39%). This allows the use of commonly
   available 1% resistors.

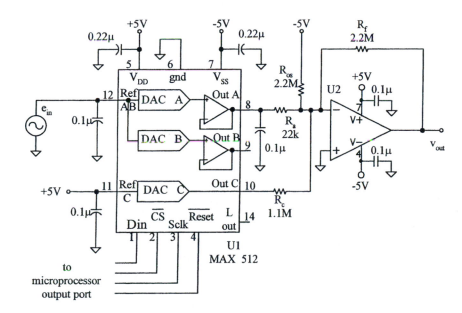

Figure 3-20 Schematic for the solution of Example 3-7

There are four resistors to pick around the inverting summer op amp. They are all interrelated and must be selected based on the circuit's specifications while keeping in mind the converter's limitations. As the output from DAC C swings from 0V to 5V, v_{out} must change from +5V to -5V. This means that there is a gain of -2 for the offset from DAC C.

$$-2 = -\frac{R_f}{R_c} \Rightarrow R_c = \frac{R_f}{2}$$

At the bottom, DAC C outputs 0V, and v_{out} must be +5V. This is caused by the -5V into R_{os}. When acting alone, this voltage has a gain of -1.

$$-1 = -\frac{R_f}{R_{os}} \Rightarrow R_{os} = R_f$$

Finally, the AC part of the circuit is to have a gain of -100.

$$-100 = -\frac{R_f}{R_i} \Rightarrow R_i = \frac{R_f}{100}$$

The maximum current that we should draw from DAC A is about 10μA, and the output impedance from DAC C may be as large as 24kΩ. So pick the resistors as large as practical. A reasonable value is

$$R_f = 2.2M\Omega$$

which leads to

$$R_{os} = 2.2M\Omega$$
$$R_c = 1.1M\Omega$$
$$R_i = 22k\Omega$$

Summary

The microprocessor or personal computer provides the intelligence of a system. The analog circuits form the muscles. Analog switches, digital potentiometers, and digital to analog converters make the connection between the two, allowing you to build a system that is both smart and able to do something with that intelligence.

Analog switches provide the simplest control. The switches may be normally open or normally closed; single pole or double pole; single throw or double throw. A digitally controlled rotary switch is called a multiplexer. The DG211 is a quad SPST-NC switch. It may be powered from ±4.5V to ±18V, which also sets the range of the analog voltages that you may apply. Don't forget to tie its V_L pin to a logic high. The

ON resistance is no more than 175Ω, not an ideal short, and it can pass only 1mA. So keep these limitations in mind when selecting an application for this analog switch.

The digital potentiometer provides much more control. It is made of 255 resistors in a voltage divider series chain, and 256 switches, one at each node. The code that you send to it selects which resistor is tied to the wiper. Both linear (normal operation) and log taper (audio) are available. Total resistance varies among models, typically in the 10kΩ to 100kΩ range. There is a tight match among resistors within an IC, providing accurate operation as a potentiometer, but between two ICs of the same model, there may be as much as 20% variation in resistance. This means there is wide uncertainty in the value of an IC when used as a rheostat. The wiper (i.e., the analog switches) has a resistance that may be as high as 1kΩ for some models. Maximum wiper current may be a low as 1mA, depending on the IC you choose. Usually, power is +5V, though there are some +3V models and some ±5V ICs available. Remember, this power not only sets the logic levels, but also limits the analog voltages that may be applied. Serial or parallel communications from the microprocessor to the digital potentiometer may be supported. Some even have built-in EEPROM, to allow the potentiometer to remember its position when power is removed.

The DG1267 comes in three bulk resistance models, allows ±5V power (and inputs), and communicates serially with its controller. Bulk resistance varies as much as 20% from one IC to the next, and wiper resistance may be as high as 1kΩ. So care must be taken to assure accurate performance. Select an application that uses the device as a potentiometer (voltage divider) rather than a rheostat (variable resistance) and that has little current flow through the wiper. Or consider the application. Perhaps you do not need to know the resistance of the potentiometer but can measure the effect as you vary it. For these applications, allow the microprocessor to monitor the effect, and vary the potentiometer until the desired effect is obtained. A final scheme is to initiate a calibration cycle periodically. Connect a known input to the potentiometer, and other electronics, and then measure the result. Knowing the setting of the pot, you can calculate its resistance.

Digital to analog converters use an R-2R ladder configuration. This allows them to require only 2n resistors for an n-bit converter, rather than the 2n resistors needed by a digital potentiometer. The converters may be voltage out or current out and may require a fixed input reference voltage. Or they may allow that voltage to vary widely (one-quadrant), or even to reverse polarity (two-quadrant or four-quadrant). Both parallel and serial communications versions are available.

The AD7524 is an 8-bit, current output, multiplying DAC, with parallel control. It requires that its outputs be at ground potential, so you must follow the AD7524 with an inverting op amp, driving the output current into virtual ground. Its bulk resistance also may vary by 20%. So the techniques described for the digital potentiometers may be needed with this IC as well.

The MAX512 is a triple, 8-bit, voltage out, multiplying DAC with serial control. Serious decoupling is needed on all power pins, inputs, and outputs to assure stable operation. Powered from ±5V, the input (and output) range is +3.5V to -5V for DAC A and DAC B and only +5V to 0V for DAC C. It does not suffer from the bulk resistance variation of digital potentiometers and current output digital to analog converters. However, the input and output impedance, its 8-bit resolution, and the accuracy of the circuitry before and after the converter must all be considered in order to produce a reliable design.

Problems

3-1   Refer to manufacturers' and distributors' catalogs, data books, or World Wide Web sites. Find one of each of the following types of analog switches. Complete a table with a row for each of the following switch types and with columns for model number (e.g., DG211), manufacturer, cost, and source of information.

SPST-NO   SPST-NC   SPDT   DPST-NO   DPST-NC

DPDT   Mux 1-of-4   Mux 1-of-8   Mux 2-of-8   Mux 2-of-16

3-2   Refer to manufacturers' and distributors' catalogs, data books, or World Wide Web sites. Submit a hard copy of the full specifications of the DG211. On those specifications, mark the following:
   a. minimum and maximum supply voltages.
   b. range of analog voltages without damage to the IC.
   c. maximum input for ±15V supplies recommended by the manufacturer to obtain rated performance.
   d. worst case ON resistance.
   e. recommended maximum current through the switch.
   f. maximum switch capacitance.
   g. current through the switch when the switch is OFF.
   h. worst case time to change states (off to on; on to off).

3-3   Repeat Problem 3-2 for the one-of-eight analog multiplexer you found in Problem 3-1.

3-4   Refer to manufacturers' and distributors' catalogs, data books, or World Wide Web sites. Obtain a hard copy of the full specifications for the Dallas Semiconductor DS1267. On those specifications, mark the following:
   a. minimum and maximum supply voltages.
   b. range of analog input voltages.

    c. range of variation in the 100kΩ body resistance between ICs. Answer in both % and Ω.

    d. range of variation among the 256 resistors within a particular IC, assuming that $R_{total} = 100$kΩ. Answer in both % and Ω.

    e. worst case wiper resistance.

    f. recommended maximum wiper current.

3-5 Draw the schematic of the DS1267-100 used as a **single** potentiometer, with the two potentiometers connected in series and the S_{out} pin correctly used as the wiper of the combined potentiometers.

3-6 For the circuit in Problem 3-5, draw a timing diagram for the three digital inputs, to properly set the cascaded potentiometer

    a. to 40% up from the bottom of the lower resistor. This is actually at 20% of the total cascaded potentiometer. Be sure to get S_{out} connected correctly.

    b. to 30% down from the top of the upper resistor. This is actually at 85% of the total cascaded potentiometer. Be sure to get S_{out} connected correctly.

3-7 Design a circuit using a DS1267 and two 555 timers, configured as one shots. Assume that the second 555 triggers the first 555, and the potentiometers in the DS1267 set the delays until the outputs of the 555s go low. Select components so that the maximum delay is 20μs. Shape the falling edge into a sharp pulse to trigger the second 555 one shot. Use the falling edge of the second 555 to trigger the first one shot. Be sure to shape that falling edge into a sharp pulse to trigger the first 555.

What code must you send to the DS1267 to

    a. produce a 50kHz signal with a 30% duty cycle out of the first 555?

    b. change the duty cycle to 85% without changing the frequency?

3-8 Between the **OUT1** pin and the R_{FB} pin of the AD7524 is a resistor whose value is closely matched to the R-2R ladder resistors. This internal resistor is used as shown in Figure 3-21. Instead of flowing out of **OUT1**, the current produced by the converter flows through R_{FB}. The op amp still keeps its inverting input pin at virtual ground. The internal resistor takes the place of an external feedback resistor, causing a voltage at v_{out}.

    a. Assuming that R is at its lowest value (10kΩ - 20%), calculate the current through R_{FB}, and v_{out}, given data of 128.

    b. Assuming that R is at its highest value (10kΩ + 20%), calculate the current through R_{FB}, and v_{out}, given data of 128.

    c. Compare the results from parts a and b. Discuss one advantage and one disadvantage of using R_{FB} rather than providing an external feedback resistor for U2.

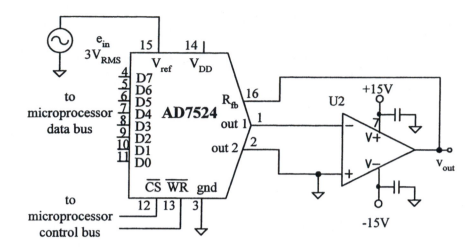

Figure 3-21 AD7524 converter using its internal feedback resistor

3-9   Refer to manufacturers' and distributors' catalogs, data books, or World Wide Web sites. Obtain a hard copy of the full specifications for the Maxim MAX512. On those specifications, mark the following:
a. minimum and maximum supply voltages.
b. range of analog input (REF) voltages.
c. range of each output voltage, assuming ±5V supplies.
d. lowest input impedance of each REF input.
e. output impedance of each output.

3-10   Figure 3-22 is an abbreviated version of the inverted R-2R ladder network used within the MAX512. Using **superposition**, calculate the output voltage.

3-11   If the RefAB input is to be driven from a voltage divider consisting of 100Ω resistors, what must you do to assure that the DAC does not load down the signal? Draw a schematic showing the source, the voltage divider, your circuit, and the DAC.

3-12   It is necessary to drive a signal into a 100Ω load. The size of the signal is to be controlled by the MAX 512, but it may be as large as 10V. Draw a schematic that includes a 1V input source, the MAX 512, your circuit, and the load.

3-13   Carefully draw the timing diagram for the three digital control signals to set OutA to 1.00V given that RefAB is 3.00V. Shut down the other two outputs.

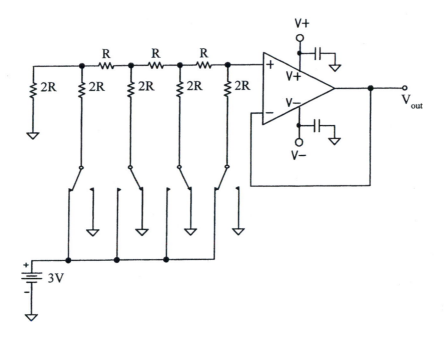

Figure 3-22 MAX512 inverted R-2R ladder network for Problem 3-10

Digital to Analog Converter Lab Exercise

Purpose

During this lab exercise you will design and evaluate the performance of a multiplying digital to analog converter combined with an external amplifier. Accuracy and frequency response of both the converter and the external amplifier will be evaluated.

A. Initial Setup

    1. Design a circuit using a MAX512 and an op amp (with appropriate feedback resistors) to input a sine wave of $1.000V_{RMS}$, $0V_{DC}$, 1kHz, and to output a signal that is adjustable in $20mV_{RMS}$ steps to $5.12V_{RMS}$.

    2. Have the lab instructor verify your design.

    3. Build the circuit.

4. Apply $\pm V_{supply}$ to the converter and to the op amp. (These may not be the same value. Think this through carefully.)

5. Apply the input signal, and verify that it is correct.

6. Pull $\overline{RST}$ low. What should happen to the output of the converter? What should happen to the output of the amplifier? Display each on the oscilloscope. When they are undistorted sine waves, measure and record their amplitudes.

7. Drive the Din, $\overline{CS}$, and SCLK lines to shut down DAC C and DAC B and to drive DAC A to half scale. Remember to use a debounced signal for SCLK and $\overline{CS}$.

8. Verify that the OutA is an undistorted sine wave of $0.500V_{RMS}$, and that the output of the amplifier is also undistorted and is $2.56V_{RMS}$. You will need to measure the amplitudes on the digital multimeter to get adequate accuracy and resolution.

Do _not_ continue until your circuit is working correctly.

B. Circuit Accuracy
   1. Calculate the codes needed to obtain the AC outputs listed in Table 3-1 from the **circuit.** Enter these into the table now.

   2. Leave the input signal at $1.000V_{RMS}$. Be sure to monitor this input with the digital multimeter and readjust it as needed. Complete Table 3-1. Make all measurements with the digital multimeter, but monitor the wave shape at the output of both the converter and the op amp with the oscilloscope to assure that the signal is not distorted.

Table 3-1 Converter and amplifier accuracy

$V_{out\ desired}$ V_{RMS}	Code 16-bit	$V_{out\ DAC}$ V_{RMS}	$Error_{DAC}$ V_{RMS}	$V_{out\ DAC}$ V_{DC}	$V_{out\ amp}$ V_{RMS}	$Error_{amp}$ V_{RMS}	$V_{out\ amp}$ V_{DC}
0.000							
1.000							
2.000							
3.000							
3.020							
4.000							
5.000							

C. Frequency Response
  1. Determine the frequency response of your circuit, both converter and amplifier, by altering the input frequency as indicated in Table 3-2. Since the frequencies will exceed the range of your digital multimeter, make the measurements with the oscilloscope. Be sure that the input is adjusted to be constant.

Table 3-2 Converter and amplifier frequency response

Frequency Hz	$V_{out\ converter}$ V_{RMS}	$V_{out\ amp}$ V_{RMS}
100		
300		
1000		
3000		
10,000		
30,000		
100,000		
300,000		

  2. Accurately locate the frequency at which the output of the converter has fallen to 0.707 of its value at 100Hz. This is the converter's high frequency cut-off.

  3. Accurately locate the frequency at which the output of the entire circuit has fallen to 0.707 of its value at 100Hz. This is the entire circuit's high frequency cut-off.

D. Analysis and Conclusions
  1. List possible causes of any errors (in the converter and/or the amplifier) in excess of $50mV_{RMS}$. Suggest ways to minimize the effect of each cause you listed.

  2. List possible causes of V_{DC} other than 0V (in the converter and/or the amplifier). Suggest ways to minimize the effect of each cause you listed.

  3. Create a semilog graph of the frequency response of the converter and of the overall circuit. Discuss causes of the roll-off.

4

Power Supplies and Integrated Circuit Regulators

The single requirement common to all phases of electronics is the need for a supply of DC power. With increases in the specifications, performance, and level of complexity of the circuits in virtually every field of electronics has come more demand on the power supplies. It is unusual to find a system today operating from an unregulated power supply. Battery power is common with run times stretching from days to months. To meet the demand for stable, protected voltages over a wide range of load current demands, input voltages, temperatures, and efficiency, analog integrated circuits have been widely incorporated into regulated power supplies. In this chapter you will find the characteristics, schematics, analysis, design, and limitations of practical power supplies with integrated circuit regulators. Discrete, op amp based, three-terminal, switched capacitor, and switching regulators will all be covered.

Objectives

After studying this chapter, you should be able to do the following:
1. Draw the schematic and explain the operation of a power supply using an op amp based regulator.
2. Draw the schematic of a power supply using a three-terminal regulator.
3. Given the required performance specifications, calculate all required component values needed to build a three-terminal regulated power supply.
4. Draw the schematics and describe the operation and precautions of each of the following variations on the basic three-terminal regulator:
   a. dual power supplies.
   b. adjustable output voltage.
5. Explain the difference between a dual and a dual tracking regulator.
6. Draw the schematic and explain the operation, applications, and limitations of a switched capacitor voltage inverter.
7. Explain the principle of a pulse width modulated switching power supply and why it is more efficient than a linear regulator.
8. Discuss the nonideal characteristics of the switching regulator's diode, inductor, capacitor, transistor, and pulse width modulator IC.
9. For a given application select the appropriate topology, IC, and other components, calculating all component values, including all currents, voltages, and power dissipation, within the circuit.

4-1   Full Wave Rectified, Capacitive Filtered Power Supplies

The purpose of a DC power supply is to convert the widely available AC power to DC, normally for bias power. This is done in several stages, as shown in Figure 4-1.

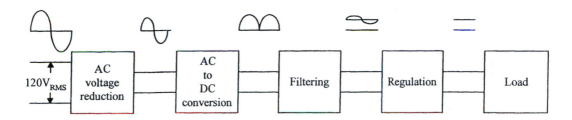

Figure 4-1 DC power supply block diagram

The first block, AC reduction, is often handled with a step-down power transformer. Here the $120V_{RMS}$, AC input is reduced to potentials in the range of $1.5V_{RMS}$ to $28V_{RMS}$. The AC to DC conversion is most efficiently done with a full wave bridge rectifier. A wide variety of epoxy bridges are available, with the four diodes already interwired in a single, easy-to-use, inexpensive package. The output of the bridge is indeed DC. Current always flows in the same direction. However, the output voltage from the rectifier varies drastically in every cycle. To smooth out this variation, a filter is needed. Although many complex filter configurations can be built, a simple, single shunt capacitor works very nicely for most applications (especially when followed by an IC regulator). The characteristics of these first three blocks of a regulated power supply will be discussed in this section. You will see the relationships between circuit performance and load current under light and heavy load.

4-1.1   Light Load Capacitive Filtering

The schematic for the first three blocks of a power supply is shown in Figure 4-2.
        As the incoming voltage from the rectifier rises, the capacitor charges to the peak of that voltage.

$$v_P = v_{sec\,RMS} \times 1.414 - 1.4V$$

The $-1.4V$ term accounts for losses across the diodes. When the rectified signal passes peak and begins to fall, the diodes turn off, and the capacitor begins to discharge slowly through the load. This continues until the incoming wave begins to recharge the capacitor. The minimum level to which the capacitor's voltage falls is called v_{min} and depends on the size of the capacitor, v_P, and the load current. The distance between the peak and the minimum output voltage is the peak-to-peak ripple voltage.

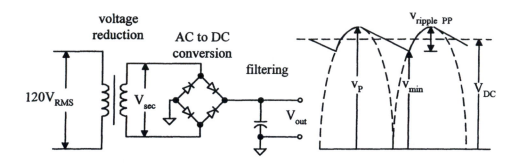

Figure 4-2 Unregulated DC power supply schematic and output waveform

$$V_{ripple\,PP} = v_P - v_{min}$$

Light load conditions cause this ripple voltage to be less than 10% of the output DC voltage. Under this condition, the capacitor discharges slightly, and almost linearly. This approximation greatly simplifies the analysis of the circuit. For **light loads**,

$$v_{min} = v_P - \frac{I_{DC}}{fC}$$

where

    I_{DC} = the load current
    f = 120 Hz for a full wave rectifier
    C = the filter's capacitance

The DC value of the output from the filter lies halfway between v_P and v_{min}, or

$$V_{DC} = v_P - \tfrac{1}{2}v_{ripple\,PP}$$

$$V_{DC} = \frac{v_P + v_{min}}{2}$$

With this information, you can now analyze the operation of a capacitively filtered power supply under light load.

Example 4-1

For the circuit in Figure 4-2, with

$$V_{secondary} = 12.6V_{RMS}$$
$$C = 1000\mu F$$
$$I_{load\ DC} = 100mA$$

assuming light load, calculate the following voltages from the filter:

a. V_P
b. V_{min}
c. $V_{ripple\ PP}$
d. V_{DC}

Solution

a.
$$v_P = 1.414 \times v_{sec\ RMS} - 1.4V$$

$$= 1.414 \times 12.6V_{RMS} - 1.4V$$

$$= 16.4V_P$$

b. The minimum voltage depends on the peak voltage, the load current, and the capacitance.

$$v_{min} = v_P - \frac{I_{DC}}{fC}$$

$$= 16.4V_P - \frac{0.1A}{120Hz \times 1000\mu F}$$

$$= 15.6V$$

c. The peak-to-peak ripple voltage is the difference between the peak voltage and the minimum voltage.

$$v_{ripple\ PP} = v_P - v_{min}$$

$$= 16.4V_P - 15.6V$$

$$= 0.8V_{DC}$$

d. The average output voltage, the DC value of the output, lies halfway between the peak and the minimum.

$$V_{DC} = \frac{v_P + v_{min}}{2}$$

$$= \frac{16.4 V_P + 15.6 V}{2}$$

$$= 16.0 V_{DC}$$

4-1.2　Rectifier Surge Current

When power is first applied, the filter capacitor is fully discharged. It presents a short circuit load to the rectifiers. Over the first cycle, as the capacitor charges, there will be a surge of current through the rectifiers. This surge is nonrepetitive, occurring only once each time the supply is turned on. Its value is

$$I_{surge} = \frac{v_P}{R_s}$$

The series limiting resistance, R_s, consists of

$$R_s = R_{transformer\ secondary} + R_{diode} + ESR_{capacitor}$$

This series resistance forms a charging time constant with the filter capacitor of

$$\tau = R_s \times C_{filter}$$

To assure that the rectifiers are not damaged, verify that

1. the diodes' ratings are adequate.

$$I_{FSM} > I_{surge}$$

2. the capacitor can acquire significant charge during the first cycle.

$$\tau < 8.3 ms$$

Ratings for typical diodes and bridges are listed in Table 4-1.

This surge occurs when the power supply is first turned on; it is nonrepetitive and destructive if you have undersized the diode. So, simply hooking up a circuit and trying to measure the surge current at start-up is a bad idea. Instead, simulate the circuit. Do not forget to include resistors in the simulation for $R_{transformer\ secondary}$ and $ESR_{capacitor}$.

Table 4-1 Diode and bridge rectifier forward and surge current limits *(Copyright of Motorola, used by permission)*

$I_{forward\,(DC)}$ A	I_{FSM} A	Series
	Diodes	
1	30	1N4000
3	100	MR500
6	400	MR750
20	400	MR2000S
25	600	MR2500S
	Bridges	
1	45	MDA100A
2	60	MDA200
4	100	MDA970
35	400	MDA3500

Should the simulation show a surge current that is larger than your diode can tolerate, add a single resistor in series with the rectifier. Size it to drop the surge current below the diode's I_{FSM}. Once the capacitor has charged to steady state, this resistor should be removed automatically. This can be done with a timer and a relay. When power is first applied, the timer delays for a fraction of a second, then energizes. Its normally open contacts are across the resistor, shorting out the resistor when they close. Or, you could use a thermistor instead of the resistor. Most thermistors have a negative temperature coefficient. When power is first applied, the thermistor is cold and has a large resistance. As it warms up, its resistance falls.

4-1.3   Heavy Load Capacitive Filtering

To continue operating in a light load condition as load currents increase, capacitance must be increased, and the regulator IC (to be added) will have to dissipate more power. This results in power supplies that are bulkier and more expensive than necessary. However, if you allow the ripple voltage to increase beyond 10%, the capacitor can be made smaller, and the required IC heat dissipation falls. Unfortunately, for such high ripple conditions, the light load equations of the previous section no longer accurately calculate v_{min}.

A rigorous mathematical analysis of the capacitively filtered power supply under high ripple (heavy load), driving a regulator, requires the solution of a linear function on the discharge side of the equation and a sinusoidal function on the charge side. Classically, these types of equations are unsolvable. Also, that does not account for the nonideal characteristics of the diodes. However, simulation produces results easily.

Example 4-2

For the circuit in Figure 4-3, display the voltage across the capacitor and across the load. Determine v_{min} of each. The components to the right of the dotted line form a simple regulator, establishing 1.0V across the load resistor. This is independent of the input voltage from the capacitor or the size of the resistor, across a wide range of input voltages and load currents. Since most capacitive filters drive regulator circuits of some form, this is a more realistic way to load the filter than with a simple resistor.

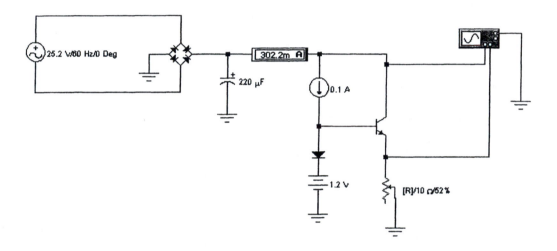

Figure 4-3 Schematic for Example 4-2

Solution

Computer simulations using *Electronics Workbench* were run. The schematic in Figure 4-3 was captured from the screen of that program. For a given transformer voltage and capacitor value, the potentiometer was adjusted until the DC current displayed on the ammeter was at 0.1A. (The value of the potentiometer can be increased by pressing the capital **R** key. Pressing the lower-case **r** key lowers the value of the potentiometer.) When the current was stable at the desired value, the simulation was paused. By double clicking on the oscilloscope icon, its face was enlarged. Its controls were then manipulated and its cursor moved until it lined up with a minimum. The result is shown in Figure

4-4. Channel A is across the capacitor, and channel B is across the load. The minimum voltage across the capacitor is 25.6V. Channel B indicates a steady 1.0V across the load, so the regulator is working properly.

Figures 4-5, 4-6, 4-7, and 4-8 were produced using this approach for each of four transformers, five different capacitor values for each transformer, and currents at 0.1A increments up to 1.5A for each capacitor. The results were plotted, and then judiciously smoothed.

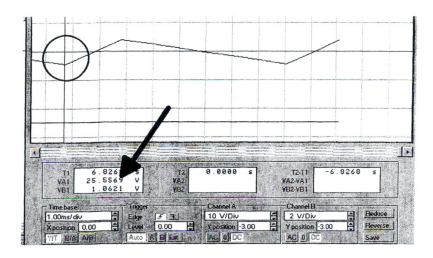

Figure 4-4 *Electronics Workbench* oscilloscope display

Using Figure 4-4, a circuit with a $25.2V_{RMS}$ transformer secondary, a 220μF capacitor, and 300mA load has a minimum voltage of 25.6V. However, using the light load equations

$$V_{min} = v_{pk} - \frac{I_{DC}}{fC}$$

$$= \left(25.2V_{RMS}\sqrt{2} - 1.4V\right) - \frac{0.3A}{120Hz \times 220\mu F}$$

$$= 22.9V$$

This is over 10% error. As the loads rise to more realistic values of a half ampere or more, this error becomes far too large to justify the simple light load equation's use in most practical cases.

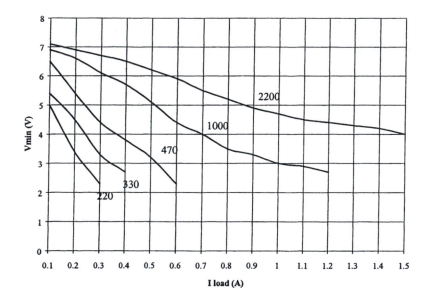

Figure 4-5 Minimum voltage from a $6.3V_{RMS}$ secondary

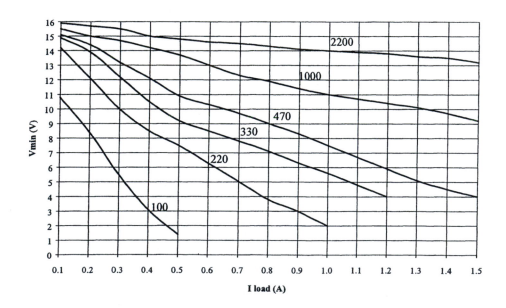

Figure 4-6 Minimum voltage from a $12.6V_{RMS}$ secondary

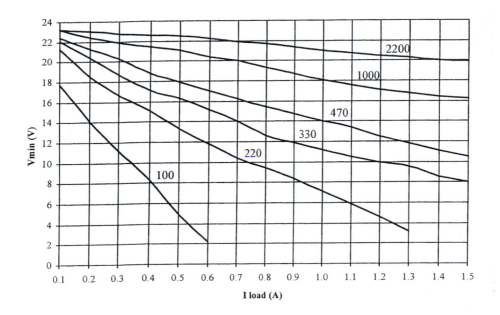

Figure 4-7 Minimum voltage from an $18V_{RMS}$ secondary

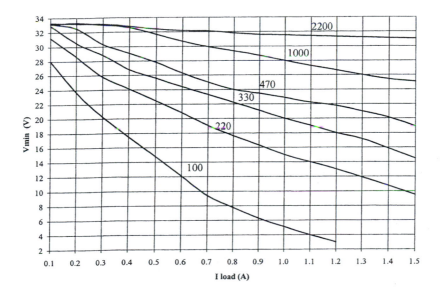

Figure 4-8 Minimum voltage from a $25.2V_{RMS}$ secondary

4-2 Simple Op Amp Regulator

You can see from the heavy load curves that the output voltage from a simple capacitive filtered power supply varies quite a bit as you change the current going to the load. In most applications this could cause major problems. The power supply must maintain a constant output voltage, regardless of how much load current is required. This calls for a regulator.

The simplest regulator consists of a series resistor and a zener diode in parallel with the load, as shown in Figure 4-9. When biased into the reverse breakdown region, the voltage across the zener diode (and therefore across the load) varies only a small amount even with relatively large changes in zener and load currents.

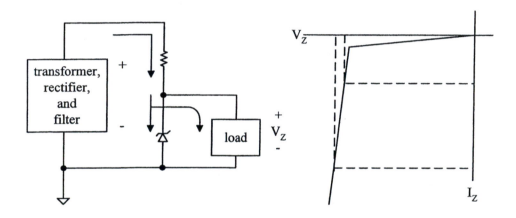

Figure 4-9 Zener regulator and characteristics

For simple applications this may be adequate, but often the change in zener voltage (and therefore load voltage) is too large. Better regulation is required. Look at Figure 4-10. A noninverting amplifier buffers the voltage from the zener diode. The op amp places a negligible load on the zener, allowing it to operate at a **single, fixed** point. This results in a very stable output from the zener diode. Also, the zener no longer must absorb large swings in current. So you may replace it with a low power, highly accurate reference diode. To alter the output voltage, you simply adjust R_f.

$$V_{load} = V_z\left(1 + \frac{R_f}{R_i}\right)$$

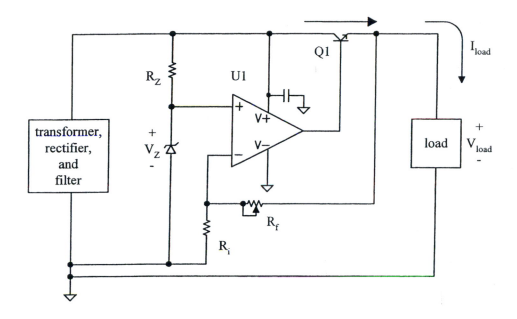

Figure 4-10 Op amp buffered, pass transistor boosted regulator

To bias the transistor, the op amp's output voltage must exceed this load voltage by about 0.7V.

$$V_{\text{out op amp}} = V_{\text{load}} + 0.7V$$

Two words of caution about the input voltage from the filter. First, its peak value must be kept below the maximum supply voltage for the op amp. (Op amps with rather high supply voltage ratings are available.) Second, the voltage from the filter must always be large enough to keep the op amp out of saturation. Usually this is about 2V above the output (load) voltage. This is also necessary to keep the zener diode in its breakdown region.

$$v_P \leq V_{\text{max supply op amp}}$$

$$v_{\text{min}} \geq V_{\text{load}} + 2V$$

Also, notice that the minimum output voltage is V_z (when R_f is shorted). So be sure to pick a zener voltage smaller than the lowest output that you want. Finally, be careful to make R_f the potentiometer. Making R_i a potentiometer and running the wiper all the way down to ground will remove all negative feedback. The output voltage will be driven to $+V_{SAT}$ of the op amp.

The op amp only has to provide the base current. This is then multiplied by β of the transistor to produce the load current. The β of the transistor is determined by

$$\beta \geq \frac{I_{load}}{I_{out\,op\,amp}}$$

The transistor must be able to dissipate the power provided by the supply but not delivered to the load.

$$P_Q = (V_{DC} - V_{load})I_{load}$$

where V_{DC} is the average voltage output from the filter.

Example 4-3

Design a regulated power supply to meet the following specifications. Use a $25.2V_{RMS}$ transformer, a full wave bridge rectifier, and a 220µF filter capacitor.

V_{load}: adjustable 5V to 15V.
I_{load}: 0mA to 500mA.

Solution

1. Filter output characteristics

$$v_P = 1.414 \times 25.2V_{RMS} - 1.4V = 34.2V$$

$$v_{min} = 22.5V \quad \text{(See Figure 4-8.)}$$

$$V_{ripple\,PP} = v_P - v_{min}$$

$$V_{DC} = \frac{v_P + v_{min}}{2} = 28.4V$$

2. Regulator design

a. Since the zener diode determines the minimum output voltage, select a 1N748. It has a zener voltage of 3.9V at 20mA. This will allow adjustment of R_f to bring the output voltage below the minimum specified.

The resistor, R_s, sets the current into the zener. It must be small enough to allow adequate current to keep the zener in its breakdown region, but large enough to keep the diode below the maximum of 100mA.

When the input from the filter is at v_{min}, minimum zener current will flow.

$$R_S \leq \frac{22.5V - 3.9V}{20mA} = 930\Omega$$

When the input from the filter is at v_P, maximum zener current will flow. This must be limited to less than 100mA.

$$R_S > \frac{34.2V - 3.9V}{100mA} = 303\Omega$$

Since $303\Omega < R_s < 930\Omega$, pick $R_S = 680\Omega$.

b. Op amp selection
The peak voltage from the filter is

$$v_P = 34.2V_P$$

which must be less than the op amp's maximum supply voltage. For the 741C the maximum supply voltage is 36V, so that op amp may be used.

c. Gain design
With an input of 3.9V from the zener diode, and with a maximum output of 15V specified, the maximum gain is

$$A = \frac{15V}{3.9V} = 1 + \frac{R_f}{R_i}$$

Solving for R_f gives

$$R_f = (3.85 - 1) \times R_i$$

Picking $R_i = 3.3k\Omega$ gives $R_f = 9.4k\Omega$. So use a 10kΩ potentiometer for R_f.

d. Transistor specifications
The 741C op amp will output at least 2mA before it begins to limit current. This is the transistor's base current. The transistor must be able to output 500mA:

$$\beta_{min} = \frac{500mA}{2mA} = 250$$

The worst case power dissipation for the transistor occurs when the output voltage has been dropped to its minimum level (because this puts most of the voltage across the transistor) and the maximum load current is being demanded.

$$P_Q = (V_{DC} - V_{load}) \times I_{load} = (28.4V - 3.9V) \times 0.5A$$

$$P_Q = 12.3W$$

Select an NPN transistor that has a β of 250 or more (this may require a Darlington configuration) and that has a power dissipating ability of 15W.

4-3 Three-Terminal Regulators

Many integrated circuit voltage regulators are available. Because of its simplicity and easy applications, the three-terminal regulator is very popular. The schematic symbol of a three-terminal regulator is shown in Figure 4-11.

These units are designed to output a fixed voltage at currents normally at or below 1.5A. The ICs find major applications as on-card regulators. This means that an unregulated power supply voltage is provided to each printed circuit board in a system, and a three-terminal regulator on each printed circuit board regulates the voltage on that card. This distribution of regulators throughout the equipment has several advantages over a single, large, centralized voltage regulator. Several three-terminal regulators each outputting 1A are cheaper to build, or buy, than one regulator outputting several amperes. The current from a centralized regulator must flow through a significant amount of series resistance and inductance to reach the board that uses it. This seriously affects the value of the load voltage, as illustrated in Figure 4-12.

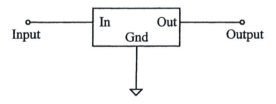

Figure 4-11 Three-terminal regulator IC schematic symbol

$$V_{load} = 10V - I_{load} \times 1\Omega$$

For $I_{load} = 1A$,

$$V_{load} = 9V$$

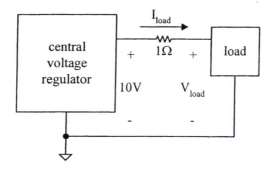

Figure 4-12 Load of high current central voltage regulator

For I_{load} = 5A,

$$V_{load} = 5V$$

Also, any change in one load connected to the centralized voltage regulator will change the input voltage to several other physically close loads. This is referred to as **coupling** and is highly undesirable. A failure on one card brings down the entire system, and deciding which board has failed may be tedious. The on-card, three-terminal regulator eliminates all of these problems.

4-3.1 Characteristics

Five characteristics of the three-terminal regulators are of key importance.

1. V_{out}

     The regulated output voltage for three-terminal regulators is fixed at the value specified by the manufacturer for the particular model you are using. Should a different output voltage be required, either use a different model, or add a voltage divider network, to be discussed in a following section. Even three-terminal regulators labeled as **adjustable** regulate at a fixed voltage, usually 1.2V. You then add a voltage divider to "amplify" this 1.2V, just as you did with the zener voltage in the op amp regulator.

2. $V_{dropout}$

     The unregulated input voltage must, at all times, be larger than the regulated output voltage. The amount that the input must exceed the regulated output is called the dropout voltage. For the older ICs this is typically 2V, but many new configurations advertise **low dropout**, as low as 0.2V. For example:

$$\left| e_{in\,min} \right| \geq \left| V_{reg\,out} \right| + V_{drop\,out}$$

$$V_{reg\,out} = 5V \qquad\qquad V_{dropout} = 2V \qquad\qquad e_{in\,min} = 7V$$

$$V_{reg\,out} = -5V \qquad\qquad V_{drop\,out} = 0.2V \qquad\qquad e_{in\,min} = -5.2V$$

The low dropout regulators allow the input ripple to drop closer to the regulated output. This allows you to select a smaller filter capacitor or, if the input is from a battery, to discharge quite a bit more before the regulator quits working.

3. $I_{out\,max}$

The output or load current may vary anywhere from zero to this rated maximum. This is the **short circuit** current limit. Any demand in excess of this short circuit limit causes the output voltage to drop as the current is held constant. Be sure you select an IC whose $I_{out\,max}$ is only a little above the maximum required load current. This protects circuit components should a short circuit occur. Picking a 1.5A rated IC when you need only 400mA means that should a failure occur, 1.5A will flow through the connectors, the printed circuit board traces, the failed component, and anything else in series with that path. Much collateral damage may occur. Selecting a 500mA rated regulator limits failure current to only 25% above that used during normal operations. No other parts may be damaged, and the repair should be easy to find and complete.

However, if adequate provisions are not made to insure the removal of heat from the regulator package, the unit will go into thermal shutdown at load currents much below $I_{out\,max}$. At what output current thermal shutdown occurs depends on the input voltage, output voltage, ambient temperature, output current, and heat sinking. The IC can provide rated current **only** if adequate heat sinking is used.

4. $T_{J\,max}$

The IC has a temperature sensor built in. When the wafer becomes too hot (usually 125°C to 150°C), the unit will turn off. The output current and voltage drop and remain low until the IC cools significantly. The regulator then turns back on. Under a continuous overload, the IC may cycle on and off continuously.

5. Ripple rejection dB

The internal negative feedback of the regulator allows it to hold the output constant, even with large swings in input voltage. This is termed **ripple rejection** and is expressed in dB. Typical values are 50dB to 80dB depending on the model and load voltage. The variation in the regulated output, caused by input ripple is

$$V_{\text{ripple out}} = V_{\text{ripple in}} \times 10^{-\frac{dB - 20\log V_{\text{out}}}{20}}$$

Example 4-4

For an input ripple of $12V_{PP}$ (a big value), an output of $5V_{DC}$, and 50dB ripple rejection, calculate the ripple at the load.

Solution

$$V_{\text{ripple out}} = 12V_{PP} \times 10^{-\frac{50 - 20\log 5V}{20}}$$

$$= 12V_{PP} \times 10^{-1.8}$$

$$= 0.19V_{PP}$$

So the point is that the IC will reduce the input variation from the capacitor. There is no need to select a large filter capacitor to reduce the voltage variation at the load. The IC's ripple rejection will do that for you. Instead, pick the filter capacitor to be sure that the regulator **always** has adequate input voltage.

$$\left| e_{\text{in min}} \right| \geq \left| V_{\text{reg out}} \right| + V_{\text{dropout}}$$

4-3.2 Regulator Circuit Components

What, then, is necessary to use a three-terminal voltage regulator? First you must insure that the input voltage exceeds the regulated output voltage by V_{dropout} at the rated maximum load current. However, the input voltage may not exceed the rated maximum input voltage under no-load conditions.

$$\left(V_{\text{out}} + V_{\text{dropout}} \right) < e_{\text{in actual}} \Big|_{I_{\text{load max}}} < e_{\text{in max rated}} \Big|_{I_{\text{load}} = 0}$$

Second, input and output capacitors may be needed in addition to the normal power supply filter capacitors. If the regulator is more than 5cm (2 in), as the electron flows from the main filter capacitor, a capacitor must be placed next to the regulator and connected between its input and ground. It should be at least a 0.22µF ceramic or film capacitor, a 2µF solid tantalum capacitor, or a 25µF aluminum electrolytic capacitor. Also, a 0.1µF ceramic or film should be located next to the regulator and connected between the output and ground. This will improve high frequency and transient (pulse) operation. Negative regulators require an additional 1µF tantalum or 25µF electrolytic capacitor at the output.

The final requirement to have a working three-terminal IC regulator is a heat sink. Although in many applications a heat sink may not be necessary, it is always wise to verify that with a calculation.

4-3.3 Heat Sinks

Any semiconductor device that carries current will dissipate power. That power is normally dissipated in the form of heat. The heat is generated at the wafer (junction) and must flow through the package to the surrounding air. The packaging material presents a certain opposition to this flow of heat (thermal resistance, Θ_{JA}). The temperature at the junction will rise. Should the junction temperature become too high, the semiconductor will be damaged (or the IC may undergo thermal shutdown).

$$T_J = T_A + \Theta_{JA} P$$

where

T_J = the temperature of the junction
T_A = the ambient temperature right next to the case
Θ_{JA} = the thermal resistance from the junction to ambient
P = the power dissipated by the device

It is necessary to keep T_J below the maximum specified by the manufacturer of the device being used. This can be done by minimizing the power dissipated through careful circuit design, by decreasing the ambient temperature (T_A) with a fan, or by decreasing the thermal resistance (Θ_{JA}) with a heat sink. We will discuss the last option (reducing Θ_{JA}) in detail. Rearranging the junction temperature equation from above gives

$$\Theta_{JA\,max} = \frac{T_{J\,max} - T_A}{P}$$

When a heat sink is used, the junction to ambient thermal resistance consists of three terms.

$$\Theta_{JA} = \Theta_{JC} + \Theta_{CS} + \Theta_{SA}$$

where

Θ_{JC} = the thermal resistance junction to case, a specification of the IC
Θ_{CS} = the thermal resistance case to sink. This depends on how well you mount the heat sink to the case. With thermal grease it may be as low as 0.5°C/W. If you have to use a mica insulating wafer, Θ_{CS} is typically 2°C/W.
Θ_{SA} = the thermal resistance of the heat sink

Example 4-5

Select a three-terminal regulator and heat sink needed to provide:

$$V_{out} = 5V_{DC} \qquad I_{load} \le 0.7A_{DC} \qquad E_{in} = 15V_{DC} \qquad T_A = 60°C$$

Solution

1. A review of manufacturers' specifications on the World Wide Web, or in data books or catalogs, reveals that the LM309K and the LM340-5K and T meet the output voltage and current requirements. These are rated at 1.5A. No values closer to 0.7A were found. Also, higher rated currents would provide even less short circuit current protection.

2. Calculate the maximum allowable Θ_{JA} under normal operating conditions.

$$P = (E_{in} - V_{out})I_{out}$$

$$= (15V_{DC} - 5V_{DC}) \times 0.7A_{DC}$$

$$= 7W$$

$$\Theta_{JA\,max} = \frac{T_{J\,max} - T_A}{P}$$

$$= \frac{150°C - 60°C}{7W}$$

$$= 12.9°C/W$$

This is for the LM340-5. The LM309 has a $T_{J\,max} = 125°C/W$. This part is rejected because it requires a lower Θ_{JA}, leading to a larger heat sink (one with a lower thermal resistance).

3. Can this part be used without a heat sink? If the specified Θ_{JA} is below the calculated $\Theta_{JA\,max}$, then a heat sink is not needed.

Is $\qquad \Theta_{JA\,spec} < \Theta_{JA\,max\,calculated}$?

For the T package (TO220), $\Theta_{JA\,spec} = 50°C/W$, too large.

For the K package (TO3), $\Theta_{JA\,spec} = 35°C/W$, also too big.

So a heat sink is needed.

4. It's now time to calculate the thermal resistance of the heat sink.

$$\Theta_{SA\,max} = \Theta_{JA\,max} - \Theta_{JC} - \Theta_{CS}$$

$\Theta_{JA\,max}$ was just calculated. Any higher resistance will cause the part to get too hot.

$Q_{JC} = 3°\,C/W$ This is a specification.

$Q_{CS} = 2°\,C/W$ This is set by how you mount the heat sink to the IC package.

$$\Theta_{SA} = 12.9°\,C/W - 3°\,C/W - 2°\,C/W$$

$$= 7.9°\,C/W$$

This is the **largest** thermal resistance that the heat sink may have. Any larger resistance would cause the regulator to overheat.

Example 4-6

Design a power supply using a three-terminal regulator that will output $5V_{DC}$ at currents up to $500mA_{DC}$ with an ambient temperature of 30°C. Select the transformer, filter capacitor, and heat sink if needed.

Solution

1. Transformer

   Initially select a $12.6V_{RMS}$ secondary transformer and an LM340-5T regulator. These are both inexpensive and commonly available.

   $$v_P = 1.414 \times 12.6V_{RMS} - 1.4V = 16.4V_P$$

   This is well below the maximum input voltage for the LM340-5.

2. v_{min}

   $$v_{min} = V_{load} + V_{dropout}$$

   $$= 5V + 2V = 7V$$

3. C_{filter}

   To select a filter capacitance, you want to pick the **smallest** capacitor that will meet the v_{min} requirements. The smaller the capacitor, the less it costs, the smaller physically it is, and the less power the regulator IC must dissipate. From Figure 4-6, a $220\mu F$ capacitor gives $v_{min} = 7.5V$. A $100\mu F$ capacitor is too small.

   $$C_{filter} = 220\mu F$$

4. $E_{in\ ave}$

$$E_{in\ ave} = 16.4V - \frac{16.4V - 7.5V}{2} = 12.0V_{DC}$$

5. Heat sinking

$$P_{IC} = (E_{in\ DC} - V_{load})I_{load}$$

$$= (12V_{DC} - 5V_{DC}) \times 0.5A_{DC} = 3.5W$$

$$\Theta_{JA\ max} = \frac{T_{J\ max} - T_A}{P}$$

$$= \frac{150°C - 30°C}{3.5W} = 34°\,C/W$$

$$\Theta_{SA\ max} = \Theta_{JA\ max} - \Theta_{JC} - \Theta_{CS}$$

$$= 34°\,C/W - 3°\,C/W - 2°\,C/W$$

$$= 29°\,C/W$$

A TO220 heat sink with a thermal resistance of **less** than 29°C/W must be **properly** mounted to the LM340-5T. The full schematic is shown in Figure 4-13.

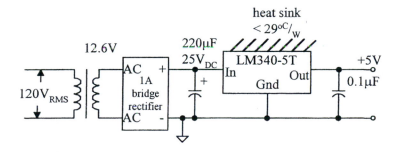

Figure 4-13 Schematic for Example 4-6

4-3.4 Adjustable Voltage Supplies

The precise output voltage you need may not be available in a standard three-terminal regulator, or you may want to vary the voltage. The LM117, 217, and 317 are part of a series of adjustable output three-terminal regulator ICs. Their performance has been optimized for adjustable operation. The output voltage can be changed over a range of 1.2V to 40V, at load currents up to 1.5A (with proper heat sinking). The basic schematic is in Figure 4-14.

This series of ICs regulates the voltage between its output and adjust pins to 1.2V. Since there is no ground connection, virtually all internal IC biasing current flows out of the output pin. To assure regulation, there must be at least a 10mA load. This is easily done by picking R1 at or a little below 120Ω.

The output voltage is set by this current and R2.

$$I_{R1} = \frac{1.2V}{R1}$$

The current from the IC's adjust pin is negligible, so

$$I_{R2} = I_{R1}$$

$$V_{R2} = I_{R2} \times R2$$

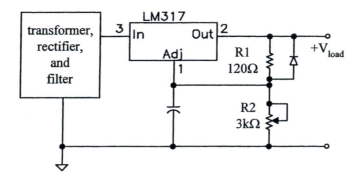

Figure 4-14 Adjustable voltage regulator

The output voltage is the sum of the voltages across R1 and R2.

$$V_{load} = 1.2V_{DC} + V_{R2}$$

The capacitor shown in Figure 4-14 is optional, but it does improve the ripple rejection of the IC.

Adjustable output power supply design requires a little thought when selecting the filter capacitor and the heat sink. The filter capacitor must provide a large enough v_{min} to insure that the regulator will work at the highest output voltage. However, highest voltage is dropped **across** the regulator IC when the output is at its minimum voltage. So the regulator's maximum power dissipation comes at minimum output voltage. Heat sink selection must be done when the output is adjusted to its minimum level. These principles are illustrated in Example 4-7.

Example 4-7

Design a power supply that will output a voltage adjustable from 1.2V to 15V at 0.6A. The ambient temperature is 40°C.

Solution

1. Schematic

    The unregulated schematic of Figure 4-2 will be used.

    $$v_{min} = V_{out\,reg} + 2V$$

    $$= 15V + 2V = 17V$$

    Looking at Figures 4-5 through 4-8, note that an $18V_{RMS}$ transformer is needed. A 470μF capacitor will provide just enough v_{min}. However, there will be some variation in capacitor value from one unit to another in manufacturing. A more prudent design is to select the 1000μF capacitor.

2. Adjustment resistors

    Select R1 = 82Ω. This gives

    $$I_{R2} = \frac{1.2V}{82\Omega} = 14.6mA$$

    To output $15V_{DC}$,

    $$V_{R2} = 15V_{Dc} - 1.2V_{DC} = 13.8V_{DC}$$

    So $$R2 = \frac{13.8V_{DC}}{14.6mA_{DC}} = 945\Omega$$

    Select R2 = 1kΩ.

3. Heat sink

$$v_P = 1.414 \times 18V_{RMS} - 1.4V = 24.1V_P$$

$$v_{min} = 20.5V \text{ (from Figure 4-7)}$$

$$V_{DC} = \frac{24.1V + 20.5V}{2} = 22.3V$$

This is the average voltage at the input of the regulator. The worst case power dissipation comes when the output is adjusted to its lowest value.

$$\Theta_{JA\,max} = \frac{T_J - T_A}{(V_{DC} - V_{load})I_{load}}$$

$$= \frac{150° C - 40° C}{(22.3V - 1.2V)0.6A} = 8.7° C/W$$

$$\Theta_{SA\,max} = \Theta_{JA\,max} - \Theta_{JC} - \Theta_{CS}$$

$$= 8.7° C/W - 3° C/W - 2° C/W = 3.7° C/W$$

4-3.5 Dual Voltage Supplies

Many discrete and integrated circuits require bipolar (dual or ±V) supplies. This requirement can be easily met with two three-terminal regulators as shown in Figure 4-15.

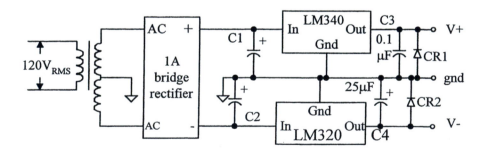

Figure 4-15 Simple dual power supply

Opposite phase AC is provided by the transformer's secondary and a grounded center tap. The single full wave bridge rectifier turns these into positive and negative DC voltages (with respect to the grounded center tap). Filtering is provided by C1 and C2. Be careful to get the electrolytic capacitors' polarities correct!

The LM340 provides regulation of the positive voltage, while the LM320 regulates the negative voltage. **Warning:** The LM320 has a **different pin configuration** than the LM340. The case of the LM320 is **not** ground. Take care when you mount the negative regulator.

These two regulators may not turn on simultaneously. If one regulator turns on before the other, the output of the slower regulator will be driven toward the potential of the faster. This may damage the slower regulator as it turns on. Diodes CR1 and CR2 prevent these reverse polarities on start up. But, be sure that you do not accidentally reverse them.

Many applications require several different voltage power supplies. One solution is to build several independent regulators. However, very often it is important that all of these supply voltages track. That is, if one supply is adjusted up 20%, it is best if **all** of the supply voltages go up the same amount. You can do this by adding an op amp to the adjustable three-terminal regulator of Figure 4-14.

The circuit in Figure 4-16 provides current limiting and thermal shutdown for the negative as well as the positive output voltage. The positive regulated voltage is produced with an LM317 adjustable regulator IC, as it was in Figure 4-14.

The regulated positive voltage is always 1.2V more positive than the voltage across R3 (on the adjust pin). This regulated output voltage is used as the input of an inverting amplifier. The op amp, U3, is the input stage of this amplifier, and the LM337 negative voltage regulator is the power output stage. Consequently, the negative regulated voltage, V-, is an amplified and inverted version of the positive regulated voltage, V+. Current limiting and thermal shutdown are independently provided by the LM317 for the positive output and the by the LM337 for the negative output. The op amp assures that the negative output voltage tracks the positive output by driving the LM337's adjust pin to 1.2V above the required output. Since the LM337 is inside the op amp's negative feedback loop, this +1.2V offset appears between the op amp and the regulator, not at the regulator's output.

To provide digital control, replace R3 with either a digital potentiometer or a voltage output digital to analog converter. Be sure that whichever you choose is able to sink the 10mA which flows down through R2.

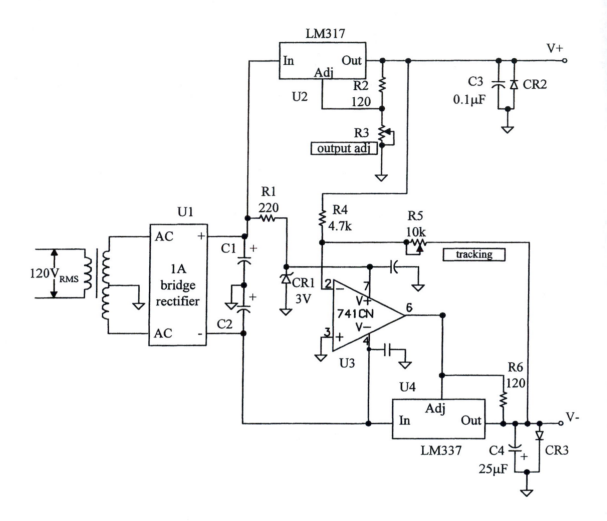

Figure 4-16 Tracking dual regulator

Example 4-8

Given:

R3 = 380Ω   R5 = 11.3kΩ

calculate:

I_{R3}, V_{R3}, V^+, $V_{U3\,INV}$, I_{R4}, I_{R5}, V^-, $V_{U3\,out}$

Solution

$$I_{R3} = I_{R2} = \frac{1.2\,V}{120\Omega} = 10\text{mA}_{DC}$$

$$V_{R3} = I_{R3} \times R3 = 10\text{mA}_{DC} \times 380\Omega = 3.8\,V_{DC}$$

$$V^+ = 1.2\,V_{DC} + V_{R3} = 1.2\,V_{DC} + 3.8\,V_{DC} = 5.0\,V_{DC}$$

$$V_{U3\,INV} = 0\,V\,(\text{virtual ground})$$

$$I_{R4} = \frac{V^+ - V_{R3\,INV}}{R4} = \frac{5.0\,V - 0\,V}{4.7\text{k}\Omega} = 1.06\text{mA}_{DC}$$

$$I_{R5} = I_{R4} = 1.06\text{mA}_{DC}$$

$$V^- = V_{R5} = -I_{R5} \times R5 = -1.06\text{mA}_{DC} \times 11.3\text{k}\Omega = -12.0\,V_{DC}$$

$$V_{U3\,out} = V^- + 1.2\,V_{DC} = -12.0\,V_{DC} + 1.2\,V_{DC} = -10.8\,V_{DC}$$

Example 4-8 illustrates a powerful technique. The 1.2V offset and any nonlinearity or inaccuracy associated with the LM337 are eliminated by including it **inside** the negative feedback loop of the op amp. The output of the circuit is the point where feedback resistors are connected, **not** the output of the op amp. The op amp alters its output voltage in whatever way necessary to assure that there is negligible difference in potential between its input pins. This assures that the gain equation is implemented, even if there are nonlinear devices with offset within the feedback loop.

$$V^- = -\frac{R_f}{R_i}\,V^+$$

You may use this trick of tucking the power stages inside the op amp's negative feedback loop with transistors, power amplifier ICs, or cards, or even with entire racks of commercial, DC coupled electronics.

4-4 Switched Capacitor Voltage Conversion

Properly biased CMOS transistors pass current in either direction. This unique feature allows for the production of an electronically controlled, bidirectional switch. The 7660, and second-generation parts, combine four of these switches with an oscillator and the required voltage level translators into a single, inexpensive, easy-to-use package. By adding two external capacitors, you can precisely invert or split a voltage, even when there is no negative power supply.

The traditional use of the 7660 is to generate a negative power supply when only a positive supply is available. Only slight loading and ripple are noticed for negative supply currents up to $10mA_{DC}$. Second-generation look-alikes are able to provide up to $100mA_{DC}$. For low power op amps this is plenty of drive. So, for that circuit which is mostly digital, requiring a hefty $+5V_{DC}$ power supply, you do not have to build a negative power supply to bias the analog ICs. The 7660 family can invert the $+5V_{DC}$, giving $-5V_{DC}$. Or in a battery powered application running from a single $1.5V_{DC}$ cell, or $3V_{DC}$, or a $9V_{DC}$ battery, you can use the 7660 to generate the negative supply, rather than requiring more batteries or trying to split the single supply (and operate over a reduced range).

If the current drawn from the switched capacitor circuit is negligible, then the voltage inversion is accurate to a few parts per million, **independent** of the value of the components used. Properly connected, these switch capacitors can produce a very accurate

$$v_{out} = kE_{in}$$

where

$$k = ..., \frac{1}{8}, \frac{1}{4}, \frac{1}{2}, -1, -2, -4, ...$$

In fact, this technique of voltage division and inversion is far more accurate, and more stable, than traditional resistor voltage division or resistor op amp voltage inversion.

A simplified block diagram and schematic symbol of the 7660 are shown in Figure 4-17.

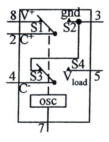

Figure 4-17 7660 schematic

Four CMOS transistors are configured as a double-pole double-throw switch (S1-S4). The oscillator is provided to throw the switch back and forth. Not shown are a regulator, a level translator, to assure that the CMOS transistors are properly driven, and other support logic.

A voltage inverter is shown in Figure 4-18. Inexpensive electrolytic capacitors are connected between the C^+ and C^- and from the V_{load} pin to ground. Ground and power are applied to their pins. On the first half cycle, with the switches to the left, as shown in Figure 4-18(a), the pump capacitor C_P is charged to E_{in}. When the oscillator throws the switches to the right [see Figure 4-18(b)], the positive end of C_P is connected to ground, while the capacitor's negative end is tied to V_{load} and to the reservoir capacitor. So the output is now negative with respect to ground. Since the two capacitors are in parallel, the charge equalizes. If the capacitors are equal, the voltage left after charges equalize is $\frac{1}{2}E_{in}$. The switches are again thrown to the left, and the pump capacitor is charged to E_{in}. When the switches transfer to the right, C_P again parallels C_R, and charge is passed to C_R. For equal capacitors, this time V_{load} settles at $\frac{3}{4}E_{in}$. The cycle repeats. Each time more charge is pumped from the input and C_P to C_R and the output. At the 10kHz oscillator frequency, very little time is required to fully charge C_R to the negative complement of the input voltage.

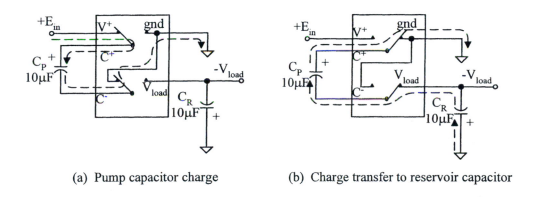

(a) Pump capacitor charge   (b) Charge transfer to reservoir capacitor

Figure 4-18 Voltage inverter

This conversion is nearly perfect, errors being introduced only by the ON resistance of the switches compared to the impedance of the two capacitors at the oscillator's 10kHz frequency. However, when you begin to draw current from the negative output, that reservoir capacitor will discharge during the cycle when it is disconnected from the pump capacitor. Also, when the pump capacitor is connected to the reservoir capacitor, it is connected to the load. So the pump capacitor is discharging during that part of the cycle. This loading effect is illustrated in Figure 4-19.

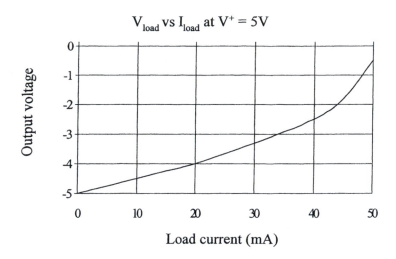

Figure 4-19 Loading effect on the output of the 7660 voltage inverter
(Courtesy of National Semiconductor)

The amount of load current also affects the output ripple voltage. Although the manufacturer recommends 10µF capacitors, increasing the capacitors decreases the ripple. This relationship between voltage and capacitance is neither simple nor linear. Second generations of the switched capacitor voltage inverter are used just as the 7660 is used. They are often pin compatible. But they are capable of outputting up to 100mA. So look carefully at your load current requirements and the ICs available before making a final selection.

Increased negative voltage is available by connecting two inverters in **series**. Look at Figure 4-20. The first 7660 inverts +5V, producing an output of −5V at its pin 5. The second 7660 has its input connected to ground, and its ground connected to −5V from the first 7660. During the first part of the cycle, the input (ground) is connected to the positive side of C_{P2}, and −5V is connected to that capacitor's negative end. This charges C_{P2} to 5V. During the second part of the cycle, the positive side of C_{P2} is tied to pin 3, −5V. This places the −5V from the first 7660 at pin 3 and the −5V charge on C_{P2} in series. The negative side C_{P2} is connected to the output. With −5V at the input and −5V across the pump capacitor, C_{R2} charges to −10V.

The circuit in Figure 4-21 outputs $-15V_{DC}$ from a $+5V_{DC}$ input. The second stage's V^+ pin is tied to $+5V_{DC}$ instead of being tied to ground as it was in Figure 4-20. During the first cycle, that capacitor has $+5V_{DC}$ on one side and the $-5V_{DC}$, from the first

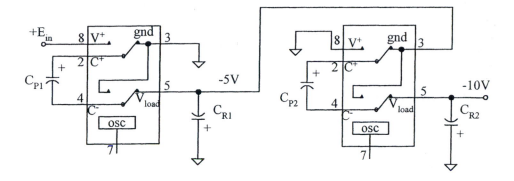

Figure 4-20　Voltage doubling with two 7660s
(*Courtesy of National Semiconductor*)

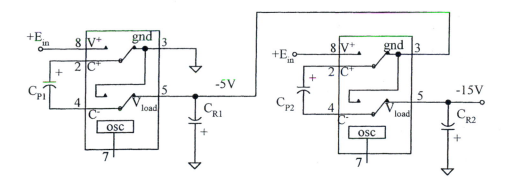

Figure 4-21　$-15V_{DC}$ from $+5V_{DC}$ *(Courtesy of National Semiconductor)*

stage, on the other side. As a result, this second-stage pump capacitor is charged to $10V_{DC}$. During the second part of the cycle, this $10V_{DC}$ is placed in series with the $-5V_{DC}$ on the IC's ground pin. The result is a $-15V_{DC}$ output. This is as large a potential difference as the 7660 can tolerate.

Figure 4-22 illustrates how you can **split** a voltage. This will generate a very precise split for light loads. When the 7660 connects the two capacitors in series, the charge flows from the source, E_{in}, and establishes voltage V_{CP} across C_P and voltage V_{CR} across C_R. The same charge that flows from the source exists on **each** capacitor. That is, for each electron that flows from the source to the lower plate of C_R, one is forced from C_R's upper plate and goes to C_P's lower plate. This, in turn, sends an electron from the upper plate of C_P to the positive side of the source. So you can write

$$Q_{series} = Q1 = Q2 = Q$$

Since the two capacitors are in series,

$$E_{in} = V_{CP} + V_{CR}$$

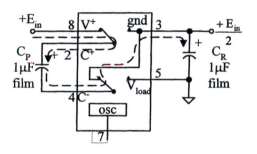
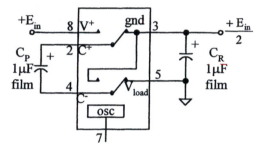

(a) Capacitors connected in series   (b) Capacitors connected in parallel

Figure 4-22 Precision supply splitter (*Courtesy of National Semiconductor*)

The fundamental relationship of voltage and charge for a capacitor is

$$V = \frac{Q}{C}$$

Apply this to each capacitor.

$$E_{in} = \frac{Q}{C_P} + \frac{Q}{C_R}$$

A little algebra yields

$$E_{in} = Q\left(\frac{1}{C_P} + \frac{1}{C_R}\right)$$

$$E_{in} = \frac{Q(C_R + C_P)}{C_R\ C_P}$$

Solve this for Q_{series}.

$$Q_{series} = E_{in} \frac{C_P C_R}{C_P + C_R}$$

This is the charge stored on **each** of the two capacitors. So the total charge stored is

$$Q_{total} = 2E_{in} \frac{C_P C_R}{C_P + C_R}$$

On the second half of the cycle, the 7660 switches the capacitors into parallel. Assuming that no charge leaks off or is added, the total charge is the same. The charge just rearranges itself so that

$$V = V_{CP} = V_{CR}$$

But

$$V = \frac{Q_{total}}{C_{total}}$$

Since C_P and C_R are in parallel, their capacitances add.

$$V = \frac{Q_{total}}{C_P + C_R}$$

Substituting the equation for total charge found several steps above,

$$V = E_{in} \frac{2C_P C_R}{\left(C_P + C_R\right)^2}$$

Example 4-9
Using two capacitors that are 10% different and two resistors that are also 10% different, compare the accuracy of a voltage splitter using a 7660 and the capacitors to a voltage divider using the two resistors.

Solution
For the resistor voltage divider,

$$R1 = R$$

$$R2 = 0.9R$$

$$V_{out} = E_{in} \frac{R1}{R1 + R2} = E_{in} \frac{R}{1.9R}$$

$$V_{out} = 0.52632E_{in}$$

Since the idea is to divide the voltage in half, the resistor divider with 10% error in one resistor produces a 5.26% error.

For the 7660 voltage splitter

$$C_P = C$$

$$C_R = 0.9C$$

$$V_{out} = E_{in} \frac{2C_P C_R}{(C_P + C_R)^2} = E_{in} \frac{2C \times 0.9C}{(C + 0.9C)^2}$$

$$V_{out} = E_{in} \frac{1.8C^2}{(1.9C)^2} = E_{in} \frac{1.8C^2}{3.61C^2}$$

$$V_{out} = 0.49861 E_{in}$$

The 10% difference in the capacitor values produces only a 0.28% error! The switched capacitor voltage splitter is over 18 times more accurate than the voltage divider.

This accuracy can be extended by cascading the ICs to divide by 2, 4, 8 ... . Equally accurate amplifiers with precise gains of 2, 4, 8, ... can be made by using the dividers as the negative feedback of a noninverting amplifier. If the signals are going to be AC, however, you will have to use a switched capacitor IC with separate positive and negative power pins. These are available for signal **processing.**

4-5 Switching Power Supplies

The regulated power supplies discussed so far are often referred to as **linear** supplies, since the pass transistors operate in their linear region. This mode of operation has several disadvantages. The power step-down transformer is the largest, most expensive component of the supply, due to the relatively low line frequency (60Hz). The low line frequency also requires large filter capacitors to decrease the ripple. Finally, operating the pass transistor in the active region forces it to dissipate quite a bit of power. These three factors combine to lower the efficiency of the power supply, while increasing its bulk and cost.

Switching power supplies overcome these problems. Switchers rely on pulse width modulation to control the average value of the output voltage, as illustrated in Figure 4-23. The average value of a repetitive pulse waveform depends on the area under the curve. Varying the duty cycle (the ON time per period) changes the average value of the voltage proportionally. For rectangular (switching) wave forms this is easily calculated.

$$V_{DC} = \frac{t_{on}}{T_{period}} v_P$$

Or, defining the duty cycle,

$$D = \frac{t_{on}}{T_{period}}$$

gives

$$V_{DC} = D v_P$$

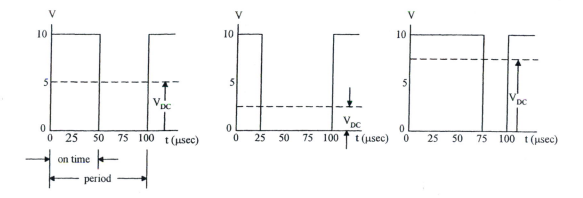

Figure 4-23 Pulse width modulation and average value

Changing the pulse width (t_{on}) while keeping the period (T_{period}) constant allows you to control the signal's average value (V_{DC}). This technique is used in switch mode power supplies to control the output voltage, compensating for changes in load current or unregulated input voltage.

4-5.1 The Buck Regulator

The block diagram of the buck regulator, a simple switching regulated power supply, is shown in Figure 4-24. The input line voltage (typically $120V_{RMS}$ at 60 Hz) is rectified and capacitively filtered. Or, this raw power may come from a battery or solar cell. This widely varying signal is applied to the switching transistor array, which is driven hard on and off by the pulse width modulator IC. The output is a constant, high frequency (higher than 20kHz) rectangular wave with a variable pulse width t_{on}. The high frequency pulse width modulated waveform is applied to a rectifier and LC filter, which smooths out the variations, outputting a smooth DC voltage. This DC is equal to the average value of the pulse width modulated wave. Should some variation in the load cause the output to try to increase, the pulse width modulator IC senses this and reduces the pulse width to the switching transistor(s). This reduces the pulse width to the high frequency rectifier and filter, lowering the output voltage.

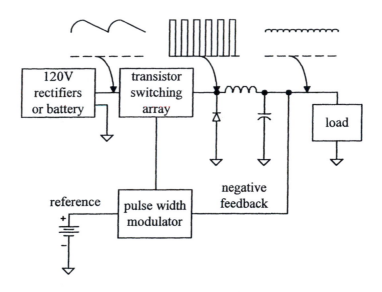

Figure 4-24 Buck regulator block diagram

When the transistor is driven on by the pulse width modulator, it acts as a short between the unregulated input and the filter. This reverse biases the diode. Current begins to increase through the inductor, swelling its magnetic field. (Remember, an inductor opposes a change in the current through it.) Part of this increasing current goes to charge the output filter capacitor, increasing the voltage to the load. This part of the cycle is shown in Figure 4-25.

During the second half of the cycle, the pulse width modulator drives the transistor off. With no more current from the transistor switch, the magnetic field of the inductor begins to collapse, causing the voltage across the inductor to switch polarity. The inductor is now acting as a generator, sourcing current as the field falls ,cutting the coils of wire. This reversal of polarity places a negative on the cathode of the diode, turning it on. When the diode is on, the left side of the inductor is clamped at a few tenths of a volt negative, placing the inductor (more or less) in parallel with the output capacitor and load. As the field in the inductor collapses, the output voltage begins to drop. The output capacitor then discharges, releasing current to the load in an effort to reduce variations in output voltage. This is illustrated in Figure 4-26.

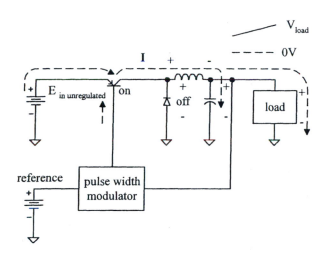

Figure 4-25 Buck regulator's charge cycle

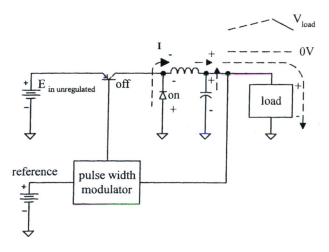

Figure 4-26 Buck regulator's discharge cycle

The four parameters of major importance for the diode(s) in the switching section of the regulator are forward current (peak and average), speed, forward bias voltage drop, and reverse bias voltage breakdown. The speed of conventional diodes is limited primarily by the reverse recovery time. That is the length of time it takes to remove the minority current carriers from near the junction when reverse bias is applied. The depletion region is then established, stopping current flow. Reverse recovery time is directly proportional to forward (load) current. The longer reverse recovery time, the longer it takes the diode to turn off, and the more power the switching components must dissipate. Efficiency drops.

There are two types of diodes whose speeds and forward current ratings are adequate for switching regulators. One is made with a process that adds gold during the diffusion phase of fabrication. These are called gold-diffused, fast recovery diodes. The Schottky diode is the second high speed power diode. There is no semiconductor junction. The diode is formed entirely from N-type material, with one metal-to-semiconductor connection formed in such a way as to allow current flow in only one direction. Rectification is achieved without minority carriers, so the inherent reverse recovery time is eliminated.

The forward bias voltage drop across the Schottky diode is approximately half that of a conventional silicon diode. This means that when biased on, the Schottky diode dissipates half the power that a gold-diffused, high speed power diode dissipates. However, the reverse voltage breakdown rating of Schottky diodes is on the order of 50V. So they may be used only in lower voltage applications.

For buck regulators, a diode must be selected that has a continuous forward current rating of at least

$$I_{on} \geq 1.21 I_{load\ max}$$

Of course, this rating must be larger than the regulator's short circuit current limit. The maximum reverse voltage rating (PIV) for the diode must be at least

$$PIV \geq 1.25 E_{in\ max}$$

The inductor in the filter stores energy in its electromagnetic field when the transistor is on, slowly releasing it during the transistor's off time. Its purpose is to smooth out the large variations in **current** to the load by the on/off, all-or-nothing nature of the switching transistor and diode.

The proper value of the inductance is derived below. When the transistor is on, the diode is off, and the output voltage V_{load} does not change significantly during this short time. So the voltage across the inductor is

$$V_{inductor} = E_{in} - V_{load}$$

But the voltage across the inductor is also related to its current by

$$V_{inductor} = L\frac{di}{dt}$$

Since you are dealing with a definite time interval and a linear rise of current, this becomes

$$V_{inductor} = L\frac{\Delta I}{\Delta t}$$

where

ΔI = the amount you allow the inductor's current to **change**. For design purposes, this is often about 25% of $I_{load\,DC}$.

Δt = the time that the transistor is on

$$\Delta t = DT = \frac{D}{f}$$

where

T = the period of the rectangular wave
f = the frequency of the rectangular wave
D = the duty cycle (a fraction) and

$$D = \frac{t_{on}}{T} = \frac{V_{load}}{E_{in}}$$

So the time interval that the inductor is charging is

$$\Delta t = \frac{D}{f} = \frac{V_{load}}{E_{in}f}$$

Substitute this into the equations for the voltage across the inductor.

$$V_{inductor} = E_{in} - V_{load} = L\frac{\Delta I}{\Delta t}$$

$$E_{in} - V_{load} = \frac{L\,\Delta I}{\dfrac{V_{load}}{E_{in}f}} = \frac{L\,\Delta I\,E_{in}\,f}{V_{load}}$$

Solving for L yields

$$L = \frac{(E_{in} - V_{load})V_{load}}{\Delta I\,E_{in}\,f}$$

Operating an inductor beyond its maximum rated current causes its core to saturate. When saturation is reached, increases in current can no longer cause an increase in the magnetic field. So the inductor can store no more energy; it can no longer oppose a change in current. Effectively, the inductance approaches zero. The only opposition the inductor then offers is its DC resistance (which has intentionally been kept as small as possible). The semiconductor switch may be damaged. Carefully review your inductor's saturation characteristics. Assure that the peak current is below its maximum level.

$$I_{peak} = I_{load} + \frac{\Delta I}{2} < I_{inductor\ saturation}$$

The way that the inductor is built strongly affects its nonideal behavior. Inductors made by wrapping wire onto a ferrite, toroidally shaped core dissipate heat well. So, as they dissipate power, their value remains relatively constant. Also this configuration contains the magnetic fields within its ring-shaped core, producing low electromagnetic interference (EMI) to contaminate surrounding circuits. Powdered iron, toroid core inductors also produce low EMI and tolerate larger peak currents than ferrite core inductors. Ferrite bobbin-core inductors are inexpensive and withstand high peak currents, but they generate considerable EMI and heat up easily.

The inductor and the output capacitor not only reduce the ripple in the output voltage, but are critical in assuring the overall stability of the regulator circuit. The capacitor reduces changes in the output voltage. But this output voltage is fed back to the pulse width modulator, which alters the width of the pulse, and therefore the current to the inductor, which alters the voltage to the capacitor, ... An analysis using full, closed loop Laplace domain, and poles and zeros is needed. The result is very sensitive to the characteristics of the pulse width modulator as well. Generally, the manufacturer of each pwm IC will provide you with guidance. Failing that, a reasonable starting point is

$$C \approx 500 \frac{\mu F}{A} \times I_{load}$$

The ripple voltage at the output contains two components. One consists of sharp spikes that occur when the transistor switches. These are present at the output because of equivalent series inductance associated with the output path. This is largely a function of the parasitic inductance in the main output path. Also the output capacitor itself has equivalent series inductance. The higher the voltage rating of the capacitor, the lower its ESL. You can reduce these spikes, then, by choosing a capacitor with the lowest ESL **at the pwm's frequency**, and by carefully laying out the circuit to reduce the length of the leads to the capacitor, and from the capacitor to the circuit's single point ground, and back to the pulse width modulator IC.

The second component to the output ripple voltage is triangular. The variation in current from the inductor, ΔI, passes through the output capacitor to ground. However, each capacitor has some series resistance. So ΔI flowing through that ESR produces an output ripple voltage.

$$V_{\text{load ripple PP}} = \Delta I \times \text{ESR}$$

Purchasing a capacitor with low ESR, then, will lower the output ripple voltage.

Film capacitors have much lower ESL and ESR than do electrolytic capacitors, but they do not have high enough capacitance to be used alone. So, often a 0.1µF film capacitor is placed **immediately** beside, and in parallel with, the larger electrolytic capacitor. This lowers the ESR and ESL of the pair while keeping the capacitance high.

The voltage rating of the electrolytic capacitor should be at least 25% above the expected output voltage. The higher the voltage rating, the lower the ESR. So using a 50V capacitor even though a 15V capacitor is all that is needed will lower the output ripple voltage. Finally, the capacitor's current rating, **at the pwm's frequency**, should be at least 50% greater than the ripple current (ΔI) that you expect to pass to it from the inductor.

Proper ground connections are critical to a low noise switching power supply. At current levels of several amperes, even 0.1Ω of resistance along a ground path generates several hundred millivolts of noise. Establish a single point ground at the ground pin of the pwm IC. The length of the leads to the input capacitor, the Schottky diode, and the output capacitor are critical. Keep these paths **short**. A ground plane is a good idea and can be implemented on either a protoboard or a printed circuit board. But be sure to remove the ground plane from under the inductor, or it will degrade the effects of the magnetic fields the inductor is trying to establish. Be sure to provide separate ground returns to the single point ground (at the pwm IC's ground pin) for the ground plane, any low level analog signals, digital ground, and high level analog (relays, speakers, and motors).

4-5.2 Continuous and Discontinuous Operation

The discussion in the previous section assumes that current **always** flows through the inductor, either building up when the switch is on or discharging when the switch is off. This is called **continuous** operation. Look again at the buck regulator, shown in Figure 4-27.

The charge current is shown in the schematic by solid lines, while discharge current is dashed. At time a, the pulse width modulator turns the transistor switch ON by pulling its base low. The transistor saturates, dropping about 0.2V. The rest of the $E_{\text{in unreg}}$ is passed to node A. Current increases (more or less linearly) through the inductor. Some of the current goes to charge the output capacitor, but most goes to the load. At time b, the pulse width modulator turns the switch OFF by driving its base high. The transistor goes off. The inductor's magnetic flux begins to collapse. This flips the polarity of the voltage across the inductor, placing a negative at node A. The Schottky diode turns ON, completing the discharge path and placing -0.2V at node A. As the flux collapses, current through the inductor falls (more or less linearly). At time c the cycle repeats.

There are two key points. First the peak-to-peak variation in the current through the inductor, ΔI, is set by the input and output voltages, the pulse width modulator frequency and the size of the inductor. It is **independent** of the current drawn by the load (I_{DC}). Second, this I_{DC} is the load current and sits halfway between the peaks of ΔI.

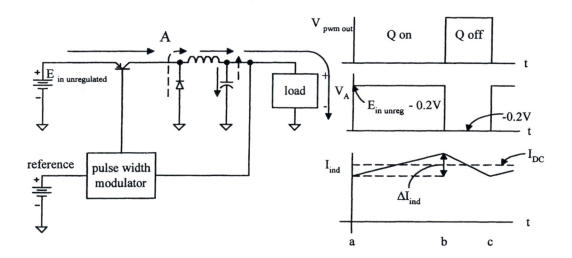

Figure 4-27 Buck regulator

$$L = \frac{\left(E_{in} - V_{load}\right)V_{load}}{\Delta I \; E_{in} \; f}$$

$$\Delta I = \frac{\left(E_{in} - V_{load}\right)V_{load}}{L E_{in} \; f}$$

Assuming that the pulse width modulator does its job properly, the ΔI wave rises and falls as more or less current is demanded by the load. But, its peak-to-peak amplitude does not change, as illustrated in Figure 4-28.

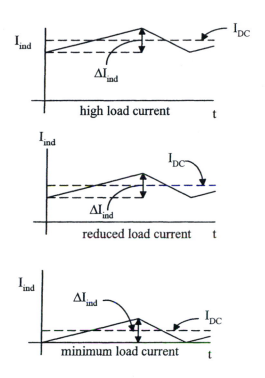

Figure 4-28 Effects of varying load current

The **minimum** current the load can draw and still keep the regulator in the continuous mode is

$$I_{load\,min} = \frac{\Delta I}{2} \quad \text{with } \Delta I \text{ set as indicated above}$$

A load **lighter** than this will force the regulator into the discontinuous mode. To prevent this, it is not unusual to see a minimum load current specified, and a resistor wired across the output terminals, to draw $I_{load\,min}$ even if the external load is completely removed.

Reducing the load current below $I_{load\,min}$ causes the inductor's magnetic field to completely collapse at some point while the transistor is on. The resulting waveforms are shown in Figure 4-29. Between times **a** and **b**, the inductor charges, a little. Since the amount of energy stored in the magnetic field is directly dependent on the current through the coil, and since that current is low, it takes only a small amount of time until the field fully collapses. When the field completely defluxes, the inductor loses the energy it stored. At time b, the switch turns off, and the magnetic field loses its ability to oppose a change in current. It becomes just a wire. The Schottky diode goes off. Point A rises to the output voltage, because the inductor no longer drops a voltage. This loss of flux **may** be accompanied by severe ringing, induced in the stray circuit inductance by the turn-off of the Schottky diode. Depending on layout, these spikes may be considerably larger than the input voltage. Since the inductor is defluxed, it passes this ringing to the output capacitor. Any voltage that the capacitor cannot short out (because of its own ESR and ESL) goes to the load as high frequency, high voltage noise. Whenever possible, it is advisable to stay out of discontinuous operation. For a switching regulator, too little load current results in a marked increase in noise at the output.

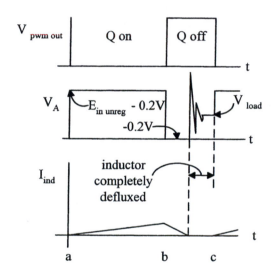

Figure 4-29 Discontinuous buck regulator operation

4-5.3 Integrated Circuit Buck Regulator

In the buck regulator, the most critical and most demanding element to design is the switch. The type of device, its rise time, delay time, fall time, storage time, parasitic capacitance, saturation voltage, gain, maximum ratings, power dissipation, and the derating of all of these for temperature, as well as base or gate drive circuitry, must be considered. Even so, the pulse width modulator is more sophisticated. It must provide a fixed frequency clock, drive the switch on and off in response to variations on the input or output, and provide short circuit limiting and some form of external logic control.

Also, the quality of the output voltage can be no better than the voltage reference used as the input. These requirements are severe enough to require the full attention of an engineering team.

Several manufacturers have made your job easier by producing a series of integrated circuits that combine these functions into a single chip. The National Semiconductor LM1575/2575 and LM1576/2576 Simple Switcher™ are part of a family of ICs that provides the switch, pulse width modulator, and reference within a single IC. The block diagram of the LM1575/LM2575 is shown in Figure 4-30.

The LM1575/2575 series provides at least 1A, while the LM1576/2576's upper current limit is at least 3A. Each series consists of four members, one which regulates at 5V (LM2575-5), one at 12V, one at 15V, and one whose output is adjustable. At 25°C these voltages are stable and accurate to ±1%. The clock is generated internally, without the need for external components. It is fixed at 52kHz, well above the audio range. Current is also monitored internally. Should a short circuit occur, the switch is turned off, on a cycle-by-cycle basis, until the short is removed. Also, the temperature of the IC is monitored, and the switch is disabled if that temperature exceeds 150°C. However, since the switching regulators built with these ICs are typically 80% efficient, many applications will not need a heat sink. Finally, placing the ON/OFF pin (pin 5) 1.6V above ground will turn the switch off and will lower quiescent current draw to about 100μA. This input is compatible with both TTL and CMOS logic.

Look again at Figure 4-30. The internal reference for the adjustable version is 1.23V. The output voltage is divided by R_f and R_i and is then fed back to the negative input (pin 4). Although this signal is applied to a noninverting input, it actually is negative feedback because of the NOR gate further down the line. So this is configured just like a noninverting op amp amplifier. The IC alters its output pulse width as necessary to drive the negative feedback voltage into alignment with its noninverting input (the 1.23V reference). V_{load} is driven to a voltage which makes the feedback voltage, the voltage across R_i, equal to 1.23V.

$$V_{R_i} = 1.23V = \frac{R_i}{R_i + R_f} V_{load}$$

Solving for V_{load} yields

$$V_{load} = 1.23V \frac{R_i + R_f}{R_i}$$

$$= 1.23V \left(1 + \frac{R_f}{R_i}\right)$$

So you can view the IC, diode, and filter as a sophisticated power op amp set up as a noninverting amplifier. The reference voltage (1.23V) is the input. Varying R_f, then,

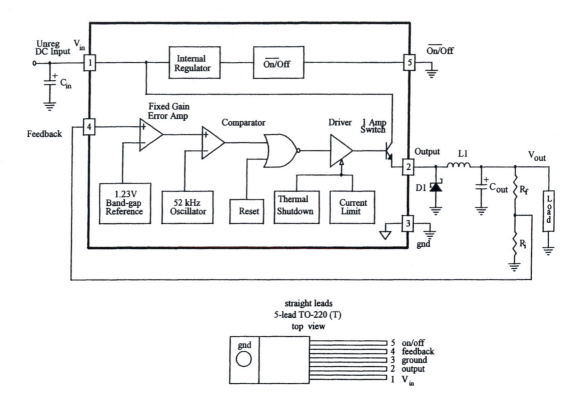

Figure 4-30 LM1575/2575 Simple Switcher™
(Courtesy of National Semiconductor)

directly changes the output voltage. In fact, you can replace R_f with a digitally controlled potentiometer and allow a microprocessor to control the output voltage.

In order to assure closed loop stability, the manufacturer recommends that the main filter capacitor

$$C_{out} > 7.785 \times 10^{-9} \frac{E_{in}}{V_{load} L}$$

However, good transient response and output ripple voltage reduction may require a capacitor several times larger than needed to assure stability.

The major advantage of the switching regulator is that it dissipates very little power itself. So less power is needed from the unregulated supply, allowing smaller, less expensive components there or prolonged battery life. Also, since the regulator is dissipating little power, little heat is generated, often eliminating the need for a cooling fan or even for a regulator heat sink.

There are two parts to the LM2575's power dissipation. The internal electronics must be provided with adequate current and voltage. The power that they dissipate is

$$P_{internal} = E_{in}I_{quiescent}$$

For the worst case

$$I_{quiescent} = 12mA_{DC}$$

If the input voltage varies, be sure to use $E_{in\ max}$.

The other element that dissipates power is the switch. When on, the transistor is saturated, producing a low voltage drop. Figure 4-31 allows you to determine that saturation voltage.

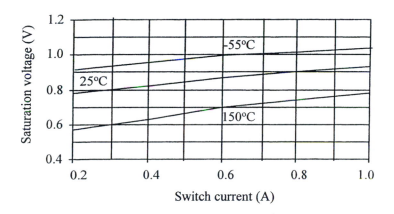

Figure 4-31 LM2575 switch saturation voltage
(Courtesy of National Semiconductor)

So, when on, the switch dissipates

$$P_{on} = V_{sat}I_{load}$$

However, the switch is on for only part of the cycle. This percentage is the duty cycle, D.

$$P_{switch} = DV_{sat}I_{load}$$

Combining this with the quiescent power that the IC uses gives the total power dissipated by the LM2575.

$$P_{LM2575} = E_{in}I_{quiescent} + DV_{sat}I_{load}$$

Now that the power dissipated by the IC can be determined, you can apply the heat sink calculations of the earlier section. The thermal resistances of the LM2575 in the TO220 package are:

$$\Theta_{JA} = 65°C/W$$
$$\Theta_{JC} = 2°C/W$$

Example 4-10

Determine what heat sink (if any) is needed for the LM2575 to provide:

$V_{load} = 5V$ $I_{load} \leq 0.7A$ $E_{in} = 15V$ $T_A = 60°C$

Solution

$$D = \frac{V_{load}}{E_{in}} = \frac{5V}{15V} = 0.33$$

From Figure 4-31, T_A is between the 25°C and the 150°C lines, closer to 25°C. At $I_{switch} = 0.7A$, $V_{sat} = 0.8V$.

$$P_{LM2575} = E_{in}I_{quiescent} + DV_{sat}I_{load}$$

$$= 15V \times 12mA + 0.33 \times 0.8V \times 0.7A$$

$$= 0.367W$$

Without a heat sink, the IC will heat up to

$$T_J = T_A + \Theta_{JA}P$$

$$T_J = 60°C + 65°C/W \times 0.367W$$

$$= 83.7°C$$

Since the IC will not go into thermal shutdown until the junction temperature reaches 150°C, a heat sink is **not** needed. The linear regulator in Example 4-6 had to dissipate (waste) 7W, compared to 0.37W for the switcher. This 7W has to be produced, requiring a larger transformer, rectifier, and filter, and the heat has to be removed, requiring a significant heat sink, added space, and added heat around the circuit.

The circuit in Figure 4-32 uses the LM2575 buck regulator to generate a **negative** voltage. It simply misleads the IC about where ground is located. The inductor, output capacitor and peak switch, and inductor currents are all higher than those predicted by the simple buck equations.

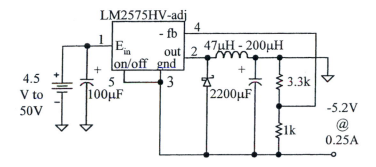

Figure 4-32 Inverting switching regulator (*Courtesy of National Semiconductor*)

4-5.4 Boost Regulator

The same circuit elements that were used to build the step-down buck regulator can be reconfigured to make a step-up boost regulator. Look at Figure 4-33. The boost circuit contains an inductor, switch, diode, capacitor, feedback voltage divider, voltage reference, and pulse width modulator, just as the buck regulator (Figure 4-25) does. Each plays a similar role, but their new arrangement allows the output to fly up above the input when the switch is turned off.

The cycle begins when the pulse width modulator outputs a high, turning the transistor ON. This grounds the node between the inductor and the diode. With ground on its anode and V_{load} held on its cathode by the capacitor, the diode is reverse biased and goes OFF. The current through the inductor ramps up, with energy being stored in its electromagnetic field. With the unregulated input voltage (E_{in}) on one side of the inductor and ground on the other,

$$V_{inductor} = E_{in} = L\frac{di}{dt} = L\frac{\Delta I}{\Delta t}$$

where

ΔI = peak-to-peak ripple current through the inductor

Δt = time that the transistor is ON

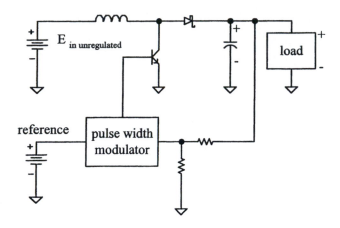

Figure 4-33 Switching boost regulator

$$\Delta t = DT$$

where

D = duty cycle
T = period of the switching signal

Combining these equations gives the relationship between the inductor and its peak-to-peak ripple current.

$$v_L = L\frac{di}{dt} = L\frac{\Delta I}{\Delta t}$$

$$L = E_{in}\frac{DT}{\Delta I}$$

So, if you are designing a boost regulator, pick the peak-to-peak ripple current to be about 25% of the load current. Then determine the size of the inductor.

If you are analyzing a given regulator, then the inductor size allows you to calculate the amount of peak-to-peak ripple current through it.

$$\Delta I = E_{in}\frac{DT}{L}$$

As with all good things, you can't get something for nothing. The boosted output voltage comes at the cost of increased **input** current. Ignoring the losses across the transistor, the diode, and the inductor's resistance, the power delivered to the load must be the same as that drawn from the source.

$$P_{out} = P_{in}$$

$$V_{load} \times I_{load} = E_{in} \times I_{in}$$

$$I_{in} = \frac{V_{load} I_{load}}{E_{in}}$$

Look at the schematic again. The inductor is directly connected in series to the input source. So the current from the source is the average current through the inductor.

$$I_{ind\,ave} = \frac{V_{load} I_{load}}{E_{in}}$$

The peak current through the inductor is its other key specification. That is,

$$I_{pk\,ind} = I_{ind\,ave} + \frac{\Delta I}{2}$$

The inductor is charging through the transistor, so during this charge time the currents through the inductor and through the transistor are the same. Look at Figure 4-34.

The wave shape of the current through the transistor is almost a rectangle with an amplitude of $I_{L\,ave}$ and a duty cycle of D. So

$$I_{Q\,ave} \approx D I_{L\,ave}$$

$$I_{Q\,RMS} \approx \sqrt{D}\; I_{L\,ave}$$

When the transistor turns OFF, the voltage across the inductor reverses polarity. This places a positive at the junction of the inductor and the diode, forward biasing the diode. Current then flows through the diode and into the load. This wave shape is also almost a rectangle.

$$I_{CR\,ave} \approx (1 - D) I_{L\,ave}$$

Since $I_{CR\,ave} = I_{load}$ and $I_{L\,ave} = I_{input}$,

$$I_{load\,ave} \approx (1 - D) I_{input\,ave}$$

Or
$$I_{in} = \frac{I_{load}}{1 - D}$$

But we just determined, using input and output power, that

$$I_{in} = \frac{V_{load} I_{load}}{E_{in}}$$

Equating these two expressions gives

$$\frac{I_{load}}{1 - D} = \frac{V_{load} I_{load}}{E_{in}}$$

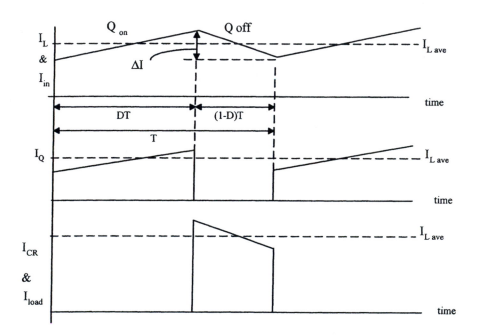

Figure 4-34 Boost regulator currents

Divide each side by I_{load}.

$$\frac{1}{1 - D} = \frac{V_{load}}{E_{in}}$$

Solve for V_{load}.

$$V_{load} = \frac{E_{in}}{1 - D}$$

and

$$D = 1 - \frac{E_{in}}{V_{load}}$$

Look closely at the equation for V_{load}. D is the duty cycle and therefore a fraction. So, the denominator is less than 1. This means that V_{load} must be **greater** than E_{in}! That's right. It's a **boost** regulator.

To see where this extra voltage comes from, look at Figure 4-35. In part a, the transistor is ON, the voltage at node A is almost 0V, the diode is OFF, and the inductor is storing energy, with voltage across it, positive to negative. The capacitor discharges, providing current to the load, trying to hold its voltage at V_{load}.

When the transistor goes OFF, the magnetic field within the inductor begins to collapse, in an effort to oppose the change in current. This means that the polarity of the voltage across the inductor must flip, becoming negative to positive. It is now a generator, as shown in part b of Figure 4-35. The voltage at node A now jumps to E_{in} + V_L. This forward biases the diode, sending voltage and current to the load (and recharging the output capacitor).

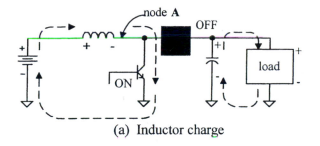

(a) Inductor charge

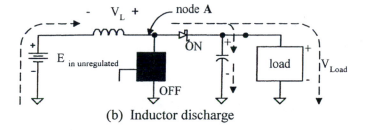

(b) Inductor discharge

Figure 4-35 Inductor flyback leads to the boost effect.

The duty cycle, which controls the output voltage, is set by the pulse width modulator in response to its clock, a voltage reference, and the voltage fed back to it from the output. National Semiconductor makes the LM1577/2577 series of boost voltage regulator ICs. These functions and several others are provided by the IC. The block diagram is shown in Figure 4-36. It is considerably different from that of the LM2575 buck regulator. The buck regulator IC (LM2575) must apply and remove E_{in} to the inductor. The boost IC (LM2577) applies and removes ground to the inductor. The LM2577 uses current feedback rather than voltage feedback (as used by the LM2575). This provides better regulation for a boost regulator but requires that you add an external compensation resistor, R_C, and capacitor, C_C. Along with C_{out}, they assure that the closed loop feedback performance of the regulator circuit is stable. The LM2577 can output voltages up to 60V from an input between 3.5V and 40V. But the input must be at least

$$E_{in} \geq 0.1V_{load}$$

Short circuit current limiting to the load is not provided. But the IC does control the current through its internal switch each cycle. Also, the temperature is monitored, and the transistor is held off if the junction temperature exceeds 125°C. The internally generated clock runs at the same 52kHz that the LM2575 uses. The reference voltage is 1.23V. For an adjustable output, you provide an external voltage divider. There are fixed voltage models available as well. For these, the feedback voltage divider is provided internally. There is also a soft start feature. This turns the pulse width modulator on slowly, allowing the switch to be turned on longer and longer each cycle. Without this feature the switch would initially turn on for 90% of the time as it tries to charge up a fully discharged L and C_{out}. Excessive current would be drawn from the input supply.

The values of R_C, C_{out}, and C_C necessary to assure stable operation are derived by a full Laplace domain closed loop stability analysis. That is certainly beyond the scope of this text. These values depend on the pulse width modulator's gain, I_{load}, and L. The manufacturer of the LM2577 provides the following guidance:

"First, calculate the maximum value of R_C,

$$R_C \leq \frac{750 I_{load\,max} (V_{load})^2}{(E_{in\,min})^2}$$

Select a resistor less than or equal to this value, and it should be no greater than 3kΩ.

"Calculate the minimum value for C_{out} using the following two equations.

$$C_{out} \geq \frac{0.19 L R_C I_{load\,max}}{E_{in\,min} V_{load}}$$

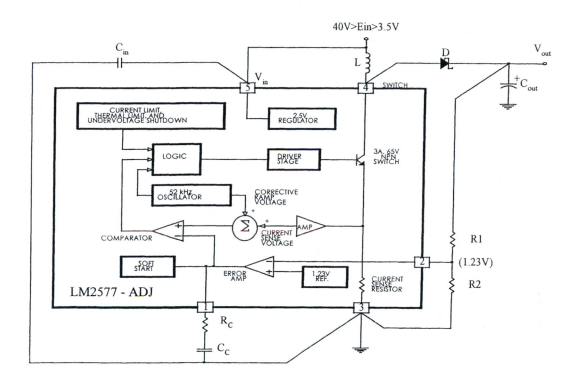

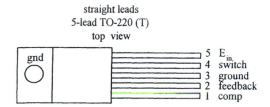

Figure 4-36 LM1577/2577-ADJ *(Courtesy of National Semiconductor)*

and

$$C_{out} \geq \frac{E_{in\,min} R_C \left[E_{in\,min} + \left(3.74 \times 10^5 L\right) \right]}{487{,}800 \left(V_{load}\right)^3}$$

The larger of these two is the minimum value that ensures stability.
"Calculate the minimum value of C_C.

$$C_C \geq \frac{58.5 \left(V_{load} \right)^2 C_{out}}{R_C^2 E_{in\,min}}$$

These values for R_C and C_C are valid only if C_{out} has a low ESR.

$$ESR_{C_{out}} \leq \frac{8.7 \times 10^{-3} E_{in}}{I_{load\,max}}$$

The compensation capacitor is also part of the soft start circuitry. ... The soft start circuit requires that $C_C \geq 0.22 \mu F$."

Ignoring the losses during the time while the switch is changing from one state to the other, power dissipated by the LM2577 consists of three terms.

$$P_{LM2577} = P_Q + P_{drive} + P_{IC\,bias}$$

First is the power that the transistor inside the LM2577 dissipates.

$$P_Q = I_{Q\,RMS}^2 R_{on}$$

$$I_{Q\,RMS} \approx \sqrt{D}\ I_{L\,ave}$$

$$I_{L\,ave} = \frac{V_{load} I_{load}}{E_{in}}$$

R_{on} is the on resistance of the power switch. It depends on current and temperature. Look at Figure 4-37.

This main switching transistor has a β of 50. So the current that drives the switch is $1/50$ of the main transistor and inductor current.

$$P_{drive} = I_{drive} E_{drive}$$

$$I_{drive} = \frac{I_{Q\,RMS}}{50}$$

The rest of the IC also dissipates a little power.

$$P_{IC} = I_{supply} E_{in}$$

I_{supply} depends on switch $I_{Q\,RMS}$ and on temperature, as illustrated in Figure 4-38. Notice that the graph is made with the switch duty cycle at 50%. This approximation should be adequately accurate for other values of D.

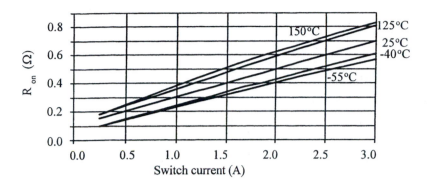

Figure 4-37 LM2577 on resistance *(Courtesy of National Semiconductor)*

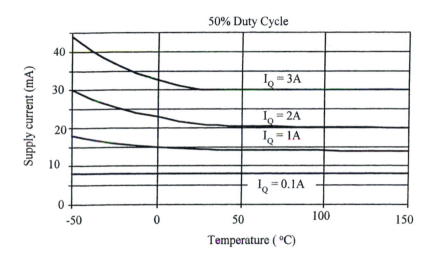

Figure 4-38 LM1577/2577 supply current versus temperature
(Courtesy of National Semiconductor)

Example 4-11

Design a switching regulator circuit that meets the following:

$E_{in} = 5V$ $V_{load} = 12V$ $I_{load\,max} = 800mA$ $\Delta I = 25\%\ I_{load\,max}$
$\Delta V_{load\,pp} = 0.05V$ $T_A = 25°C$ $\Theta_{JA} = 65°C/W$

Calculate each of the following parameters:

duty cycle (D), $I_{ind\,ave}$, $I_{Q\,RMS,}$, $I_{ind\,P}$, L, R_C, C_C, C_{out}, ESR, P_{LM2577}, T_J, heat sink if needed

Solution

$$D = 1 - \frac{E_{in}}{V_{out}} = 1 - \frac{5V}{12V} = 0.58 = 58\%$$

$$I_{ind\,ave} = I_{in} = \frac{I_{load}\,V_{out}}{E_{in}} = \frac{800mA \times 12V}{5V} = 1.92A_{DC}$$

$$I_{Q\,RMS} \approx \sqrt{D}\ I_{L\,ave} = \sqrt{0.58} \times 1.92A = 1.46A_{RMS}$$

$$\Delta I = 0.25I_{ind\,ave} = 0.25 \times 1.92A_{DC} = 0.48A_{PP}$$

$$I_{ind\,P} = I_{ind\,ave} + \frac{\Delta I}{2} = 1.92A_{DC} + \frac{0.48A_{PP}}{2} = 2.16A_P$$

$$L = \frac{E_{in}D}{\Delta If} = \frac{5V \times 0.58}{0.48A \times 52kHz} = 116\mu H \qquad \text{Pick } L = 100\mu H.$$

$$R_C \leq \frac{750I_{load}\,V_{load}^2}{E_{in}^2} = \frac{750 \times 800mA \times (12V)^2}{(5V)^2} = 3.5k\Omega$$

Pick $R_C = 2.2k\Omega$.

$$C_{out} \geq \frac{0.19LR_CI_{load}}{E_{in}\,V_{load}} = \frac{0.19 \times 100\mu H \times 2.2k\Omega \times 800mA}{5V \times 12V} = 557\mu F$$

and

$$C_{out} \geq \frac{E_{in}R_C\left[E_{in} + \left(3.74 \times 10^5\,L\right)\right]}{487,800\,V_{load}^3}$$

$$C_{out} \geq \frac{5V \times 2.2k\Omega \times \left[5V + \left(3.74 \times 10^5 \times 100\mu H\right)\right]}{487,800(12V)^3} = 553\mu F$$

Pick $C_{out} = 680\mu F$.

$$ESR \leq \frac{\Delta V_{load\,pp}}{\Delta I} = \frac{0.05V_{PP}}{480mA_{PP}} = 0.104\Omega$$

and

$$\text{ESR} < \frac{8.7 \times 10^{-3}\, E_{in}}{I_{load}} = \frac{8.7 \times 10^{-3} \times 5V}{800\text{mA}} = 0.054\Omega$$

Pick a C_{out} that has an ESR$<0.05\Omega$ at 52kHz.

$$C_C \geq \frac{58.5 V_{load}^2 C_{out}}{R_C^2 E_{in}} = \frac{58.5 \times (12V)^2 \times 680\mu F}{(2.2\text{k}\Omega)^2 \times 5V} = 0.237\mu F$$

Pick $C_c = 0.33\mu F$.

$$P_Q = I_{Q\,RMS}^2 R_{on} = (1.46A)^2 \times 0.4\Omega = 853\text{mW}$$

$$I_{drive} = \frac{I_{Q\,RMS}}{50} = \frac{1.46A}{50} = 29.2\text{mA}$$

$$P_{drive} = I_{drive} E_{drive} = 29.2\text{mA} \times 5V = 146\text{mW}$$

$$P_{IC} = I_{supply} I_{in} = 17.5\text{mA} \times 5V = 87.5\text{mW}$$

$$P_{LM2577} = 853\text{mW} + 146\text{mW} + 87.5\text{mW} = 1.09\text{W}$$

$$T_J = T_A + \Theta_{JA} P = 25^\circ C + 65^\circ \text{C/W} \times 1.09\text{W} = 95.9^\circ C$$

This is below the 125°C maximum junction temperature, so a heat sink is not needed.

A dual output regulator can be produced with this same circuit topology, by replacing the inductor with a switching transformer and adding the necessary rectification and filtering. The circuit shown in Figure 4-39 produces $\pm 15 V_{DC}$ at 225mA_{DC} from a single $+5V_{DC}$ input. The transformer is selected from Table 4-2 and Table 4-3. The compensation components and the output filter are calculated just as for the simple boost regulator. However, each equation must be slightly altered to account for the transformer and the dual output.

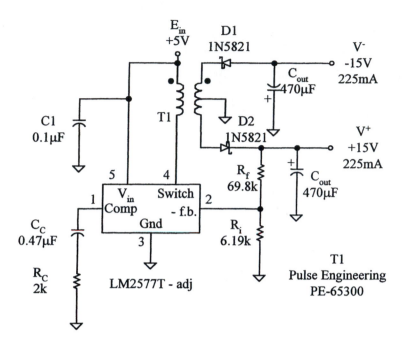

Figure 4-39 ±15V regulator *(Courtesy of National Semiconductor)*

Table 4-2 Dual output regulator transformer selection
(Courtesy of National Semiconductor)

Type	L_P μH	Input Volts	Output Volts	$I_{output\ max}$ mA
1	100	5	±10	325
	N=1	5	±12	275
		5	±15	225
2	200	10	±10	700
	N=0.5	10	±12	575
		10	±15	500
		12	±10	800
		12	±12	700
		12	±15	575
3	250	15	±10	900
	N=0.5	15	±12	825
		15	±15	700

Table 4-3 Transformer manufacturers' part numbers
(Courtesy of National Semiconductor)

Type	AIE	Pulse	Renco
1	326-0637	PE-65300	RL-2580
2	330-0202	PE-65301	RL-2581
3	330-0203	PE-65302	RL2582

The parameters associated with the dual output regulator in Figure 4-39 are listed in Table 4-4.

4-5.5 Other Switching Regulator Topologies

The buck and the boost regulators of the previous sections are certainly among the most popular ways to build switching regulators. They are simple, effective, inexpensive, and often most of the electronics are contained in a single integrated circuit.

However, these simple schemes also have several disadvantages. The buck can only step the voltage down and the boost can only step the voltage up. If your application requires several outputs, some above and some below the input DC voltage, you would have to build several separate boost and buck regulators. If the input voltage may be above or below the regulated output, neither a buck nor boost, alone will work. A more complicated, less efficient series combination of boost followed by buck may be required. These two regulators require DC input voltage and furnish no isolation between the input power and the regulated output voltage. So neither works well directly from the AC line voltage supplied by the power company. A bulky, heavy, expensive, low frequency transformer, rectifier, and low frequency filter are needed to produce the isolated, unregulated DC input to the buck or boost. Finally, practical output power is typically less than 50W.

The forward converter is shown in Figure 4-40. The input DC voltage may come directly from the line voltage through a rectifier and filter. A low frequency isolation transformer is not required. Isolation is provided by the high frequency transformer, T1. It is much smaller, lighter, and potentially less expensive than a 60Hz transformer.

There may be as many secondaries as you need. Each is isolated from the primary and from each other. So you may create either polarity output by selecting which output to tie to ground. The output connected to R3 is considered the **master** output. Its voltage is sampled and used to control the pulse width of Q1. Should that output voltage begin to decrease, the pulse width modulator IC senses it and increases the time that Q1 is on. This increases the energy coupled into the secondary, bringing the output voltage back to its regulated value.

Table 4-4 Dual switching regulator parameter calculations
(Courtesy of National Semiconductor)

transformer turns ratio	N	# secondary turns / # primary turns
transformer efficiency	η	typically 0.95
load currents	ΣI_{load}	$|+I_{load}| + |-I_{load}|$
duty cycle	D	$\dfrac{V_{load}}{NE_{in} + V_{load}}$
primary current ripple	ΔI_P	$\dfrac{D(V_{in} - V_{sat})}{25,000\, L_P}$
peak primary current	$I_{P\ (diode)}$	$\dfrac{N}{\eta} \times \dfrac{\Sigma I_{load}}{1-D} + \dfrac{\Delta I_P}{2}$
diode reverse voltage	$V_{piv\ (diode)}$	$V_{load} + N(E_{in} - V_{sat})$
average diode current	$I_{D\ ave}$	I_{load}
peak diode current	$I_{P\ (diode)}$	$\dfrac{I_{load}}{1-D} + \dfrac{\Delta I_P}{2}$
short circuit diode current		$\approx \dfrac{6\,A}{N}$
R_C		$\leq \dfrac{750\,\Sigma I_{loadmax}(V_{load} + E_{in\,min}N)^2}{(E_{in\,min})^2}$
C_{out}		$\geq \dfrac{0.19\,L R_C\,\Sigma I_{load\,max}}{E_{in\,min}V_{out}}$
		$\geq \dfrac{E_{in\,min}R_C N^2\,[E_{in\,min} + (3.74 \times 10^5 L)]}{487,800\,(V_{load})^2\,(V_{load} + E_{in\,min}N)}$
C_C		$\geq \dfrac{58.5\,(V_{load})C_{out}\,(V_{load} + E_{in\,min}N)}{R_C^2 E_{in\,min}N}$
ESR		$\leq \dfrac{V_{loadPP}}{\Delta I}$
		$\leq \dfrac{8.7 \times 10^{-3}\,E_{in}\,V_{load}N}{\Sigma I_{loadmax}[V_{load} + (E_{in\,min}N)]}$
P_{LM2575}		$0.25\,\Omega\left(\dfrac{N\Sigma I_{load}}{1-D}\right)^2 + \dfrac{NI_{load}D}{50(1-D)}V_{in}$

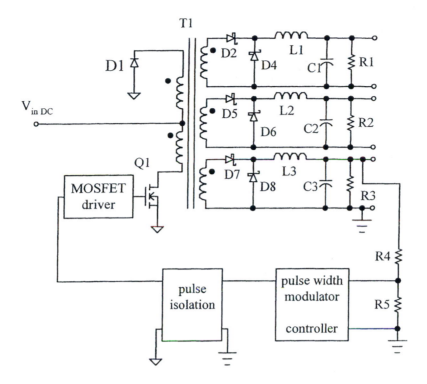

Figure 4-40 Forward converter

The other outputs are called **slaves**. Changes in their load current are sensed as a change in the secondary magnetic flux. This causes a variation in the master output, which is then corrected by the pulse width modulator controller. Typical regulation of a slave output is ≈8%.

The pulse from the controller must be coupled back to the transistor driver. To maintain the isolation between the primary and the secondary, use another set of windings on the main transformer, a separate transformer, or an optocoupler.

The n-channel enhancement mode MOSFET is a good choice for the transistor switch. These transistors are fast, have very low on resistance, and are inexpensive. Assuming that the regulator is 80% efficient,

$$P_{out} = 0.8P_{in}$$

or

$$P_{in} = 1.25P_{out}$$

$$P_{in} = I_{in\,ave} \times V_{in\,DC\,min}$$

It is critical to assure that the circuit always operates in the continuous mode. So, it is reasonable to assume that at the minimum input voltage, the controller will be producing its largest duty cycle, 0.8. But, since there are two sides of the primary, one driven by the transistor, and the other by the flyback diode, D1,

$$D \approx \frac{0.8}{2} = 0.4$$

So

$$I_{in\,ave} = 0.4 I_{peak\,in}$$

$$P_{in} = 0.4 I_{peak\,in} V_{in\,DC\,min}$$

$$P_{in} = 1.25 P_{out}$$

$$0.4 I_{peak\,in} V_{in\,DC\,min} = 1.25 P_{out}$$

$$I_{peak\,in} = \frac{3.13 P_{out}}{V_{in\,DC\,min}} = \frac{3.13 \sum I_{load} V_{load}}{V_{in\,DC\,min}}$$

The maximum voltage across Q1 occurs when it is off and D1 goes on. This grounds the top of the transformer, dropping $V_{in\,DC}$ across that part of the primary. Since the top and the bottom halves of the transformer are identical, a similar voltage drop is coupled across the lower half of the primary, with the undotted end at $V_{in\,DC}$ above the dotted end. This drop across the primary must be added to the $V_{in\,DC}$ at the junction of the two primaries. Allowing a 30% safety factor, the result is

$$V_{Q1\,max} = 1.3 \times 2 V_{in\,DC\,max} = 2.6 V_{in\,DC\,max}$$

To turn Q1 on quickly, a lot of charge must be moved onto its gate's parasitic capacitance. This requires a driver that can provide an ampere or more of current, for less than a microsecond. A special purpose MOSFET driver IC is often used.

The flyback diode must handle the same current and the same maximum voltage that Q1 handles. It also must be as fast as Q1, turning on or off in fractions of a microsecond. A signal diode cannot handle the current, and regular rectifier diodes are too slow. Typically the peak inverse voltage rating of the Schottky diode is too small. A gold-diffused diode is usually required.

The diodes in the secondary must be equally fast. Their peak inverse voltage ratings, however, depend on the output voltages and the transformer turns ratios. The inductor and the capacitor are selected just as they were in the buck regulator. Remember, though, that the input to these filters is the **secondary** voltage, which is set by the transformer turns ratio. The output resistor is selected to assure that there is always 10% to 20% of the maximum load current flowing, keeping the inductor in the linear region of its hysteresis curve.

The biggest disadvantage of the forward converter is the requirement that the transistor withstand more than twice the primary peak voltage. For equipment powered from 220V_{RMS}, this is over 600V. The **half bridge** topology of Figure 4-41 requires that each primary transistor tolerate only the peak voltage, half of that required by the forward converter.

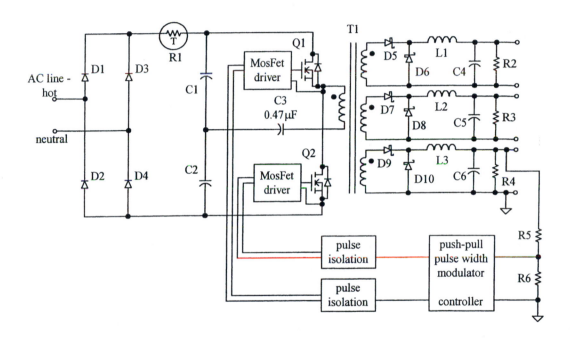

Figure 4-41 Half bridge switching regulator

During the positive half cycle of the line voltage, the capacitors C1 and C2 charge through D1 and D4. On the negative half cycle, D3 and D2 turn on to charge the series capacitors C1 and C2. The thermistor R1 is needed when the power supply is first turned on. When the power is first applied, the capacitors are fully discharged, presenting a dead short across the line. Without the thermistor, this would either damage the rectifier diodes or blow any line fuse you may have. Initially the thermistor is cool, offering a relatively high resistance. This limits the in-rush current. After many cycles, the capacitors charge up, requiring less current each cycle. During this charge time, the thermistor warms up. Since the thermistor has a negative temperature coefficient, its resistance drops. By the time the capacitors are charged, the thermistor's resistance is negligible. This puts $168V_{DC}$ between the top and the bottom rails, $84V_{DC}$ across each capacitor.

The transistors are turned on during alternate half cycles by the push-pull controller. Either transistor may be on, or both may be off. They are never both on at the same time. When Q1 is turned on, capacitor C1 provides current through Q1, down through the primary, and back to C1. This is the positive switching half cycle. During the negative switching half cycle, Q2 is on. Current flows from C2, through C3, up through the transformer primary, and then down through Q2, back to C2.

If the two main capacitors are not charged to exactly the same voltage, more current will flow one half cycle than the other. On the average, this results in a DC voltage across the primary and a DC current through it. That DC current will alter the magnetic characteristics of the transformer and eventually cause the supply's failure. Capacitor C3 blocks this DC current, preventing the failure.

When transistor Q1 is on, the voltage on its source is that at the top rail. This voltage is also on the drain of Q2, which is off. The source of Q2 is tied to the bottom rail. So each transistor will experience the full rail-to-rail voltage during part of the cycle that it is off. When powered directly from the $120V_{RMS}$ line, this is about 168V.

Maximum forward current for each transistor is the same as in the forward converter, from Figure 4-40.

$$I_{peak\,in} = \frac{3.13P_{out}}{V_{in\,DC\,min}} = \frac{3.13\sum I_{load}V_{load}}{V_{in\,DC\,min}}$$

Driving the transistors is a little more complicated than in the forward converter. To turn an n-channel enhancement MOSFET hard on requires 4V to 10V (depending on the transistor) between the gate and the source. But, the source is **not** tied to circuit common. It is tied to the transformer. When the transistor goes on, this voltage will be very close to the input rails (84V on each side). So the MOSFET drive must place the gate above the rail. That is why there are two lines coming from the isolation. One is the reference and is tied to the transistor's source. When the transistor goes on, and its source voltage jumps to 84V, the gate to source remains constant, keeping the transistor on.

The two transistors must be driven on in alternate half cycles. The duty cycle during that **half** cycle is varied by the controller to regulate the master output. Both transistors must **never** be on at the same time. During a full cycle, the push-pull controller turns both transistors off, turns Q1 on for a variable time, turns both transistors off, and then turns Q2 on for a variable time. Such push-pull pulse width controllers are available as an integrated circuit.

With proper attention to detail, over 500W may be delivered to a variety of loads at a mix of voltage levels and polarities.

Summary

This chapter has described the use of analog integrated circuits as power supply regulators. A simple power supply uses a step-down transformer, bridge rectifier, filter capacitor, and an op amp based regulator. Under light load (<10% ripple), there is a group of equations that allows you to analyze or design a filtered supply. With higher ripple, a set of curves or simulation should be used.

A zener diode can be used to provide a degree of regulation, but the addition of an op amp significantly improves regulation and even allows you to adjust the regulated output. A series (pass) transistor can be added to the op amp to boost the load current. You must be careful to consider the transistor's power dissipation and heat sink requirements.

There are several series of three-terminal IC regulators. These combine the reference zener, error op amp, and pass transistor with thermal shutdown, short circuit limiting, and high ripple rejection. Filter capacitance should be kept as low as possible while assuring that the input, unregulated voltage is 2V above the regulated output at all times. This procedure lowers cost, size, and power dissipation requirements.

Although most three-terminal regulators provide a single fixed voltage, the ground potential can be floated to allow you to adjust the output voltage. The LM317 and LM337 are optimized to work as adjustable regulators. Combined with an op amp, these regulator ICs can be used to produce a circuit in which multiple outputs track a single input voltage.

A switched capacitor voltage converter, such as the 7660, generates a negative voltage when only a positive supply is provided. This is done by charging an external pump capacitor to the positive supply and then reversing that capacitor's connections when it is switched to the output capacitor. A series connection of the ICs allows you to double or triple as well as invert the input voltage. Arranging the capacitors so that they are charged in series and then placed in parallel provides a split in voltage that is far more accurate than that obtained using resistive voltage division.

Linear regulators are highly inefficient because there is significant voltage across their pass element while significant current flows through them. Switching regulators, however, have the pass elements either hard on (very low voltage drop) or hard off (no current). Their efficiencies may exceed 90%. The output amplitude is adjusted by altering the width of the switching pulse. Two popular configurations are the buck (step-down) and the boost (step-up). Circuit design has been simplified with the LM2575 for buck circuits and the LM2577 for boost configurations. In addition to the power switching IC, an inductor is selected to limit current variation. If that inductor is discharged fully every cycle, the operation is called discontinuous. Although regulation may continue, it is not unusual for considerable noise to enter the system. You can keep the filter's operation in the continuous mode, by increasing the size of the inductor. This assures that it never fully discharges, and it reduces ripple and noise. To complete each supply, a main filter capacitor must be added in parallel with the load.

This lowers the ripple voltage and assures loop stability. A modification of the boost circuit produces a dual output.

Both the buck and the boost regulators are limited to output power in the tens of watts. Neither isolates the input common from that of the output. Producing multiple outputs of either polarity is unusual. Both the forward converter and the bridge switching regulator solve all of these limitations. A transformer isolates the input from the output. It also allows whatever combination of output voltages you need. However, in addition to the pulse width controller IC, isolation and a transistor driver are needed. The bridge configuration requires half the input voltage rating for the transistors, compared to the forward converter. This means that you can drive the bridge configuration directly from the $120V_{RMS}$ line, with proper precautions. Hundreds of watts can be delivered to the load.

These first four chapters have assumed ideal op amp characteristics. Under many circumstances this assumption works adequately. However, as the input voltage drops into the millivolt range, source impedance rises above $100k\Omega$, signal frequency rises above 10kHz, or rise times fall into the submicrosecond range, the actual performance of the transistors within the IC op amp noticeably alter the circuit's performance. In Chapter 5 you will see how the DC and AC characteristics of the op amp affect its performance. Limitations and compensation techniques will be discussed. You will also see how to select the op amp that is right for your particular application.

Problems

4-1   For the circuit in Figure 4-42, let $v_{sec} = 6.3V_{RMS}$, $C = 1500\mu F$, and $I_{load} = 73mA$. Calculate the following for the load voltage: v_p, v_{min}, $v_{ripple\ pp}$, and V_{DC}.

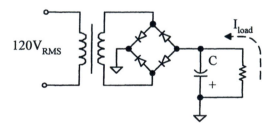

Figure 4-42 Schematic for Problem 4-1

4-2   Draw the schematic and determine the component values for a capacitive filtered, unregulated power supply which meets or exceeds the following:

$$I_{load} = 120mA_{DC}$$
$$V_{load\ p} < 9V_P$$
$$V_{load\ min} > 5V$$
$$V_{load\ ripple\ pp} < 2V_{PP}$$

4-3   For the circuit in Figure 4-42, let $v_{sec} = 12.6V_{RMS}$, $C = 470\mu F$, and $I_{load} = 500mA_{DC}$. Determine the following load voltages: v_p, v_{min}, $v_{ripple\ pp}$, and V_{DC}.

4-4   Calculate the values of the components necessary to build a capacitive filtered, unregulated power supply. Specify the smallest standard components that meet the following:

$$I_{load} = 1A_{DC}$$
$$V_{load\ min} \geq 7V$$
$$V_{load\ ripple\ pp} \leq 8V$$
$$V_{load\ p} \leq 30V$$

4-5   For the circuit in Figure 4-43, calculate the following:

$$V_{load\ DC}, \quad V_{out\ op\ amp}, \quad I_{out\ op\ amp}, \quad V_{Q1\ collector}, \quad P_{transistor}$$

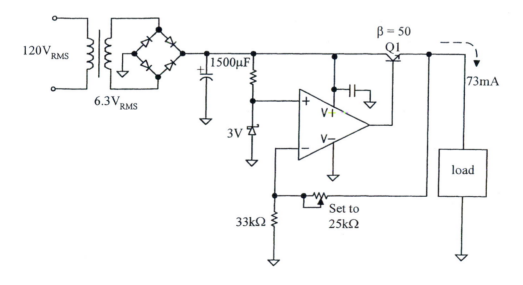

Figure 4-43 Schematic for Problem 4-5

4-6   Determine the component values necessary to build a $+5V_{DC}$, op amp/zener/transistor-type regulator for a load current of 120mA. Draw the complete power supply schematic.

4-7   For the circuit in Figure 4-44, $I_{load} = 50mA$. Do the following:
   a. Calculate $V_{capacitor\ DC}$ and T_J.
   b. If T_J is less than the maximum allowable T_J, assume a higher I_{load} and repeat step a.
   c. If T_J is above the maximum allowable T_J, assume a lower I_{load} and repeat step a.
   d. Continue steps b and c until you find a T_J that is within 10% of $T_{J\ max}$. This is I_{load} that causes thermal shutdown.

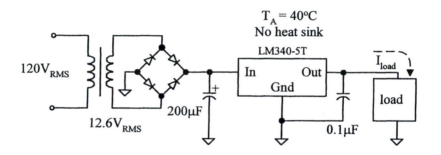

Figure 4-44 Schematic for Problem 4-7

4-8   Design a solid state power supply, using a full wave bridge rectifier and a three-terminal integrated regulator, to meet or exceed the following:

$$V_{load} = 8V_{DC}\ regulated \qquad\qquad I_{load} \leq 800mA_{DC}$$
$$V_{ripple\ load} \leq 0.1V_{pp} \qquad\qquad\qquad T_A \leq 60°C$$

   a. Draw the schematic diagram.
   b. Determine the AC input to the rectifier.
   c. Determine the allowable input minimum voltage to the regulator.
   d. Determine the input filter capacitance required.
   e. Determine the appropriate DC input voltage to the regulator.
   f. Select the diode ratings.
   g. Select the regulator and heat sink (if necessary)

4-9   Repeat Problem 4-8 using an LM317T with an output adjustable from 3V to 15V.

4-10   Calculate all of the output voltages for the circuit in Figure 4-45.

4-11   Design a tracking regulator using an LM317 and op amps to output +5V, ±12V, all adjustable ±10% with a single adjustment. Select a heat sink (if necessary) for I_{load} = 100mA$_{DC}$, E_{in} = ±18V$_{DC}$, and T_A = 30°C.

4-12   Design a power supply using an LM2575-ADJ to meet the following (these specifications are similar to those for the linear regulator in Problem 4-8):

V_{load} = 8V$_{DC}$ $E_{in\,min}$ = 12V$_{DC}$ I_{load} = 800mA$_{DC}$

$\Delta I \leq$ 200mA$_{PP}$ $V_{load\,ripple} \leq$ 0.1V$_{pp}$ T_A = 60°C

Determine all components and the heat sink, if needed.

4-13   Repeat Problem 4-12 using a boost regulator with E_{in} = 5V$_{DC}$.

4-14   Compare the three designs you have produced in Problems 4-8, 4-12, and 4-13. Discuss ease of design, size, cost, efficiency, noise, availability of parts, and any other factors that will influence your choice of a circuit topology.

4-15   Review manufacturers' literature to find an integrated circuit pulse width modulator controller that could be used in the forward converter of Figure 4-40.

4-16   Explain the reason for the pulse isolation in Figure 4-40, and find parts to implement it.

4-17   Explain how the pulse width modulator controller used in the bridge converter of Figure 4-41 is different from the simpler controller used in the forward converter of Figure 4-40. Find an IC that will work in the bridge converter.

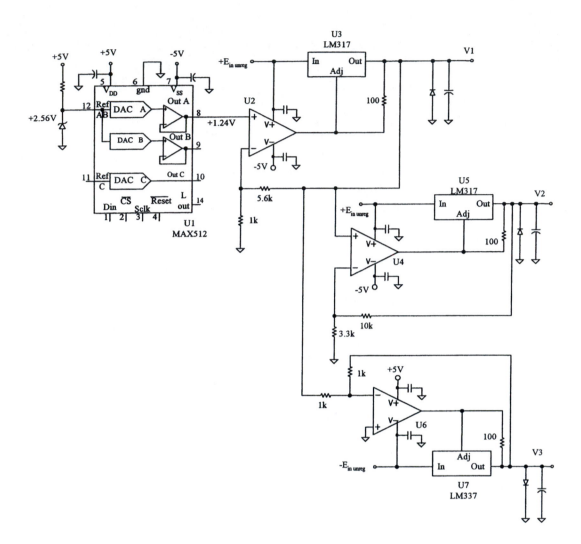

Figure 4-45 Schematic for Problem 4-10

Dual Tracking Regulator Lab Exercise

Purpose

Examine the performance of a dual tracking regulator with adjustable three-terminal regulators and an op amp.

A. Circuit Construction

1. Refer to the circuit in Figure 4-16. Use a $25.2V_{RMS}$ center tapped transformer. Set C1 and C2 to 330μF. Build just the unregulated part of the circuit; the transformer, rectifier and the two capacitors. Apply power.

2. Verify that the voltage across each 330μF capacitor is approximately $18V_{DC}$.

3. Build the remainder of the circuit, without the 500Ω load. The 3V zener may be replaced with two forward biased LEDs. Be careful of the pinouts for the LM317 and the LM337. They are different. Also, neither tab is ground. Both are electrically hot.

4. Set the R3 potentiometer to a short. Set the R5 potentiometer to 4.7kΩ.

5. Apply power. Verify the following DC voltage measurements with respect to ground: $V^+ = 1.2V_{DC}$, $V_{U2\ adj\ pin} = 0V$, $V_{U4\ adj\ pin} = 0V$, $V^- = -12V_{DC}$

6. Adjust R3 to set $V^+ = 8V_{DC}$.

7. Adjust R5 to set V^- to –8V. Do not alter that potentiometer during the remainder of the experiment.

B. Tracking Evaluation

1. Without the load resistor, adjust the R3 potentiometer to sweep V^+ in 1V steps from 5V to 15V. At each voltage, record V^+, V^-, $V_{U4\ adj\ pin}$, and $V_{U4\ adj\ pin}$.

2. Connect the 500Ω, 5W load with a series DC ammeter as shown in Figure 4-16. Display V^+ on one channel of the oscilloscope and the voltage across C1 on the other channel of the oscilloscope. Assure that v_{min} is not violated. Repeat step B1 with $I_{load} = 100mA_{DC}$. You will have to adjust the load resistor to keep $I_{load} = 100mA_{DC}$ each time you adjust R3.

C. Data Analysis

1. Plot two separate graphs, one for each data table gathered in section B.

2. On each graph, plot V^-, $V_{U4\ adj\ pin}$, and $V_{U4\ adj\ pin}$ as a function of V^+.

3. Compare the plots to the theoretical plots and to each other, and discuss the effect of load on tracking.

Switched Capacitor Voltage Converter Lab Experiment

Purpose

During this lab experiment you will investigate the performance of the 7660 as a voltage inverter, a voltage doubler, and a precision voltage splitter. You will also evaluate the effect of loading the 7660.

A. Voltage Inverter Basic Operation
1. Configure the 7660 as a simple voltage inverter. Use 10μF capacitors and no load resistor.
2. Set E_{in} to $+5V_{DC}$.
3. Measure and record the output DC voltage and the output peak-to-peak ripple.
4. Increase E_{in} to $9V_{DC}$.
5. Measure and record the output DC voltage and the output peak-to-peak ripple.

B. Loading Effect
1. Connect a load potentiometer, initially set to 9kΩ.
2. Tabulate the output V_{DC} and the output ripple V_{PP} at each of the following load currents: 1mA, 2mA, 5mA, 10mA, 20mA, 30mA, 40mA, and 50mA.

C. Loading versus Capacitance
1. Change the capacitors to 100μF.
2. Repeat the load test, completing another data table as you did in step B2.

D. Voltage Doubler
1. Configure a 7660 to output $-10V_{DC}$ with a $+5V_{DC}$ input and no load.
2. Measure E_{in}, V_{load}, and $v_{out\ ripple\ pp}$.
3. Discuss the accuracy and any sources of error.

E. Voltage Splitter
1. Configure a 7660 to output $+4.5V_{DC}$ with a $+9V_{DC}$ input, no load.
2. Measure E_{in}, V_{load}, and $v_{out\ ripple\ pp}$.
3. Discuss the accuracy and any sources of error.

5

Operational Amplifier Characteristics

In the previous chapters you have used an ideal op amp in which the DC and AC characteristics have been ignored. For temporary circuits with rather loose requirements these characteristics may be ignored. However, you must be much more careful when building prototype or production grade circuits. These circuits are expected to exhibit stable operation over long time intervals and wide variation in temperature. They may have to respond identically to DC, audio, and radio frequencies. High speed transients may need to be transmitted. Input levels may be in or below the millivolt range from signal sources with high output impedances. To meet these requirements, you must choose an op amp whose DC and AC characteristics are appropriate to the unique set of requirements for that particular circuit. In this chapter you will learn the performance limitations imposed by the op amp's DC and AC characteristics, appropriate compensation techniques, and how to configure a composite amplifier using both precise and fast op amps to obtain the best from both.

Objectives

After studying this chapter, you should be able to do the following:
1. Define each of the following and explain the effect each has on the operation of the amplifier:
   a. input offset voltage.   f. small signal rise time.
   b. bias current.   g. slew rate.
   c. offset current.   h. full power response.
   d. drift.   i. noise.
   e. unity gain bandwidth
      (gain bandwidth product)
2. Draw a schematic and explain compensation techniques for each of the above.
3. Draw a schematic and explain techniques for measuring each nonideal DC characteristic.
4. List four ways to decrease internally generated noise in an op amp.
5. Explain the application and effects of single capacitor frequency compensation.
6. Explain the difference between voltage feedback and current feedback op amps, and use each correctly.
7. List the major types of op amps and the characteristics of each group.
8. Given the details of an op amp application, select an appropriate op amp.
9. Design a composite suspended supply amplifier.
10. Use an integrator to set the DC output offset of a high speed amplifier.

5-1 DC Characteristics

The ideal op amp draws no current from the source driving it. Both the inverting and the noninverting inputs look and respond identically. The circuit does not vary with temperature. Real op amps do not work this way. Current is taken from the source into the op amp's inputs. There are slight differences in the way the two inputs respond to current and voltage. A real op amp shifts its operation with temperature. These nonideal DC characteristics and compensation techniques are described in this section.

5-1.1 Input Offset Voltage

The op amp's input is a differential amplifier. For an op amp made with bipolar transistors, a simplified schematic is given in Figure 5-1. These two transistors are made to be as identical as possible. However, even when each input is driven from similar, low impedance sources, producing identical base currents, the two base-emitter junctions do not have the same V_{be} versus I_b characteristics. The result is that $V_{be1} \neq V_{be2}$. So the voltage at the inverting input does not equal that at the noninverting input. This voltage difference between the two inputs is the **input offset voltage**.

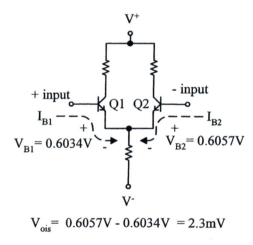

$$V_{ois} = 0.6057V - 0.6034V = 2.3mV$$

Figure 5-1 Op amp input with input offset voltage

        The effect of this is most clearly seen in the voltage follower configuration of Figure 5-2(a). Summing the voltages around the loop consisting of the noninverting input to the inverting input to V_{load} gives

$$-0.6034V + 0.6057V - V_{load} = 0$$

Solving for V_{load} yields

$$V_{load} = 0.6057V - 0.6034V = 2.3mV = V_{ios}$$

The input offset voltage has an even greater effect on the output voltage when the amplifier's gain is greater than 1. For the circuit in Figure 5-2(b)

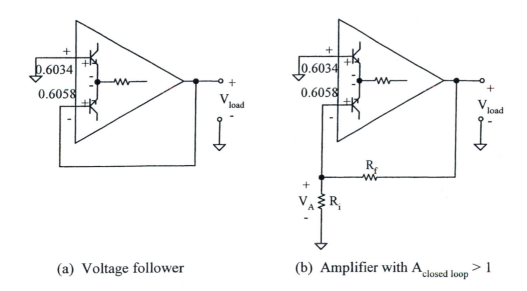

(a) Voltage follower   (b) Amplifier with $A_{closed\ loop} > 1$

Figure 5-2 Effect of input offset voltage

$$-0.6034V + 0.6057V - V_A = 0$$

$$V_A = 2.3mV = V_{ios}$$

But, by the voltage divider law,

$$V_A = \frac{R_i}{R_i + R_f} V_{load}$$

$$V_{ios} = \frac{R_i}{R_i + R_f} V_{load}$$

Solve for V_{load}.

$$V_{load} = \frac{R_i + R_f}{R_i} V_{ios}$$

$$V_{load} = \left(1 + \frac{R_f}{R_i}\right) V_{ios}$$

So, in reality, the input offset voltage of the op amp is multiplied by the noninverting gain of the amplifier circuit. To emphasize this, the input offset voltage is often modeled as a voltage source in series with an ideal (no offset voltage) op amp, as illustrated in Figure 5-3.

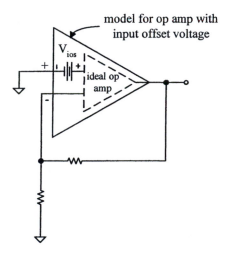

Figure 5-3 Amplifier using the model of an op amp with input offset voltage

Many op amps have offset null pins. These allow you to add an external network, including a potentiometer, to adjust the output DC voltage to zero. However, the addition of these extra components, the manufacturing step necessary to set the potentiometer, and the mechanical and thermal vulnerability of potentiometers make using these external compensation networks a bad idea. Precision op amps have become quite inexpensive, in fact, cheaper than the cost of the added parts and procedures required by an external compensation network. So, if the effect of the op amp's input offset voltage is too large, the single simplest, and best, answer is just to buy a better quality op amp.

To reduce the effect of input offset voltage, buy a better quality op amp.

5-1.2 Bias Currents

The op amp's input is a differential amplifier. It may be made of bipolar transistors or field effect transistors. In either case, these transistors must be biased, and that takes current. Look at Figure 5-4.

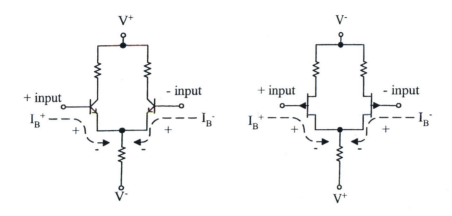

Figure 5-4 Op amp bias currents

For the bipolar transistors this bias current, supplied by the external circuit, is the base current, biasing the input transistors into their linear region. The bias current for the FET input op amps is actually the leakage current through the reverse biased gate-to-channel junction. For CMOS op amps, the bias current is orders of magnitude smaller, because of the SiO_2 insulation between the gate and the channel. But even for these transistors, some current leaks along the surface and through parasitic capacitances associated with the gate.

Manufacturers specify input bias current as the **average** of the bias current at each input.

$$I_B = \frac{I_B^+ + I_B^-}{2}$$

For the 741, a bipolar input op amp, the bias current is 500nA or less. A CMOS input op amp has currents in the fentoamperes (0.000001nA) range at room temperature.

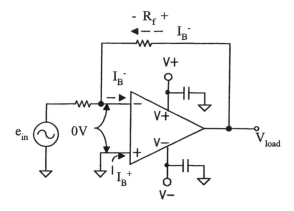

Figure 5-5 Basic amplifier with bias currents

Can such small current levels really have much effect? Figure 5-5 shows a basic inverting amplifier. When e_{in} is set to 0V, the output, V_{load}, should also be 0V (ignoring the input offset voltage). However, both the inverting and the noninverting inputs are drawing bias current from the external circuit. Current I_B^+ comes directly from ground and produces no voltage drop. Current I_B^- comes entirely through the feedback resistor (since both ends of the input resistor are at 0V, no current flows through it). Therefore, bias current I_B^- produces a voltage drop across R_f. Since the left end of R_f is held at virtual ground, the right end of R_f must be driven up by the op amp's output. Instead of an output voltage of 0V (ignoring V_{ios}),

$$V_{load} = I_B^- \times R_f$$

For the 741 and a 1MΩ feedback resistor,

$$V_{load} = 500nA \times 1M\Omega = 500mV_{DC}$$

The output is driven to 500mV because of the bias currents. In applications where signal levels are measured in millivolts, or even hundreds of millivolts, this is totally unacceptable.

This effect can be compensated for, as shown in Figure 5-6. A compensating resistor, R_{comp}, has been added between ground and the noninverting input. Current I_B^+ flowing through this resistor develops a voltage V_{NI}. Summing the voltages around the input-feedback-output loop, you obtain

$$- V_{NI} + 0 + V_f - V_{load} = 0V$$

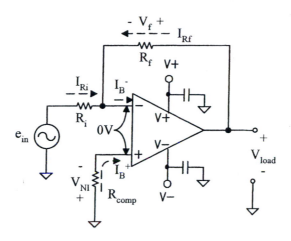

Figure 5-6 Bias current compensation

or

$$V_{load} = V_f - V_{NI}$$

If the proper size R_{comp} is selected, V_{NI} cancels V_f, and the output goes to zero. The value of R_{comp} is derived below.

$$V_{NI} = I_B^+ R_{comp}$$

or

$$I_B^+ = \frac{V_{NI}}{R_{comp}}$$

$$I_{Ri} = \frac{V_{NI}}{R_i}$$

$$I_{Rf} = \frac{V_f}{R_f}$$

For compensation, $$V_f = V_{NI}$$

$$I_{Rf} = \frac{V_{NI}}{R_f}$$

The current I_B^- is the sum of I_{Rf} and I_{Ri}.

$$I_B^- = I_{Rf} + I_{Ri}$$

$$I_B^- = \frac{V_{NI}}{R_f} + \frac{V_{NI}}{R_i}$$

$$I_B^- = V_{NI}\left(\frac{R_i + R_f}{R_f R_i}\right)$$

Assuming $\qquad I_B^- = I_B^+,$

$$V_{NI}\left(\frac{R_i + R_f}{R_f R_i}\right) = \frac{V_{NI}}{R_{comp}}$$

$$R_{comp} = \frac{R_f R_i}{R_i + R_f}$$

$$R_{comp} = R_i \,/\!/\, R_f$$

So to compensate for bias currents, the compensating resistor should equal the parallel combination of the resistors tied to the other input.

> **To compensate for bias currents, set the total resistance seen from the noninverting input to ground equal to the total resistance seen from the inverting input to ground.**

5-1.3 Offset Current

Bias current compensation works **if** the bias currents are equal. Since the input transistors cannot be made identical, there is always some small difference between I_B^+ and I_B^-. This difference is the **offset current**.

$$I_{OS} = \left| I_B^+ - I_B^- \right|$$

The absolute value sign indicates that there is no way to predict which of the bias currents is larger. Offset current for the 741 general purpose (bipolar transistor) op amp is always less than 200nA.

Even with bias current compensation, offset current produces an output voltage when the input voltage is zero. In Figure 5-7 (ignoring V_{ios}) and with $e_{in} = 0V_{DC}$,

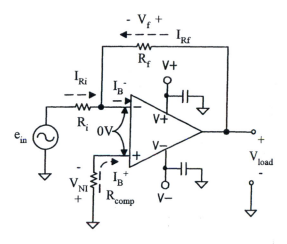

Figure 5-7 Effect of offset current

$$V_{NI} = I_B^+ R_{comp}$$

$$I_{Ri} = \frac{V_{NI}}{R_i} = \frac{I_B^+ R_{comp}}{R_i}$$

$$I_{Rf} = I_B^- - I_{Ri} = I_B^- - \frac{I_B^+ R_{comp}}{R_i}$$

Summing the input-feedback-output loop yields

$$-V_{NI} + I_{Rf} R_f - V_{load} = 0$$

$$V_{load} = I_{Rf} R_f - V_{NI}$$

$$V_{load} = I_{Rf} R_f - I_B^+ R_{comp}$$

Substituting for I_{Rf}, you obtain

$$V_{load} = \left(I_B^- - \frac{I_B^+ R_{comp}}{R_i} \right) R_f - I_B^+ R_{comp}$$

Select

$$R_{comp} = \frac{R_i R_f}{R_i + R_f}$$

After several steps of algebraic manipulation,

$$V_{load} = R_f \left(I_B^- - I_B^+ \right) = R_f I_{os}$$

So, even with bias current compensation, if the feedback resistor is 1MΩ, a 741 op amp could have an output as large as

$$V_{load} = 1M\Omega \times 2100nA = 200mV$$

with a zero input voltage.

5-1.4 Combined Effects

Input offset voltage, bias current, and offset current all occur in an op amp simultaneously. So, though you calculated the effect that each has singly on the output, in reality, the effects of these nonideal characteristics must be combined.

Example 5-1

Calculate the output voltage in the circuit in Figure 5-8(a), assuming that:

e_{in} is off (0V)
$V_{ios} = 2mV$
$I_B^+ = 110nA$
$I_B^- = 90nA$ ($I_B = 100nA,\ I_{os} = 20nA$)

These numbers are typical for the 741C op amp.

Solution

Look at the schematic as drawn in Figure 5-8(b). I_B^+ flows through the 180kΩ resistor, producing V_A.

$$V_A = -110nA \times 180k\Omega = -19.8mV$$

The negative polarity occurs because bias current flows into the op amp's input.

Between the noninverting input (point A) and the inverting input (point B) is V_{ios}.

$$V_B = -19.8mV + 2mV = -17.8mV$$

This causes a current to flow through the 100kΩ resistor.

$$I_{Ri} = \frac{17.8mV}{100k\Omega} = 178nA \quad \uparrow$$

The current flowing into the inverting input of the op amp is 90nA. So, of the 178nA coming up through R_i, 90nA goes into the op amp. The rest must go, left to right, through R_f.

$$I_{Rf} = 178nA - 90nA = 88nA \quad \rightarrow$$

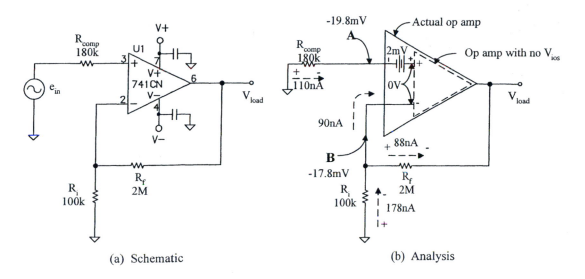

(a) Schematic   (b) Analysis

Figure 5-8 Example 5-1 manual calculations

This current flows through R_f, producing a voltage drop of

$$V_{Rf} = 88nA \times 2M\Omega = 176mV$$

Watch the polarity. The current flows left to right. So, the left end of R_f must be more positive than its right end. This voltage is added to the voltage at point B to get the load voltage.

$$V_{load} = -18mV - 176mV = -194mV$$

The simulator *Electronics Workbench*™ allows you to set many of an op amp's key parameters. Just bring the op amp onto the drawing area; then double click on it. The Opamp Properties dialog box appears. From it select the

LM741C. Then click on the Edit button. In the Edit window change the offset voltage, bias current, and offset current. These changes are shown in Figure 5-9.

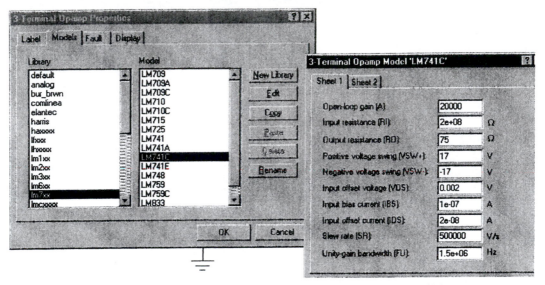

Figure 5-9 *Electronics Workbench™* op amp parameter editing

Then enter the rest of the circuit with appropriate DC voltmeters and ammeters. The result is shown in Figure 5-10. There is a little difference between the manually calculated voltages and those from the simulation, primarily at V_A. Following this difference through the rest of the parameters gives good correlation.

5-1.5 Characteristic Measurements

Direct measurement of the nonideal DC characteristics of an op amp cannot be done easily. Even the least expensive op amps have input impedances comparable to those of many digital multimeters. So measuring directly at any input pin seriously loads the circuit, altering the measurement. In addition, bias and offset currents are at or below the level measurable by many commonly available meters. However, by using the op amp to buffer the voltage produced by these input parameters, you can accurately measure the **output** voltage with no loading. A little thought then leads you to the op amp's DC characteristics.

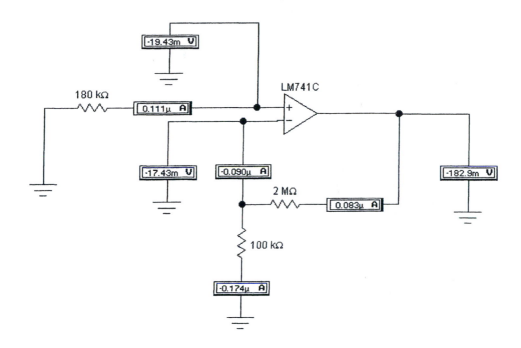

Figure 5-10 *Electronics Workbench™* simulation of Example 5-1

Measurement of the input offset voltage can be done directly with the voltage follower. Since there are no resistors, the bias current and the offset current have no effect on the output voltage. The output voltage is the input offset voltage. Of course, the voltmeter that you use must be accurate at the level of V_{ios}, often into the microvolt range.

Measurement of bias currents is shown in Figure 5-11. Bias current at one input is forced to flow through a large resistor. Since the bias current at the other input does not flow through a resistor, it produces no voltage and has no effect on the output voltage. The input offset voltage is also shown. Since each circuit is a voltage follower (as far as V_{ios} is concerned), this voltage just adds to the output. Having already determined V_{ios}, and by using large (MΩ) resistors, you can determine the bias current.

For I_B^- $$V_{load} = V_{ios} + I_B^- R_f$$

For I_B^+ $$V_{load} = V_{ios} - I_B^+ R_i$$

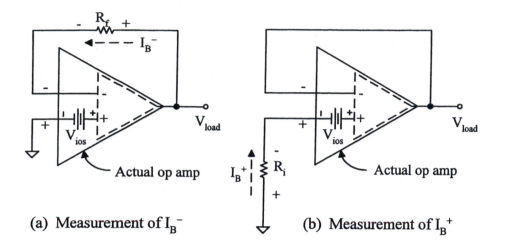

(a) Measurement of I_B^- (b) Measurement of I_B^+

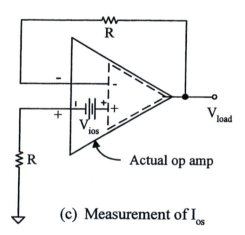

(c) Measurement of I_{os}

Figure 5-11 Measurement of bias currents

A combination of (a) and (b) allows you to measure offset current. With both resistors **exactly** equal, the effects of the bias currents cancel. Any voltage at the output is caused by offset current and input offset voltage.

$$V_{load} = V_{ios} + I_{os}R$$

Example 5-2

Use the techniques just described to "measure" V_{ios}, I_B^+, I_B^-, I_{bias}, and I_{os} of the LM741C used in Example 5-1.

Solution

The op amp model has been set up as it was in Figure 5-9, with $V_{ios} = 2mV$, $I_{bias} = 100nA$, and $I_{os} = 20nA$. The voltage follower in Figure 5-12 measures V_{ios} correctly.

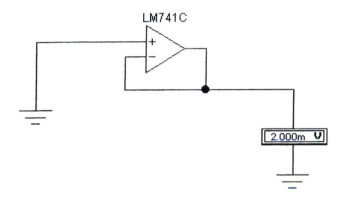

LM741C

2.000m V

Figure 5-12 Measurement of V_{ios} in *Electronics Workbench™*

I_B^+ is measured in Figure 5-13.

$$V_{load} = V_{ios} - I_B^+ R$$

$$I_B^+ = -\frac{V_{load} - V_{ios}}{R} = -\frac{-98mV - 2mV}{1M\Omega} = 100nA$$

I_B^- is measured in Figure 5-14.

$$V_{load} = V_{ios} + I_B^- R$$

$$I_B^- = \frac{V_{load} - V_{ios}}{R} = \frac{82mV - 2mV}{1M\Omega} = 80nA$$

I_{os} is measured in Figure 5-15.

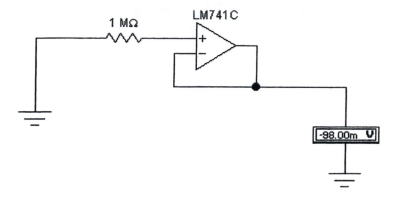

Figure 5-13 Measurement of I_B^+ in *Electronics Workbench*™

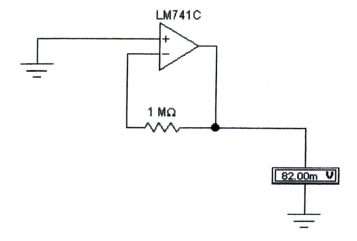

Figure 5-14 Measurement of I_B^- in *Electronics Workbench*™

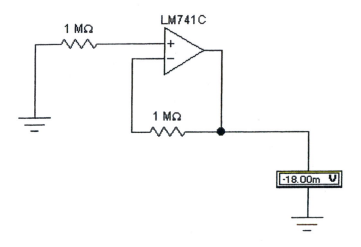

Figure 5-15 I_{os} measured in *Electronics Workbench*™

$$V_{load} = V_{ios} + I_{os}R$$

$$I_{os} = \frac{V_{load} - V_{ios}}{R} = \frac{-18mV - 2mV}{1M\Omega} = 20nA$$

5-1.6 Drift

Offset current and offset voltage change with temperature. For example, you may carefully null an amplifier with external resistors and a potentiometer shortly after turning on power, at an ambient temperature of 25°C. Later, you return to find that the temperature has risen to 35°C, and the circuit is no longer nulled. This is called **drift.**

Often, offset current drift is expressed in nA/°C, while offset voltage drift is measured in mV/°C. These indicate the **change** in offset for each degree Celsius **change** in temperature.

Example 5-3

If a noninverting amplifier with a gain of 100 is nulled at 25°C, what happens to the **output** voltage if the temperature rises to 50°C, given an input offset voltage drift of 0.15mV/°C?

Solution

$$0.15 \text{mV}/^\circ\text{C} \times \left(50^\circ\text{C} - 25^\circ\text{C}\right) = 3.75 \text{mV}$$

The **input** offset changes by 3.75mV. Since this is an input change, the output voltage changes by

$$V_{load} = V_{ios} \times A = 3.75 \text{mV} \times 100 = 375 \text{mV}$$

This could represent a major shift at the output.

The numbers specified for drift are not necessarily constant across the temperature range of an op amp. A drift of 0.15mV/°C may be 0.25mV/°C or –0.05mV/°C at 75°C. Read the specifications carefully. If a plot of offset versus temperature is given, it is far better to use that.

There are very few circuit techniques that you can use to minimize the effects of drift. Careful layout, to keep op amps away from sources of heat (power supply regulators and power transistors), and forced air cooling help stabilize the ambient temperature.

5-2 AC Characteristics

Although the op amp is used in some industrial applications as strictly a DC amplifier, many op amp applications are AC. Bias currents, offset current, input offset voltage, and drift all affect the steady state (DC) response of the op amp. How the op amp's output responds to **changes** in its input must also be considered. The effects of small signal sinusoidal ($<1V_p$) and large signal ($>1V_p$) or transient inputs on the op amp's output are presented in this section.

5-2.1 Gain Bandwidth Product

For the ideal op amp, you were told to assume that the open loop gain is arbitrarily large, and that this does not vary significantly with frequency. Neither of these assumptions is valid for a real op amp.

The frequency response of the 741C is given in Figure 5-16. The DC and very low frequency open loop gain is about 200k, which is generally large enough to be considered arbitrarily large. However, as frequency increases, the gain drops proportionally. The open loop gain falls until it reaches 1 at a frequency of 1MHz. This is a decrease of 20dB/decade. Expressed another way, **increasing** the frequency by a factor of 10 **decreases** the gain by 10. At any point along the curve, the product of gain and frequency (or bandwidth) is a constant 1MHz.

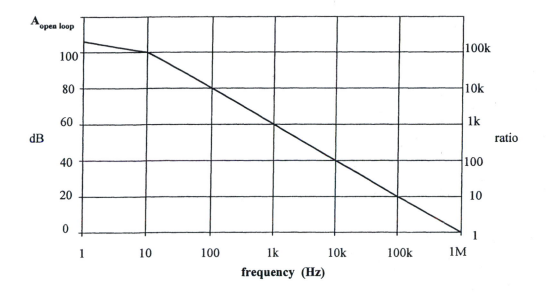

Figure 5-16 Open loop frequency response of a 741C

Example 5-4

Determine the gain and gain bandwidth product at 10Hz, 1kHz, 100kHz, and 1MHz.

Solution

For each of the frequencies, A_{OL} can be read from Figure 5-16.

At 10Hz:	$A_{OL} = 100k$	$GBW = 100k \times 10Hz = 1MHz$
At 1kHz:	$A_{OL} = 1k$	$GBW = 1k \times 1kHz = 1MHz$
At 100kHz:	$A_{OL} = 10$	$GBW = 100k \times 10Hz = 1MHz$
At 1MHz:	$A_{OL} = 1$	$GBW = 1 \times 1MHz = 1MHz$

Initially, you might think that this drastic roll-off of gain at high frequencies is a disadvantage. However, very often it is intentionally built into the op amp by installing a simple low pass RC filter, as shown in Figure 5-17.

This reduced gain ensures that high frequency noise does not become large enough, when fed back into the input through parasitic capacitance, to cause oscillations. A stable amplifier is assured this way, but only by sacrificing frequency response.

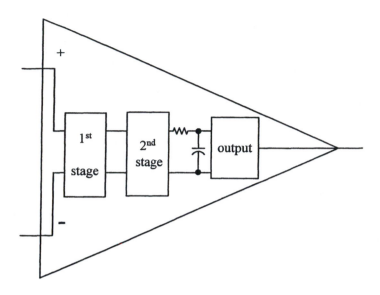

Figure 5-17 Internal frequency compensation

This reduction in open loop gain also severely limits the high frequency gain when negative feedback is used. This is illustrated in Figure 5-18 for an amplifier with a closed loop gain (A_o) of 100 (40dB). At low frequencies, the open loop gain is so much larger than the closed loop gain that the circuit's performance is not affected. But as the open loop gain falls closer to the closed loop gain, the closed loop gain begins to suffer, and falls below that predicted by ideal theory. When the open loop gain coincides with the low frequency closed loop gain (100 in this example), the actual closed loop gain has fallen to 0.707 of its low frequency level. In terms of decibels, the gain has fallen -3dB below its low frequency level. This is the high frequency cut-off, f_{-3dB}, and defines the small signal bandwidth of the amplifier.

$$f_{-3dB} = \frac{GBW}{A_O}$$

where   GBW = gain bandwidth product (1MHz of the 741C)
              A_o = circuit's closed loop DC gain (100 in Figure 5-18)

Above this high frequency cut-off, the closed loop gain closely follows the fall of the open loop gain.

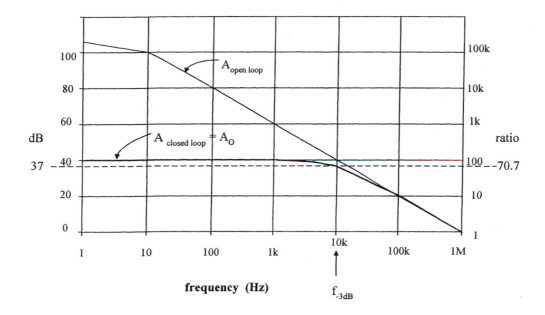

Figure 5-18 Closed loop frequency response

Example 5-5

What is the maximum closed loop gain you can get from a 741C based circuit with a cutoff of 20kHz (upper audio frequency limit)?

Solution

$$f_{-3dB} = \frac{GBW}{A_O}$$

or

$$A_O = \frac{GBW}{f_{-3dB}} = \frac{1MHz}{20kHz} = 50$$

So to use a 741C op amp in an audio amplifier, you must limit the closed loop gain to 50 or less.

Closely examine Figure 5-18 and you see that even at frequencies below f_{-3dB}, the closed loop gain falls significantly below the 100 predicted for the DC gain. At f_{-3dB} the gain has fallen 30% below its DC value. This amount of error may be hard to accept. What is the highest frequency that can be used before the gain drops 1% from its DC value? Ten percent from the low frequency value? Knowing these two frequencies may give better guidance than f_{-3dB}.

$$f_{1\%} = \tfrac{1}{7} f_{-3dB}$$

$$f_{10\%} = \tfrac{1}{2} f_{-3dB}$$

These equations tell you the frequency at which the gain has fallen 1% (or 10%) below its value at DC.

Example 5-6

Run a computer simulation of an inverting amplifier with a gain of -1000 using a 741C. Plot the frequency response from 1Hz to 1MHz. Locate f_{3dB}, $f_{10\%}$, and $f_{1\%}$. Compare these values to those obtained from theory.

Solution

The simulation is run using *Electronics Workbench*™. The schematic and the setup for the 741C are shown in Figure 5-19. The library for the 741C has been edited to set the gain bandwidth to 1MHz. The Bode Plotter has been connected to the input and output to allow the display of the amplifier's frequency response. Double clicking on the Bode Plotter in *Electronics Workbench*™ enlarges it and allows you to manipulate its cursor. This is shown in Figure 5-20(a). The cursor is set at the lowest frequency, giving a pass band gain of $A_o = 995$, not the 1000 predicted by theory.

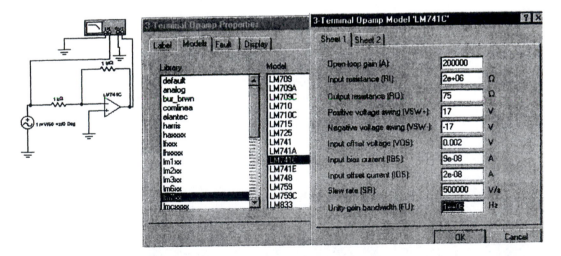

Figure 5-19 Schematic and setup for Example 5-6 using *Electronics Workbench*™

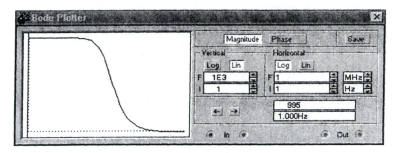

(a) A_O=995 at 1Hz

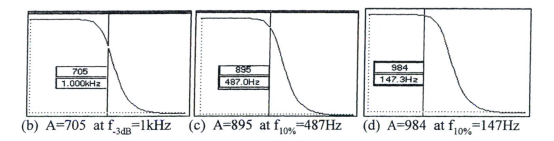

(b) A=705 at f_{-3dB}=1kHz   (c) A=895 at $f_{10\%}$=487Hz   (d) A=984 at $f_{10\%}$=147Hz

Figure 5-20 Example 5-6 frequency responses using *Electronics Workbench*^{TM}

The high frequency cutoff occurs at

$$f_{-3dB} = \frac{GBW}{A_O} = \frac{1MHz}{995} = 1005Hz$$

Moving the Bode plotter's cursor shows a simulated cut-off in Figure 5-20(b) of 1000Hz at A = 705. This is within the resolution of the simulation.

The 10% down frequency occurs at

$$f_{10\%} = \frac{f_{-3dB}}{2} = \frac{1005Hz}{2} = 503Hz$$

The simulation, shown in Figure 5-20(c), gives a gain of 895 (90% of 995) at 487Hz.

The 1% down frequency occurs at

$$f_{1\%} = \frac{f_{-3dB}}{7} = \frac{1005Hz}{7} = 144Hz$$

The simulation in Figure 5-20(d) gives a gain of 984 (99% of 995) at 147Hz.

So the frequency response can prove to be much more restrictive than the 1MHz gain bandwidth product specification may at first seem to indicate.

There are two solutions to this restriction. The simplest is to select an op amp with a higher gain bandwidth product. Careful though! Excessive gain bandwidth will allow high frequency **noise** to contaminate the output. Select an op amp with **just enough** gain bandwidth. The signal that you are processing passes properly, and high frequency noise is eliminated, since the gain rolls off as the frequency goes up.

An alternative to buying more expensive, higher frequency op amps is to build the circuit with several stages, each stage with a lower gain (and therefore higher frequency response). This is illustrated in Example 5-7.

Example 5-7

Design a circuit with a gain of $125 \pm 10\%$ over the full audio frequency band. Set the input impedance to 560Ω.

Solution

From the problem statement,

$$f_{10\%} = 20\text{kHz}$$

But

$$f_{10\%} = \frac{f_{-3\text{dB}}}{2}$$

So

$$f_{-3\text{dB}} = 2f_{10\%} = 2 \times 20\text{kHz} = 40\text{kHz}$$

You can now calculate the op amp's required gain bandwidth.

$$\text{GBW} = A_O \times f_{-3\text{dB}} = 125 \times 40\text{kHz} = 5\text{MHz}$$

a.   Single op amp solution

Since the 741C has a GBW of 1MHz, you need a different op amp. The LM318 has a GBW of 15MHz. This works well. For an inverting amplifier,

$$R_i = Z_{in} = 560\Omega$$

$$R_f = -A_O \times R_i = -(-125) \times 560\Omega = 70\text{k}\Omega$$

Use a $2\text{k}\Omega$ resistor in series with a $68\text{k}\Omega$ resistor.

b.   Two op amp solution

Another solution is to use two cascaded amplifiers, each with a gain of 11.2. Then each separate amplifier needs a

$$\text{GBW} = A_O \times f_{-3\text{dB}} = 11.2 \times 40\text{kHz} = 448\text{kHz}$$

A much slower, more stable, and less expensive amplifier may be used. In fact, an IC containing two op amps, each with GBW = 500kHz, may be much less expensive than the faster single op amp.

5-2.2 Small Signal Rise Time

For small signals ($<1V_P$), the gain bandwidth product, and therefore f_{3dB}, gives all of the speed information you need. However, measuring the open loop gain at a variety of frequencies to generate the frequency response plot of the op amp is impractical. The op amp's nonideal DC characteristics drive the output into saturation if you try to run the op amp open loop. So, another measurement is needed.

In its simplest version, the compensation network inside the op amp is an RC low pass filter. Applying a step to a series RC circuit causes the capacitor to charge exponentially.

$$v_c(t) = V_P\left(1 - e^{-\frac{t}{RC}}\right)$$

Solve this for t.

$$t = -RC \ln\left(1 - \frac{v_c}{V_P}\right)$$

This is the time it takes the capacitor to charge to v_c, where V_P is the final, peak voltage the capacitor reaches.

The time it takes to get to 10% of its final value is

$$t_{10\%} = -RC \ln\left(1 - \frac{0.1V_P}{V_P}\right) = 0.11RC$$

To get to 90% of the final value , it takes

$$t_{90\%} = -RC \ln\left(1 - \frac{0.9V_P}{V_P}\right) = 2.3RC$$

The rise time is defined as the time it takes to go from the 10% of V_P to 90% of V_P. That is,

$$t_{rise} = t_{90\%} - t_{10\%} = 2.3RC - 0.1RC = 2.2RC$$

Rearranging yields

$$RC = \frac{t_{rise}}{2.2}$$

From your studies of AC circuits, you should remember that looking across the capacitor in a simple series RC circuit forms a low pass filter. That filter has a cutoff frequency of

$$f_{-3dB} = \frac{1}{2\pi RC}$$

Rearranging gives $\qquad RC = \frac{1}{2\pi f_{-3dB}}$

Now, equate the two relationships, one from the rise time and one from the frequency response.

$$RC = \frac{1}{2\pi f_{-3dB}} = \frac{t_{rise}}{2.2}$$

Simplify. $\qquad t_{rise} = \frac{2.2}{2\pi f_{-3dB}} = \frac{0.35}{f_{-3dB}}$

Since open loop gain frequency response is so difficult to measure, it is simpler to set the op amp as a voltage follower, input a small step ($<1V_p$), and measure the amplifier's output rise time. The gain bandwidth product can then be calculated.

$$GBW = A_O \times f_H = 1 \times \frac{0.35}{t_{rise}}$$

5-2.3 Slew Rate

Gain bandwidth product and small signal rise time are characteristics of small signals, with peak voltages less than 1V. For many front-end applications this may be adequate. But most power or driver stages must output larger signals.

For larger signals, $v_p > 1V$, the op amp's speed is limited by its slew rate. Slew rate is defined as the maximum rate at which the output voltage can rise.

$$SR = \frac{dv}{dt}\Big|_{max}$$

The rate at which the voltage can change is limited primarily by the internal frequency compensation capacitor. The voltage across this capacitor cannot change instantaneously but is governed by the size of the capacitor and the amount of current available to charge the capacitor.

$$i_{capacitor} = C\frac{dv}{dt}$$

Or
$$\frac{dv}{dt} = \frac{i}{C}$$

This gives
$$SR = \frac{dv}{dt}\Big|_{max} = \frac{i_{max}}{C}$$

The 741C uses a 30pF internal capacitor and has a maximum internal capacitor charging current of about 15µA. So the slew rate of a 741C is

$$SR = \frac{dv}{dt}\Big|_{max} = \frac{i_{max}}{C} = \frac{15µA}{30pF} = 0.5V / µs$$

The units of slew rate are volts per microsecond. Independent of all other considerations (GBW, t_{rise}, A_{OL}, A_o, input, etc.), it takes at least 2µs for the output to change 1V, or 40µs for the output voltage to swing from one extreme (−10V) to the other (+10V).

Example 5-8

A square wave with negligible rise time and a peak-to-peak amplitude of $500mV_{PP}$ must be amplified to a peak-to-peak amplitude of $6V_{PP}$ with a **rise time** of 4µs or less.

a. Can a 741C be used?

b. Can a 318 be used?

Solution

Since the output has a peak amplitude greater than 1V, the slew rate must be considered. The time interval given is the time for the output signal to go from 10% to 90% of its peak-to-peak swing. So

$$\Delta V_{out} = 90\% V_{out\,PP} - 10\% V_{out\,PP} = 80\% V_{out\,PP}$$

The slew rate is

$$SR_{needed} = \frac{\Delta V}{\Delta t} = \frac{0.8 \times 6V_{PP}}{4µs} = 1.2V / µs$$

a. The 741C has a 0.5V/µs slew rate. It is too slow.

b. The 318's slew rate is 50V/µs. It is more than fast enough. The rise time using the 318 is

$$SR = \frac{\Delta V}{\Delta t}$$

or $\qquad \Delta t = \dfrac{\Delta V}{SR} = \dfrac{0.8 \times 6V_{PP}}{50V / \mu s} = 0.096\mu s = 96ns$

5-2.4 Full Power Response

Slew rate limits the response speed to all large signal wave shapes. The effects on a square wave are easily calculated, as in Example 5-8. However, deriving the slew rate limiting of a sine wave requires a bit more work.

$$SR = \frac{dv}{dt}$$

$$v(t) = V_P \sin(2\pi ft)$$

$$SR = \frac{d}{dt}\left[V_P \sin(2\pi ft)\right]$$

$$SR = 2\pi f V_P \cos(2\pi ft)$$

This is a maximum (worst case) at $t = 0$ (cosine goes to 1).

$$SR_{max} = 2\pi f V_P \cos(0) = 2\pi f V_P$$

This tells you that the maximum frequency and the peak amplitude of a sine wave are limited by the op amp's slew rate. Solve for f_{max}.

$$f_{max} = \frac{SR_{max}}{2\pi V_P}$$

This f_{max} is called the **full power response**. It is the highest frequency for a large amplitude sine wave that an op amp can output without distorting. Typically that distortion turns the sine wave into a triangle. Notice that this is entirely separate from the gain bandwidth product (which limits the output frequency because of a drop in gain). Above the full power response frequency, the op amp cannot charge the compensation capacitor fast enough to cause the output signal to swing to $V_{out\,P}$.

Example 5-9

It is necessary to amplify a $20mV_{RMS}$ sine wave, 15kHz, to $6V_{RMS} \pm 1\%$. Calculate the op amp's GBW and slew rate specifications.

Solution

The amplifier must have a gain of

$$A_O = \frac{V_{load}}{e_{in}} = \frac{6V_{RMS}}{20mV_{RMS}} = 300$$

$$f_{1\%} = 15kHz = \frac{f_{-3dB}}{7}$$

$$f_{-3dB} = 7f_{1\%} = 7 \times 15kHz = 105kHz$$

So
$$GBW = A_O \times f_{-3dB} = 300 \times 105kHz = 31.5MHz$$

This is quite high. But, don't forget the trick of using two stages. If you did that, each stage would need a gain of

$$A_{\text{single stage of 2 stage amp}} = \sqrt{300} = 17.3$$

This would mean that each stage would need an op amp with a GBW of

$$GBW_{\text{single stage of 2 stage amp}} = 17.3 \times 105kHz = 1.8MHz$$

This may be a much less expensive, and much more stable solution.

However, the slew rate must also be considered.

$$SR = 2\pi f V_P = 2\pi \times 15kHz \times \left(\sqrt{2} \times 6V_{RMS}\right) = 800kV/s = 0.8V/\mu s$$

5-2.5 AC Noise

Any unwanted signals at the output of an amplifier are often referred to as **noise.** Under this broad definition, variation in V_{load} caused by bias current, offset current, or input offset voltage are examples of DC noise. You have seen how to minimize these effects.

Alternating current noise can be divided into two major classes: that caused by external effects (or interference), and that caused by effects internal to the amplifier circuit itself. This section deals with internally generated noise.

Noise **voltage** is primarily **Johnson** or **thermal** noise. Every resistive element, inside and outside the op amp IC, generates Johnson noise.

$$e_n = \sqrt{4kTR\ \Delta f}$$

where   k = Boltzmann's constant
   T = temperature
   R = resistance of the conductor
   Δf = bandwidth of the measurement

Simplifying this for room temperature

$$e_n = 4\frac{nV}{\sqrt{Hz}} \times \sqrt{\frac{R}{1k\Omega}}$$

So a 100kΩ resistor, at room temperature, will produce a noise of

$$e_n = 4\frac{nV}{\sqrt{Hz}} \times \sqrt{\frac{100k\Omega}{1k\Omega}} = 40\frac{nV}{\sqrt{Hz}}$$

The noise **current** is often referred to as **shot noise**. The exact number of electrons flowing past a point at any instant is random. The average rate is the DC current. This random variation of the current, about I_{DC}, is the noise current. It is directly proportional to I_{DC} and to the circuit's bandwidth.

$$i_n \propto I_{DC}, \Delta f$$

To determine the overall output noise, you must consider the effects of the Johnson noise from the resistors around the op amp, the op amp's own Johnson noise, and its shot noise. These effects are shown in Figure 5-21. The source, e_{Rc}, is the Johnson noise generated by R_{comp}.

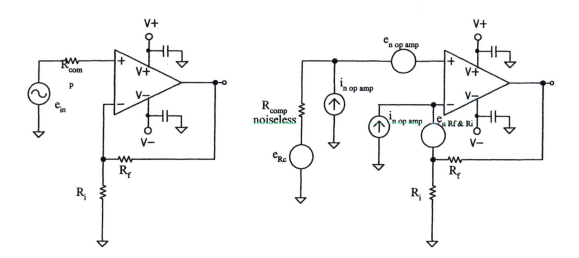

(a) Noninverting amplifier schematic   (b) Noise source

Figure 5-21 Op amp amplifier noise sources

$$e_{R_c} = 4\,\frac{nV}{\sqrt{Hz}} \times \sqrt{\frac{R_{comp}}{1k\Omega}}$$

The shot noise, $i_{n\,op\,amp}$, is a manufacturer's specification. At the noninverting input it flows through R_{comp}, producing a noise voltage.

$$i_{n\,op\,amp} \times R_{comp}$$

The op amp's Johnson noise, $e_{n\,op\,amp}$, is also a manufacturer's specification.

At the inverting input, the shot noise current flows through the parallel combination of R_f and R_i ($R_f\,//\,R_i$) to produce the noise voltage

$$i_{n\,op\,amp} \times \frac{R_f R_i}{R_f + R_i}$$

The parallel resistor combination also generates its own noise, $e_{n\,Rf\,\&\,Ri}$.

$$e_{n\,R_f\,\&\,R_i} = 4\,\frac{nV}{\sqrt{Hz}} \times \sqrt{\frac{\dfrac{R_f R_i}{R_f + R_i}}{1k\Omega}}$$

Because noise is random, none of these sources correlates with another. This means that you cannot just add them algebraically. At any instant in time they may add, or they may cancel. Statistical techniques must be applied. When determining the combined effect of three random events (a, b, and c), you must calculate their rms sum.

$$\text{total effect} = \sqrt{a^2 + b^2 + c^2}$$

You must take this same approach when determining the total input noise voltage. Each of the voltage terms above must be squared, and then added; then the square root must be taken.

$$e_{n\,total} = \sqrt{e_{Rcomp}^2 + \left(i_n R_{comp}\right)^2 + e_{n\,op\,amp}^2 + \left[i_n\left(R_f\,//R_i\right)\right]^2 + e_{n\,R_f\,\&\,R_i}^2}$$

This total **noise** voltage is an **input** voltage. To find the noise at the output, you must multiply this input voltage by the noninverting gain of the circuit.

$$e_{n\,out} = e_{n\,total}\left(1 + \frac{R_f}{R_i}\right)$$

But this does not yet give you the noise voltage at the output of the op amp. Each term of the equation above has $\sqrt{Hz}$ in its denominator. Each is sensitive to frequency. The higher the circuit's bandwidth, the larger the output noise. If the op amp stopped passing frequencies abruptly and totally at its cut-off, you could simply multiply by the square root of this frequency. However, you have seen that signals above f_{-3dB} do pass through the op amp, with a reduced gain. This tailing off effectively broadens the noise bandwidth. For simple, first order roll-off (which most op amps have), this effect increases the bandwidth by 1.57.

$$BW_n = 1.57 \times f_{-3dB}$$

$$v_{noise\,out} = e_{n\,out}\sqrt{BW_n}$$

The voltage that you finally calculate is in RMS. This is the value you would expect to read from a true RMS, indicating digital voltmeter. Since this noise is random, not sinusoidal in shape, you cannot just multiply this RMS by 1.414 to determine the peak value. The relationship is statistical in nature and is given in Table 5-1.

Table 5-1 Peak-to-Peak vs. RMS for AC noise

Source: Thomas M. Frederiksen, *Intuitive Operational Amplifiers*, New York: McGraw-Hill, 1988, pp127-145

Peak-to-Peak Amplitude	Probability of Having a Larger Amplitude (%)
2 x RMS	32.0
3 x RMS	13.0
4 x RMS	4.6
5 x RMS	1.2
6 x RMS	0.3
7 x RMS	0.05

Example 5-10

Calculate the output noise from an op amp amplifier built using a 741 with:

$$R_f = 1M\Omega \qquad R_i = 10k\Omega \qquad R_{comp} = 10k\Omega$$

$$i_{n\,op\,amp} = 0.54\frac{pA}{\sqrt{Hz}} \qquad\qquad e_{n\,op\,amp} = 20\frac{nV}{\sqrt{Hz}}$$

Solution

$$R_f\,//R_i = \frac{R_f R_i}{R_f + R_i} = \frac{1M\Omega \times 10k\Omega}{1M\Omega + 10k\Omega} = 9.9k\Omega$$

$$e_{n\,R_f\,\&\,R_i} = 4\frac{nV}{\sqrt{Hz}} \times \sqrt{\frac{9.9k\Omega}{1k\Omega}} = 12.6\frac{nV}{\sqrt{Hz}}$$

$$e_{n\,R_{comp}} = 4\,\frac{nV}{\sqrt{Hz}} \times \sqrt{\frac{10k\Omega}{1k\Omega}} = 12.6\,\frac{nV}{\sqrt{Hz}}$$

$$i_n \times R_{comp} = 0.54\,\frac{pA}{\sqrt{Hz}} \times 10k\Omega = 5.4\,\frac{nV}{\sqrt{Hz}}$$

$$i_n \times \left(R_f\,//R_i\right) = 0.54\,\frac{pA}{\sqrt{Hz}} \times 9.9k\Omega = 5.4\,\frac{nV}{\sqrt{Hz}}$$

$$e_{n\,total} = \sqrt{\left(12.6\,\frac{nV}{\sqrt{Hz}}\right)^2 + \left(12.6\,\frac{nV}{\sqrt{Hz}}\right)^2 + \left(5.4\,\frac{nV}{\sqrt{Hz}}\right)^2 + \left(5.4\,\frac{nV}{\sqrt{Hz}}\right)^2 + \left(20\,\frac{nV}{\sqrt{Hz}}\right)^2}$$

$$= 27.9\,\frac{nV}{\sqrt{Hz}}$$

$$e_{n\,out} = 27.9\,\frac{nV}{\sqrt{Hz}} \times \left(1 + \frac{1M\Omega}{10k\Omega}\right) = 2.81\,\frac{\mu V}{\sqrt{Hz}}$$

$$f_{-3dB} = \frac{GBW}{A_O} = \frac{1MHz}{101} = 9.9kHz$$

$$BW_n = 1.57 \times f_{-3dB} = 1.57 \times 9.9kHz = 15.5kHz$$

$$v_{noise\,out} = 2.81\,\frac{\mu V}{\sqrt{Hz}} \times \sqrt{15.5kHz} = 350\mu V_{RMS}$$

Almost no peak-to-peak output noise (i.e., 99.7%) will exceed

$$6 \times V_{RMS} = 6 \times 350\mu V = 2.1mV_{PP}$$

To lower the noise voltage and current present in your IC, keep the supply voltage as low as practical. This decreases the power dissipated by the IC and therefore its temperature. This also lowers the internal bias currents, lowering the noise current. Keep the circuit's bandwidth as low as possible. (Do not use a 318 if a 741 will work.) This drops the high frequency noise harmonics. Do not allow shunting capacitance around R_i. Capacitance here lowers Z_i (R_i in parallel with the capacitance), therefore raising the high frequency (noise) gain. Also, inverting summers effectively place each of their input resistors in parallel. This lowers R_i, raising the noise gain. Use inverting summers only when necessary, and then carefully limit the number of inputs. Finally, keep the input and source impedances low. This allows you to drop **both** R_f and R_i.

5-2.6 External Frequency Compensation

As you have just seen, internally generated noise can contaminate your amplifier's output. Lowering the circuit's bandwidth reduces the effect of this internally generated noise. Perhaps even more damaging is the wide variety of externally generated interference. These include any digital ICs in the circuit, microprocessors, power electronics, relays, motors, and so on. In fact, personal computers, laptops, personal electronic notepads, garage door openers, portable telephones, cellular phones, and pagers flood the environment with electrical noise. Any or all of this noise may enter your amplifier, overwhelming its output.

Although you can most certainly shield against most of these external problems, the simplest and most effective solution for both internally and externally generated noise is to restrict the bandwidth of your amplifier to just barely what is needed. This assures that your signal is properly amplified. But the gain rolls off at higher frequencies, reducing or eliminating the effect of the noise at the output. Taken to its logical conclusion, this would mean that you should stock hundreds of different model op amps, each with a different gain bandwidth product. Then you would need to carefully select the correct one for each amplifier.

Although this could be done, it is a rather impractical solution. Adding an external frequency compensation capacitor in parallel with R_f is a far better approach, as shown in Figure 5-22. At low frequencies the capacitor looks like an open. So the gain is set by R_f and R_i, as usual. As frequency increases (i.e., for the noise harmonics), the capacitive reactance decreases, shorting out R_f. This drops the gain. Assuming that the op amp itself does not roll off, R_f and C_f set the high frequency cut-off of the amplifier, $f_{-3dB \text{ amplifier}}$.

To properly do an AC derivation, you must remember that impedances are phasors.

$$\overline{A} = -\frac{\overline{Z_f}}{R_i}$$

$$\overline{A} = -\frac{\dfrac{\overline{Z_c}R_f}{\overline{Z_c} + R_f}}{R_i}$$

$$\overline{A} = -\frac{\dfrac{(X_C \angle -90°)(R_f \angle 0°)}{R_f - jX_C}}{R_i}$$

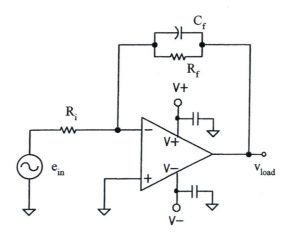

Figure 5-22 External frequency compensation for an inverting amplifier

$$\overline{A} = -\frac{\dfrac{X_C R_f \angle -90°}{R_f - jX_C}}{R_i}$$

To complete the division in the numerator, you must change the rectangular notation to polar.

$$\overline{A} = -\frac{\dfrac{X_C R_f \angle -90°}{\sqrt{R_f^2 + X_C^2} \angle \arctan\left(\dfrac{-X_C}{R_f}\right)}}{R_i}$$

At $f_{-3dB \ amplifier}$, the **magnitude** of the gain drops by -3dB, or 0.707. So, you only have to consider the magnitudes.

$$A = \frac{\dfrac{X_C R_f}{\sqrt{R_f^2 + X_C^2}}}{R_i}$$

$$A = \frac{R_f}{R_i} \times \frac{X_C}{\sqrt{R_f^2 + X_C^2}}$$

The point that you are looking for is the frequency at which the magnitude of the gain falls below its DC value by 0.707.

$$A_{f_{-3dB}} = \frac{A_O}{\sqrt{2}} = \frac{\dfrac{R_f}{R_i}}{\sqrt{2}}$$

Equate this to the general gain magnitude equation.

$$A_{f_{-3dB}} = \frac{\dfrac{R_f}{R_i}}{\sqrt{2}} = \frac{R_f}{R_i} \times \frac{X_C}{\sqrt{R_f^2 + X_C^2}}$$

$$\frac{1}{\sqrt{2}} = \frac{X_C}{\sqrt{R_f^2 + X_C^2}}$$

$$\frac{1}{2} = \frac{X_C^2}{R_f^2 + X_C^2}$$

$$R_f^2 + X_C^2 = 2X_C^2$$

$$R_f^2 = X_C^2 \quad \text{or} \quad R_f = X_C$$

$$R_f = \frac{1}{2\pi f_{-3dB} C}$$

$$f_{-3dB} = \frac{1}{2\pi R_f C_f}$$

Even though the circuit looks entirely different, the cut-off frequency is set just as it is in a simple series RC low pass filter.

Select the $f_{\text{-3dB op amp}}$ of the op amp IC to be 5 or more times above the $f_{\text{-3dB circuit}}$ set by R_f and C_f. The op amp then has all of the bandwidth needed to provide the gain at the upper frequencies, and the R_f and C_f roll off the actual performance.

In reality, you should get into the habit of **always** placing a capacitor across R_f. This is a very inexpensive way to insure that the amplifier works over the frequency range you are interested in, but it ignores all of the internally and externally generated noise above your signal's bandwidth. Also, you do not need a special op amp. This allows you to tailor whatever you have in stock to fit the unique requirements of each circuit.

5-3 Types of Op Amps

The proper selection of which op amp to use in a particular application is often the key factor that determines the success or failure of the circuit. Unfortunately, too often, whatever is on the shelf is used, whether it is inadequate or marginal or provides overkill. You should put as much effort into selecting the op amp as you do in determining the other circuit components. A wide spectrum of op amps are available, from those requiring only 3V supplies with bias currents in the 10^{-15}A range; to those that output hundreds of volts at tens of amperes; to those specialized for precision DC performance, with only kHz bandwidth; to video op amps that amplify DC to GHz.

So, with this variety, how do you go about selecting the right IC? First you must determine several **circuit** characteristics. These can be divided into three groups: speed, DC, and system. Grouped under speed considerations are:

speed
> gain
> gain accuracy
> upper frequency
> output maximum rate of change (slew rate)

To determine the op amp's DC characteristics, you must know the circuit's:
DC
> source impedance
> DC output error allowable
> temperature range

Finally, to properly support this circuit and to integrate it with the rest of the system, consider:
system
> power supply voltage(s) and current available
> maximum output voltage and current needed
> maximum input voltage
> cost
> availability

Use these items as a checklist. From the circuit's performance requirements, determine a value for each item in the list above. Then you are ready to select an op amp.

Op amps can be divided into five major groups:
> general purpose
> FET input
> CMOS
> wide bandwidth
> high voltage or current

Your analysis of the circuit parameters (as described above) usually allows you to select which group to investigate and often pinpoints ten or so op amps that may do a good job. Final decision then must be made on cost, availability, and prototype testing of engineering samples.

General purpose op amps are the lowest cost parts. They have gain bandwidths up to 8MHz, with input impedances of about 1MΩ, bias currents in the 100nA range, offset currents around 10nA, input offset voltage in the millivolts, and offset drifts around 5μV/°C. Power supply requirements are usually between ±5V to ±18V with quiescent currents of a few mA. Output voltages are restricted to about 2V below the supplies, as are the input voltages. Output current may or may not be limited. If it is, the maximum output current is around 10mA.

These parts are the usual choice. Most specifications are adequate for undemanding applications, and the price and availability are certainly right. Members of the general purpose class are the 741, 324, and 318. Applications include simple amplifiers, summers, impedance buffers, and audio frequency active filters.

The second group of op amps, in terms of improved performance, and increased cost, are the FET input op amps. The NPN input transistors of the general purpose op amps have been replaced with p-channel JFETs. Because these transistors are operated reverse biased, FET input op amps have notably higher input impedance ($10^{10}\Omega$ to $10^{13}\Omega$) with bias currents and offset currents in the 10pA to 100pA range. This is a factor of 1000 improvement over general purpose op amps. Input offset voltage is also improved to the order of 0.1mV, with drifts about 1μV/°C. Even with these improved DC characteristics, the p-channel input JFETs allow speeds comparable to or slightly better than the general purpose op amps. Power supply requirements, quiescent current, and input and output voltage swings are similar to those of the general purpose op amps, since both groups use bipolar transistor output stages.

Applications with high source impedance, such as piezoelectric crystals, accelerometers, pH probes, or electrometers, require the high input impedance of the FET input op amp. Also, circuits with high feedback resistance or with capacitor feedback benefit from their low bias currents. Typically, buffers are made from FET input op amps. The entire LF series of op amps from National Semiconductor, as well as the TL074 and TL084 from Texas Instruments, are FET input op amps.

The next level in DC performance beyond the FET input op amp is the CMOS op amp. The IC is built of complementary symmetry metal oxide semiconductors. Each of these transistors has a silicon dioxide (i.e., glass) layer between its gate and its channel. The result is an input impedance which is greater than $10^{12}\Omega$ and bias and offset currents on the order of 10fA (1 fentoampere = 0.001pA = 10^{-15}A). Input offset voltage is about 1mV, with drift around 1μV/°C.

Early CMOS op amps suffered from seriously reduced gain bandwidth and slew rate, especially when driving load resistances less than 10kΩ. But now there are available CMOS op amps with gain bandwidths of over 1MHz and slew rates faster than 1V/µs.

In addition to exceptional DC input characteristics, CMOS op amps have very low power supply voltage and current requirements. In fact, many have been optimized to work from a single supply of 3V to 16V (so beware when you try to run CMOS op amps from a split supply). This also means that even if they will work from a split supply, a maximum supply difference of 16V is ±8V, **not** ±16V. This limitation prevents you from simply replacing a general purpose op amp or FET input op amp, running from ±9V to ±15V with a CMOS op amp with a supply limit of 16V.

Because each transistor within a CMOS op amp is driven with voltage, not current, and because each has an insulator at its input, the total current drawn from the power supply is on the order of 100µA. There are series of these op amps that allow you to reduce this to 10µA with an external connection, at the sacrifice of speed. Some versions may even be put to sleep, drawing only a few microamperes, until an input arrives to awaken it.

The saturation voltage of CMOS op amps is 0.01V to 0.5V, increasing as you draw more current from the op amp. This means that when operating from a 1.5V battery, virtually the entire 1.5V is available at the output. CMOS op amps do not lose ±2V as both general purpose and FET input op amps do. Also, early CMOS op amps suffered from very restricted output current. However, many can now output 20mA.

Single supply or battery powered operations are ideal for CMOS op amps, as are those indicated for the FET input op amp requiring high input impedance or low bias currents. Maxim Integrated Products provides a wide variety of CMOS op amps. The LTC series from Linear Technology and the TLC series from Texas Instruments are CMOS, as is the LMC660 from National Semiconductor.

Another group of CMOS op amps is the chopper stabilized op amp. Most op amps have input offsets of 100µV to several millivolts. The drift for these devices is typically a few microvolts/°C. By adding a second amplifier, analog switches, an oscillator, and control logic within the IC, manufacturers of chopper stabilized op amps are able to produce devices with a few microvolts of initial input offset voltage and drift at 0.01µV/°C to 0.5µV/°C. Since these devices are CMOS, most of the other parameters are similar to those of the CMOS op amps described above. Bias currents are on the order of 10pA, with offset currents even less. Input impedance is typically $10^{12}\Omega$. Gain bandwidth and slew rate of the chopper stabilized op amps are comparable to those of the general purpose and FET input devices.

Power supply parameters are also similar to those of the other CMOS op amps. The maximum power supply voltage is usually 16V total, **not** ±16V. However, the output voltage swings within a few hundred millivolts of the supplies. Output current should be limited by restricting load resistance to above 10kΩ. Because of the additional circuitry needed to reduce the offset voltage, current drawn from the power supply is a few milliamperes.

A simplified block diagram of a chopper stabilized op amp is shown in Figure 5-23. The main amplifier is always active, driving the output in response to its inputs and the external feedback resistors (just as all op amps work). The null amplifier is switched in and out of the main circuit. Both amplifiers have a null pin which allows the offset voltage of each amplifier to be adjusted by changing the voltage on that pin.

Initially, switches A and A' are closed and B and B' are open. This shorts the null amplifier's inputs. The null amplifier's output, switched to its own offset adjust pin, drives that amp's output to zero. This voltage is stored on C_{XA}. On the earlier versions of chopper stabilized op amps, the two capacitors were external. Many of the newer models provide these capacitors internally. During this first cycle, the null amplifier sends its own offset to zero.

Next, switches A and A' are opened, and B and B' are closed. The null amp's inputs now parallel the main amplifier's inputs. With proper negative feedback applied externally, there should be no difference in potential between the main amplifier's inputs. Any existing difference is that amplifier's input offset voltage. The null amp's output, responding to that difference, drives the main amp's null pin until the difference is zero. That voltage is stored on C_{XB}. The switching occurs at several hundred hertz or faster with the amplifier repeatedly nulling the null amp and then nulling the main amp.

Although there is a 1000-fold improvement in the offset voltage and drift, several potential problems require your careful consideration. Most obvious are the two additional capacitors that may have to be added to the op amp. These must be high quality, stable and have low dielectric absorption. Polypropylene or polystyrene capacitors of about 0.1µF are recommended. Linear Technology makes chopper stabilized op amps with the capacitors internal. Each is a drop-in replacement for other op amps. Just do not forget the power supply limitations, and be sure that the offset adjust pins that other op amps might use are left open. Because you are working at such low voltages, leakage across the printed circuit board may become a noticeable problem. Thoughtful layout, rigorous cleaning and sealing, and input pin guarding are needed. Finally, beware of the chopper's noise. Specifications appear comparable with those of most op amps. But the noise voltage specs are usually limited to below 10Hz, and the chopper clock is running several hundred hertz or more. So read the data sheets carefully, and look hard at the prototype's output for glitches at the chopper's frequency. They may be over 100mV_{PP}.

Wide bandwidth op amps can be implemented two ways. Extending the classical op amp design with special care to optimize speed parameters produces an op amp with gain bandwidths as large as 1GHz and slew rates of over 1000V/µs. This increased speed is expensive, in terms of both dollars and a noticeable degradation in other specifications. Bias currents may rise to a few milliamperes, and the quiescent current that the power supply must provide goes up proportionally. Recent advances are beginning to produce traditional high speed op amps with lower current requirements and price. But these are unusual.

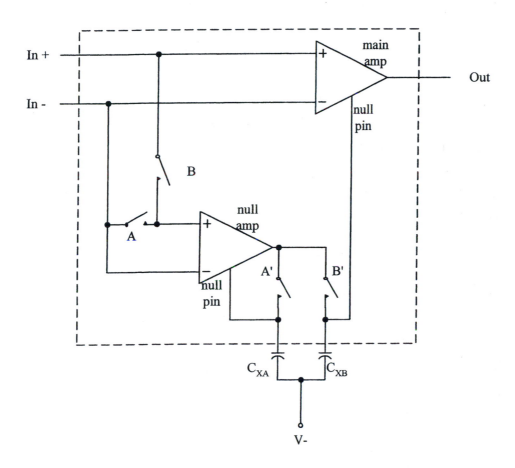

Figure 5-23 Chopper stabilized op amp diagram
(Courtesy of Texas Instruments)

The current feedback op amp alters the way that the op amp function is implemented. Figure 5-24 compares a voltage-controlled (traditional) op amp with a current feedback op amp. On the current feedback op amp, an input buffer forces the inverting input to equal the noninverting voltage. The current sourced or sunk by this buffer (i.e., current at the inverting input) is sensed and controls the output voltage through a transimpedance amplifier, Z(s).

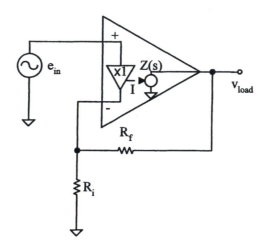

Figure 5-24 Voltage feedback and current feedback op amps
(Courtesy of Analog Devices)

An increase in input voltage causes the inverting input to be driven higher by the input buffer. This in turn puts more voltage across R1, drawing more current from the buffer. But more current from the buffer sends the output voltage up. An increase in output voltage produces current through R_f and then R_i, replacing some of the current sourced from the input buffer. This is negative feedback. An equilibrium is established, with the output voltage driven to a higher voltage, dependent on the increase in the input and the divider set up by R_f and R_i.

Although externally the current feedback op amp is used similarly to the classical voltage feedback op amp, this altered internal architecture allows much faster circuits, including a common base input stage (much faster than the traditional difference amp). A second advantage of the current feedback amplifier is the way its internal frequency compensation is accomplished. The high frequency cut-off, f_{-3dB}, is dependent on an internal capacitor and the external feedback resistor. It does not change in a simple way as you change the gain. So the concept of gain bandwidth product does not apply. The high frequency cut-off, f_{-3dB}, depends on R_f, gain, load resistance, and supply voltage. Look carefully at the manufacturer's data sheets.

The altered internal configuration requires that you select the feedback resistor, R_f, from a narrow range. This range is specified by the manufacturer but is typically between 500Ω and 2kΩ. Look carefully at the data sheets of the part you are using.

Other specifications of the current feedback op amp are comparable to those of general purpose op amps. Bias current is a few hundred nanoamperes, and input offset voltage is at or below 1mV. Input impedance is typically in the megohms range. Power supply voltage is typically ±5V to ±18V, with quiescent current needs of a few milliamperes. The output drives a 50Ω load, usually providing up to 20mA.

Except for speed and an optimal feedback resistance, these current feedback op amps look just like most general purpose op amps. But it is its very wide bandwidth that dictates special handling. At high frequencies even a small stray capacitance creates a low impedance parasitic path and a phase shift. These paths short out desired signals, but they couple unwanted rf signals into the worst places, often altering gain or producing oscillations. Select low resistance resistors to minimize the effect of parallel low impedance (parasitic capacitance) paths. Do not use IC sockets, or wire wrap or plug-in protoboards. The stray capacitance is far too large. Place all components as close to each other as possible to reduce the effects of the series inductance in the interconnections. A 22AWG wire may have inductance as large as 75nH/in. As you approach 100MHz, this inductance becomes the **dominant** element in your circuit. Consider surface mount instead of through-hole layout. A ground plane is absolutely necessary above 1MHz. Not only does it shield against interference, but it allows each signal's magnetic field to induce its own return current in the ground plane directly below. This drastically reduces the current carrier's inductance. Power supply decoupling is also critical. A 0.1μF capacitor should be used at each supply pin of each op amp. These connections must be the closest connections made to the op amp.

The op amps that you have seen so far are all designed to operate with small signals, typically ±15V at 20mA or less. Although that is fine for most signal **processing**, when it is time to drive a servomotor, relay, solenoid, lamp, heater, valve, speaker, or antenna, much more power is needed.

If your application is an audio frequency-driven into a speaker, then there are a wide variety of audio power drivers available. They are optimized to deliver watts to tens of watts of audio power to a speaker, often from a single (+3V to +12V) supply. However, they are **not** general purpose op amps and must be used strictly in accordance with their manufacturers' guidelines.

Unity gain power buffers are also available. Operating from ±15V supplies, these buffers can deliver hundreds of milliamperes to a load at remarkably low cost. The compromise, however, is that their input impedance and bias currents are poor, and the actual gain is not stable at 1 but may vary considerably. These devices are intended to be used **inside** the feedback loop of another op amp, as shown in Figure 5-25. This combination assures the excellent input characteristics of the precision, instrument quality op amp It also takes advantage of its high open loop gain to establish that the closed loop gain of the entire amplifier is $1 + R_f/R_i$. The power buffer provides the current to the load.

This technique of placing one amplifier within the negative feedback loop of another, forms a **composite** amplifier. One word of caution: the inner amplifier (the power buffer) must be considerably faster than the outer amplifier (the precision instrument quality op amp). If the inner amplifier is too slow, at higher frequencies its output lags its input. Effectively, this is a phase shift. At some frequency, this phase shift is enough to cause the negative feedback to become **positive**. The entire composite amplifier tries to break into oscillations. So select an IC for the output op amp that is barely fast enough for your application. Then assure that the inner IC is several times faster than the outer. Presto! The oscillations disappear.

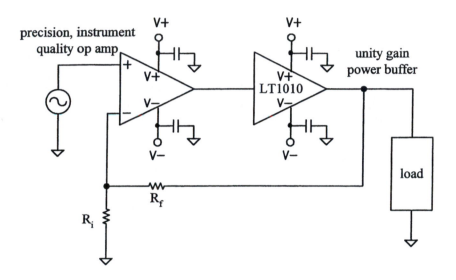

Figure 5-25 Proper application of a power buffer

But, many applications require 100V or more at 10A or more. High power op amps are available which can sink or source over 10A at voltages over 100V. Often bias current, offset voltage, drift, input impedance, open loop gain, gain bandwidth, and slew rate are comparable to those of the op amps in the general purpose class. Current limiting may be externally programmable with resistors that you connect to the IC. This allows you to set the maximum current delivered into a shorted load. Like the three-terminal IC regulators you saw in Chapter 4, many of the power op amps have internal thermal sensing and shutdown. The same calculations you did there for heat sinking must also be completed for these power op amps. Many of the parameters, including maximum allowable power dissipation, are sensitive to temperature. So, when using power op amps, you must carefully calculate the case temperature and then verify that the op amp's performance at that temperature is acceptable.

Finally, breadboarding and testing at high current and voltage levels require extra care. Mistakes can be spectacular. Follow any manufacturer's suggestions rigorously.

Begin by testing the circuit at a low supply voltage, from a supply with a low current limit. Set the load resistance to the largest allowed by spec. Do your initial test with the input voltage grounded (not open). Once zero in has produced the appropriate output, at low voltage and load, apply as large an input signal as is consistent with the low power supply that you are using. Verify that the circuit functions. Turn the input, then the power, off. Don appropriate protective eyewear. An exploding part can put out your eye. Raise the power supply to the highest level that the amplifier will need. Repeat the zero input voltage and the full input voltage tests. During these high voltage tests, carefully monitor the current that the power supply is delivering to the IC. Also keep an eye on the IC's case temperature. From previous calculations you know how high this should go, if everything is working correctly. Finally, **gradually** lower the load resistance to the rated value. (Or if you are adventurous, or foolish, you could just wire everything up, hook up the smallest load resistance and the highest supply voltage, flip the switch, and hope. If you do, stand back, way back; be sure to have a fire extinguisher handy; and don't expect to get any useful information from the molten mess on your bench.)

5-4 Composite Amplifiers

A composite amplifier uses op amps of several types, combined in such a way as to take advantage of each op amp's strengths while overcoming the weaknesses of each. Usually the op amps will, in some way, affect each other's feedback. You can build amplifiers using several inexpensive op amp ICs that outperform much more expensive single ICs.

The power amplifier in Figure 5-25 is a composite amp. Used alone, the input op amp can only drive a few milliamperes into a load. The power buffer can drive the load but has an unstable gain and high bias currents and input offset voltage. When the two are used together, the weaknesses of each are overcome. The external circuit sees the input of the precision op amp: high impedance, low bias, and offset. The load sees the power buffer, with all of its drive.

However, the key is placing the power amp **inside** the precision op amp's negative feedback. The high gain of the input op amp insures that its two input pins are at the same potential. Using fixed, stable resistors for R_f and R_i then means that the output voltage is fixed and stable, determined by those two resistors. The input op amp compensates for any variation in gain by the power buffer. Should the power buffer's gain go up, the input op amp receives a slight increase in its negative feedback voltage. This lowers its output drastically, enough to exactly compensate for the change in gain of the power buffer. Conversely, should the power buffer's gain go down, the negative feedback to the input amp drops minutely. This causes a major increase in its output, compensating for the power buffer's drop in gain. As long as the power buffer is considerably faster than the input amp, the circuit is stable. You have just built a power amplifier using two parts, one which has poor DC characteristics and the other which has low output current. Properly used, each overcomes the weakness of the other. Champagne performance on a beer budget!

5-4.1 Suspended Supply

The suspended supply amplifier in Figure 5-26 forms the core of another unusual, but powerful, amplifier. Alone this circuit can deliver over $100V_{PP}$ but suffers from such widely varying gain that it requires precision resistors to even produce a gain accurate to ±20%, not exactly a manufacturable circuit.

But before looking at the composite amp, you need to investigate the suspended supply amp itself. From simple, common, inexpensive parts you can build an amplifier with a very high voltage output. Of course, you have to provide a high voltage power supply, but those are not hard to build, or rare. The key is to remember that the op amp has no ground pin. There is no reference to ground. It performs just as well with V+ at 56V and V– at 32V (a difference of 24V) as it does from ±12V. The only constraint is that both of the input pins must be between V+ and V-.

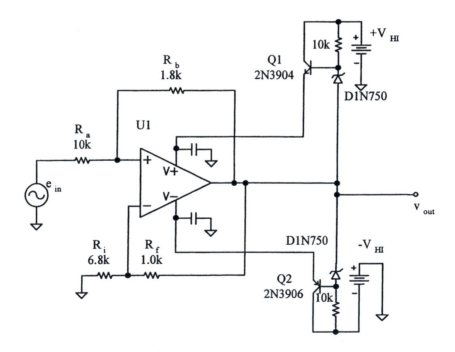

Figure 5-26 Suspended supply amplifier

Transistor Q1 and zener CR1 form a positive 12V regulator to provide V+. However, instead of referencing this regulator to ground, it is tied to V_{out}. So, V+ is about 11.4V above the output. Should the output be driven to 35V, the positive supply is pushed up to 47V. Dragging V_{out} in the opposite direction to -50V sends V+ to -38V. Transistor Q2 and zener CR2 do the same for V-, setting it about -11.4V more negative than V_{out}.

So, as the output drives up and down, its supplies are pushed and pulled along, staying ±12V on either side. In practice, the output can be driven to within about 5V of the fixed, high voltage supplies. At that point you run into the saturation voltages of the op amp and the transistor/zener regulators. So, from a ±56V supply, the output of even a 741C can be driven to $100V_{PP}$.

The other constraint is that you, somehow, keep both inputs between V+ and V-. Actually, this is not all that hard. The negative feedback through R_f and R_i forces the inverting input to equal the noninverting pin. So all you need to do is to be sure that the noninverting input is between V+ and V-. That is the purpose of R_a and R_b. They form a voltage divider between V_{out} and E_{in}. Assuming 12V zeners, and giving a little headroom (5V), pick R_a and R_b such that

$$V_{NI} = \left(V_{out} - E_{in}\right)\frac{R_a}{R_a + R_b} + E_{in} < V_{out} - 5V$$

Although this looks a little formidable, knowing the maximum out that you want sets V_{out}. Knowing the gain then tells you E_{in}. Then pick R_a, and solve for R_b.

Although this works well, it also creates **positive** feedback. This greatly complicates the gain equation.

$$A = \frac{R_i R_b + R_f R_b}{R_i R_b - R_f R_a}$$

The negative sign in the denominator is the problem. If $R_f R_a$ is larger than $R_i R_b$, this noninverting configuration can actually invert. Worse is that as $R_i R_b$ approaches $R_f R_a$, the denominator approaches zero, and the gain grows radically. To prevent this and keep the amplifier noninverting, assure that

$$R_i R_b > R_f R_a$$

Be sure to take into account the tolerance of the resistors.

$$\text{For 5\% resistors:} \quad R_i > \frac{1.22 R_f R_a}{R_b}$$

$$\text{For 1\% resistors:} \quad R_i > \frac{1.041 R_f R_a}{R_b}$$

Example 5-11

For the schematic in Figure 5-26, assume that each of the four gain setting resistors has a 5% tolerance. Determine the nominal gain and the worst case low and high gains.

Solution

The tolerance for each resistor is set using its dialog box, as shown in Figure 5-27. Then the Worst Case Analysis is enabled, set to give the lowest gain. A second run determines the highest gain. Look at Figure 5-28.

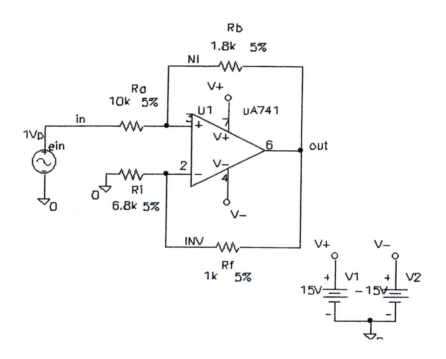

Figure 5-27 Worst case analysis setup of the suspended supply amplifier

Even though the amplifier is designed to have a gain of 6.3, using 5% resistors means that some units may be produced with gains as low as 3.4 while others come off the assembly line with gains of 488. This is **not** acceptable!

The solution is to enclose the suspended supply within the negative feedback loop of another, traditional, low voltage noninverting amplifier. This is just what was done with the high power buffer from the previous section. Look at Figure 5-29.

The suspended supply is shown as an amplifier (triangle) with a gain of 3.4 to 488. Assume an input to the output op amp of $5V_P$. Since there is negative feedback, and assuming that neither of the op amps is in saturation, there is little difference in potential between the two inputs of U1. So its noninverting input also has a $5V_P$ signal on it. That puts $5V_P$ across the 11kΩ resistor. Five volts across 11kΩ means 455µA flowing through it. This same current flows through the 100kΩ resistor. The output, v_{load}, is 455µA flowing through the series combination of 100kΩ and 11kΩ. This gives a $50V_P$ output from the suspended supply amplifier, regardless of its gain.

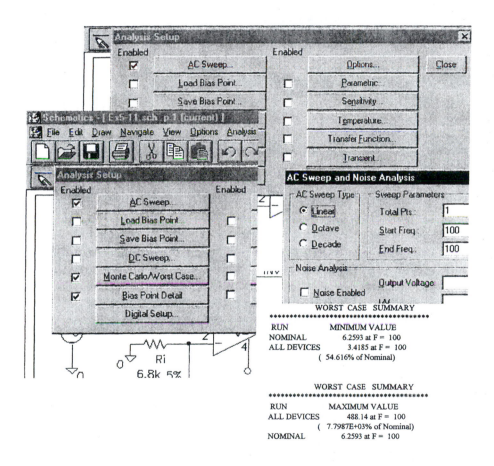

Figure 5-28 Worst case analysis sample results

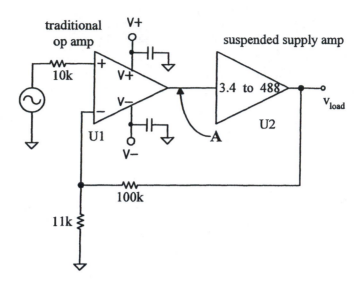

Figure 5-29 Composite amp using a suspended supply amp inside the loop

If the suspended supply amp has a gain of 448, then the traditional op amp has to produce only a 111mV$_P$ output. Should the suspended supply amplifier's gain be only 3.4, then the output from the traditional op amp, at point A, must go to 14.7V$_P$. Most general purpose or FET input op amps can reach this level operating from ±18V supplies.

Tucking the high voltage, suspended supply amplifier inside the feedback loop of a traditional low voltage op amp provides the large signal size needed (from the suspended supply amp) while overcoming the huge uncertainty in its gain.

It is critical to select an op amp for U2 that is several times faster than U1. Otherwise, the composite amplifier oscillates.

5-4.2 Integrator Feedback

The circuit in Figure 5-30 is an integrator. It look like an inverting amplifier. However, the feedback resistor has been replaced with a capacitor. This changes a lot.

To calculate the output voltage, v$_{load}$, you can still assume that there is no difference in potential between the input pins. For really precise performance, you can use a chopper stabilized op amp so that the input offset voltage is negligible. The inverting pin is at virtual ground.

So $$i_{R_i} = \frac{e_{in}}{R_i}$$

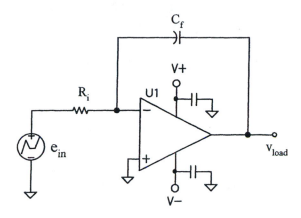

Figure 5-30 Op amp integrator

With an FET or CMOS op amp the bias currents are tiny. All of i_{Ri} goes to C_f. The voltage across C_f requires a little calculus. Remember that

$$i_C = C \frac{dv_C}{dt}$$

Solve this for v_C.

$$i_C dt = C_f dv_C$$

$$dv_C = \frac{1}{C_f} i_C dt$$

$$\int dv_C = \int \frac{1}{C_f} i_C dt$$

$$v_C = \frac{1}{C_f} \int i_C dt$$

But we know i_C in terms of e_{in}.

$$v_C = \frac{1}{C_f} \int \frac{e_{in}}{R_i} dt = \frac{1}{R_i C_f} \int e_{in} dt$$

This voltage is developed across the capacitor from positive to negative. With the left end at virtual ground, this puts v_{load} at a negative potential. It is, after all, an inverting amplifier.

$$v_{load} = -\frac{1}{R_i C_f} \int e_{in}\, dt$$

Although this is mathematically correct, and it does explain why the circuit is called an **integrator**, it may not be very intuitive. So consider another approach. If e_{in} is above the voltage at the op amp's inverting input (virtual ground in the simplest case), current flows through R_i, left to right, and charges the capacitor from positive to negative. The output, v_{load}, is driven down. If e_{in} is below the voltage at the inverting pin, current flows right to left through R_i, charging the capacitor in the opposite direction from negative to positive. Only when e_{in} exactly equals the voltage at the inverting input does the capacitor receive no current. When this happens, the capacitor **holds** its charge, neither charging nor discharging. The output voltage is constant, at whatever potential it has when e_{in} becomes equal to the voltage at the inverting pin.

For this to work correctly there are three practical limitations. The capacitor must not leak. Be sure to choose a film capacitor. There should be little bias current flowing into the op amp. The assumption is that **all** of the current through R_i goes to C_f. Bias current takes some of the current through R_i. An FET or CMOS op amp is needed. Finally, the input offset voltage must be small, since that sets voltage at the inverting input. Often a chopper stabilized op amp is selected.

Combining a fast op amp (with lots of input offset voltage and bias currents) with this integrator produces an amplifier that is fast and very precise. The schematic is shown in Figure 5-31.

The LM318 is an inexpensive, commonly available, reasonably fast op amp with input offset voltage in the millivolts range and bias currents in the 10-100nA range. You should select whichever op amp provides the gain bandwidth and slew rate performance you need. Resistors R1 and R2 set the overall gain. Since the input offset is increased by this gain, it is reasonable to expect hundreds of millivolts or more output offset.

The LTC1049 is a chopper stabilized CMOS op amp. It is fairly inexpensive, has an input offset of a few **microvolts**, and negligible bias currents and drift. But it is slow. Also, notice that it requires ±5V supplies rather than the ±15V of the high speed amp.

The voltage at U2's inverting input is its input offset voltage, a few microvolts. Assuming that the overall amplifier's output, v_{load}, has a positive offset of several hundred millivolts means that current flows from right to left through R5. Since the bias current of U2 is tiny, all of the current from R5 goes onto C1, charging it from negative to positive. That is, the output of U2 is driven negative. That negative voltage is divided and then sent to the noninverting input of U1. This negative voltage is amplified by the gain of U1, driving its output (which was hundreds of millivolts positive) down.

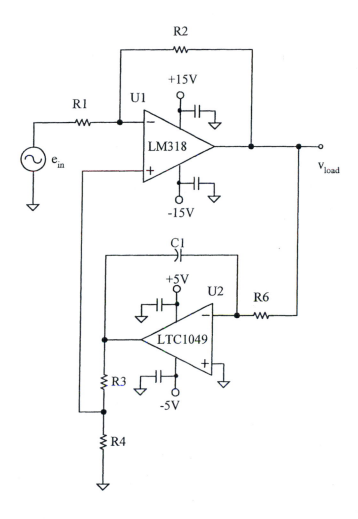

Figure 5-31 Integrator stabilized high speed amplifier

This process continues until $v_{load\ DC} = V_{ios\ U2}$, which is only a few microvolts. At that point, there is no difference in potential across R5, so no current flows through it. With no current, C1 holds the charge it has. This holds the voltage at the noninverting input of U1 that is needed to drive its output to ground. The integrator has automatically removed the DC offsets of U1. You have built an amplifier that is as fast as U1 and as precise as U2.

Several additional comments are in order. The integrator is inherently a low pass filter. You want it to respond only to DC and slow changes in that DC, not to the

AC signal at the output. So select its high frequency cut-off several orders of magnitude below your input signal's lowest possible frequency.

$$\frac{1}{2\pi R5 \times C1} << \frac{f_{lowest}}{100}$$

But be careful to use a film capacitor for C1. If necessary, make R5 very large to slow the integrator's response enough.

Resistors R3 and R4 divide the output of the integrator. This allows it to swing across its entire range (or about +4V to -4V) as it adjusts the input of U1. Select resistors that are on the same order of magnitude as R1 and R2. Set their precise value so that a 4V output from U2 produces an input into U1 that is a little larger than the worst case **input** offsets of U1. Be sure to account for the effects of input offset voltage, bias currents, and offset current.

Finally, there is no reason that the output of the amplifier has to be driven to 0V. It goes to the potential on U2's noninverting input ($\pm$ a few microvolts). So if you want an accurate offset adjust, place a voltage out digital to analog converter, a potentiometer, or some other DC control circuit driving the noninverting input pin of U2. The $V_{load\ DC}$ is driven to this voltage without altering the AC signal at all. (You may have to change the size of R3 and R4.)

Example 5-12

Determine the output offset of an LM318 op amp based amplifier with a gain of -100, without and then with integrator feedback.

Solution

The uncompensated inverting amplifier is shown in Figure 5-32.

The simulation is run in *Electronics Workbench*™. The input offset voltage of the LM318 is set at 4mV and the gain bandwidth product at 10MHz. With a gain of -100, then the output DC of 417mV and the AC voltage of 979mV are accurate. However, the DC error signal is half the size of the desired AC output! This is a **major** problem.

The amplifier with integrator feedback is shown in Figure 5-33. The LTC152 has an input offset of 0.5µV but a low gain bandwidth product. The integrator feedback configuration has reduced the amplifier's output offset to less than 1mV without affecting the AC amplitude. This is now a usable circuit.

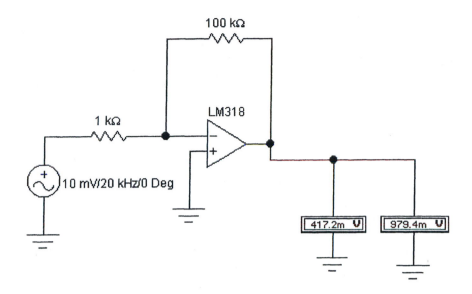

Figure 5-32 Uncompensated amplifier simulation using *Electronics Workbench*^{TM}

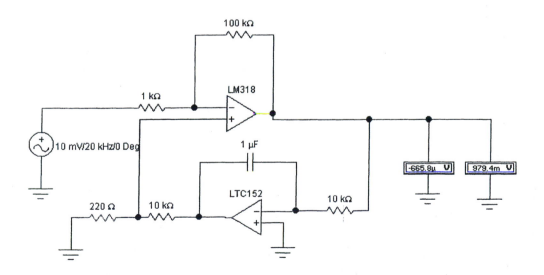

Figure 5-33 Integrator feedback simulation using *Electronics Workbench*^{TM}

Summary

This chapter has presented the nonideal DC and AC characteristics of operational amplifiers. Compensation techniques of each were also given. Bias current is the DC current necessary to flow into the inputs of an op amp to turn it on. The effects of bias currents can be eliminated by placing a resistor in the noninverting input lead which equals the resistance in the inverting input lead. Offset current is the difference between the two bias currents. Its effect on the output voltage can be minimized by keeping the value of the negative feedback resistor small. Input offset voltage is modeled as a DC source at the noninverting input. To lower the effect of input offset voltage, null pins are available. But usually it's best just to buy a better quality op amp. Drift describes the variation of offsets with temperature. Layout, power dissipation, and ambient temperature all must be considered if drift is critical.

To prevent high frequency oscillations, the gain of an op amp must be reduced at high frequency. The gain bandwidth product is the open loop gain times frequency and is a constant. So as frequency rises, the open loop gain falls off. The high frequency cut-off for a closed loop amplifier (f_{-3dB}), then, is this gain bandwidth product divided by the closed loop gain. At that frequency the closed loop gain falls to 0.707 of its DC value. For gain errors of less than 10%, you must lower the signal frequency to one-half of f_{-3dB}. One percent gain error requires that the frequency be dropped to one-seventh of f_{-3dB}.

Large signal outputs ($v_{load} > 1V_P$) are limited by the speed at which the op amp can charge its internal frequency compensation capacitor. This is defined as the slew rate and is expressed in V/µs. The limits that slew rate places on the op amp's response to a pulse is easily seen. However, slew rate limiting of a sine wave output requires a derivative calculation. The result is defined as full power frequency response. An amplifier may be gain bandwidth (small signals) or slew rate or full power limited (large signals). You should check both.

The frequency limits of an op amp vary widely between ICs of the same model and across the operating range of an IC. So relying on the internal compensation to set an amplifier's upper frequency response is a bad idea. Instead, select a much faster op amp than is needed in the circuit; then place a capacitor across the feedback resistor. This sets the high frequency cut-off precisely with the value of the feedback resistor and its parallel capacitor (both are accurate, stable, and inexpensive). But **always** limit the high frequency gain of your amplifier this way, to just barely what is needed. This reduces the effects of external and internal noise and the tendency to oscillate.

This noise is generated by both the op amp and all associated resistors. Noise currents at the inputs of the op amp flow through external resistors, also producing a noise voltage at the inputs. These noise-produced voltages must be root-mean-squared summed and then multiplied by the noninverting gain of the amplifier to determine their effect at the output. But this effect is dependent on the $\sqrt{BW_n}$. So you must multiply the output noise by this factor to determine the output noise voltage. Also, remember that since noise is random, not sinusoidal, there is a statistical relationship between RMS and peak (not 1.414).

Op amp selection must be done with the same care given to the rest of the design implementation. First you determine the gain, gain accuracy, upper frequency, slew rate, source impedance, allowable DC errors, temperature range, power supplies available, and output current and output voltage requirements. With these circuit parameters, you can limit your op amp choice to one of five categories. Then, based on cost and availability, you can select a specific op amp. The five categories are general purpose, FET input, CMOS, wide bandwidth, and high V and I. Each category has its strengths and weaknesses. Improved performance in one means a trade-off in another. Be sure to take the time to select the op amp that is best for your application.

Composite amplifiers make the most of the advantages of two different types of op amps, combining them within the feedback loop to produce an amplifier with the best characteristics of both. Placing a power buffer within the negative feedback loop of a precision op amp results in stable gain and low DC errors and high power to the load. The suspended supply amplifier allows the output of the op amp to drive the common of its own power supply. With some positive feedback, this allows you to get $100V_{PP}$ or more from a simple 741C op amp. However, the gain is unpredictable. So enclose it within the negative feedback loop of a precise, stable traditional noninverting op amp amplifier. The result is a stable, accurate, high voltage amplifier for very little money. Remember, however, that whenever you place one stage within the feedback loop of another, the inner amplifier must be **much** faster than the outer. Finally, adding an integrator feedback loop around a high speed op amp allows the integrator to drive the DC offset at the output to zero (or whatever other voltage you want).

The amplifiers discussed work well, **if** you build them correctly. That does not mean that you just wire the correct parts to the correct pins. Component position, orientation, spacing, trace size, supply runs, grounds, ground planes, and junctions are all critical to assure that a circuit that works well on paper and in a simulator can actually be made to work on a breadboard and on a printed circuit board. All of these topics are presented in the following chapter. Making your circuit work first on a protoboard and then on a PCB may appear at first to be black magic. But the key tricks and incantations are all explained.

Problems

5-1   For each of the following op amps, find a data sheet, and then determine the typical and the worst case input offset voltage:

741C, LM318, TL084, LMC660, MAX4251, LT1227, CLC440

5-2 For a noninverting amplifier configuration with $R_f = 1M\Omega$ and $R_i = 47k\Omega$, ignoring bias and offset currents, calculate the effect on the output voltage of a 3mV input offset voltage.

5-3 For an inverting amplifier with no R_{comp}, when the input source is set to $0V_{DC}$, all of I_B^- flows through R_f. None flows through R_i. Explain why.

5-4 For an inverting amplifier with $R_f = 1M\Omega$ and $R_i = 47k\Omega$, with no R_{comp}, calculate the output voltage produced by a bias current into each input of 80nA. Ignore the effect of V_{ios}.

5-5 For a noninverting amplifier with $R_f = 10k\Omega$ and $R_i = 470\Omega$, with no R_{comp}, calculate the output voltage produced by a bias current into each input of 80nA. Ignore the effect of V_{ios}.

5-6 Calculate the components needed to compensate the amplifiers of Problems 5-4 and 5-5 for bias current. Draw the schematic of each circuit with the compensation in place.

5-7 For the circuit if Figure 5-34, calculate

   a. V_{Rcomp} **b.** V_{Ri} **c.** I_{Ri} **d.** I_{Rf} **e.** V_{Rf} **f.** V_{load}

5-8 Repeat Problem 5-7 for $I_B^+ = 80nA$ and $I_B^- = 60nA$ ($I_{os} = 20nA$).

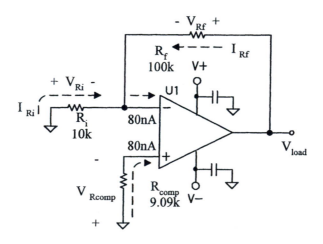

Figure 5-34 Schematic for Problems 5-7, 5-8, and 5-9

5-9   How can you modify the circuit in Figure 5-34 to reduce the effect of the offset current by a factor of 10? Repeat the calculations of Problem 5-7 to prove that the offset current effect on V_{load} has dropped by 10.

5-10   For the circuit in Figure 5-35, calculate V_{load} with $e_{in\,DC} = 0V_{DC}$.

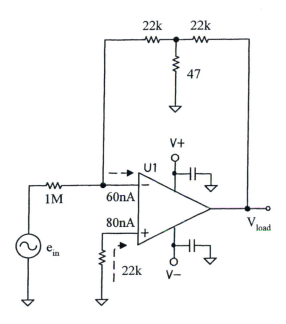

Figure 5-35 Schematic for Problem 5-10

5-11   Draw the schematic of the circuits needed to allow you to measure I_B^+ and I_B^- (two separate circuits) of a 741C with three digits of precision using a digital voltmeter on the 1.999V range. Give all component values and the output voltage to bias current conversion equation.

5-12   Draw the schematic of the circuits needed to allow you to measure the offset current of a 741C with three digits of precision using a digital voltmeter on the 1.999V range. Give all component values and the output voltage to offset current conversion equation.

5-13   The small signal rise time of the 741C is specified as typically 0.35µs. Calculate the
   a.   open loop gain bandwidth.

b. open loop gain at 20Hz, 15kHz, and 650kHz.
c. small signal bandwidth (high frequency cut-off) for a closed loop gain of 80.
d. maximum frequency for a gain of 80 with a 1% gain error.
e. maximum frequency for a gain of 80 with a 10% gain error.

5-14 What must be the open loop gain bandwidth product of an op amp used in an amplifier with a gain of 50, and a gain error of 10% at 20kHz?

5-15 How could you use several 741C op amps to build the circuit described in Problem 5-14?

5-16 A 741C is used to build a comparator. How long does it take the output pulse to rise from 0.8V to 2.4V?

5-17 Design a comparator using an op amp (to be used as the input stage of a digital counter). The output should switch from –3V to +3V in 150ns, whenever the input crosses ground going positive. Specify the op amp's minimum required slew rate.

5-18 Calculate the full power response (frequency) of a 741C that must output a $2V_{RMS}$ sine wave.

5-19 A circuit is needed to amplify a $50mV_{RMS}$, 10kHz input sine wave, outputting $3V_{RMS}$ with a gain error of 1%. Calculate the
a. closed loop gain.
b. required open loop gain bandwidth product of the op amp.
c. required slew rate of the op amp.

5-20 Calculate the output noise from an op amp built using a MAX4251 with $R_f = 1M\Omega$, $R_i = 10k\Omega$, and $R_{comp} = 10k\Omega$.

$$GBW = 3MHz \qquad i_{n\,op\,amp} = 0.5\frac{fA}{\sqrt{Hz}} \qquad e_{n\,op\,amp} = 9\frac{nV}{\sqrt{Hz}}$$

Compare your results with those for the 741 in Example 5-10.

5-21 **a.** List three ways to decrease the noise signals generated within an op amp circuit.
b. List two ways to decrease the **effect** of internally generated noise signals you cannot eliminate.

5-22 Explain why adding inputs to an inverting summer increases the effect of the noise at the circuit's output.

5-23 Design an inverting amplifier using a TL084 op amp. Set the input impedance at 10kΩ, the gain at -20, and $f_{-3dB} = 20kHz$. Draw the schematic and indicate all component values.

5-24   Referring to manufacturers' data sheets, construct a table to compare the typical values of bias current, offset current, input offset voltage, drift, gain bandwidth, slew rate, and maximum output current of the following op amps:

MAX4251,   LTC1049,   741C,   LF411,   LT1227,   LT1010, OPA512SM (Burr Brown),   CLC440 (National Semiconductor)

Add a column to your table indicating to which class each op amp belongs.

5-25   **a.** Which class of op amps has the lowest drift?
        b. Explain how such low drift is accomplished.
        c. Referring to manufacturers' data, select an op amp with a typical drift of $0.1\mu V/°C$ or less.

5-26   In precision circuits, it is often considered more important to have a circuit with low drift than a circuit with low I_{os}, and V_{ios}. Explain why.

5-27   Referring to manufacturers' data, find an op amp with specifications that meet or exceed (i.e., are better than) the following:

$V_{ios} < 5mV$ $I_{os} < 2nA$ $drift < 10\mu V/°C$ $GBW > 3MHz$

Submit a copy of the data sheet with these specifications highlighted. Also indicate **precisely** where you found these specifications.

5-28   Calculate the voltage at all five pins of the op amp, for the suspended supply amplifier in Figure 5-26. Assume that $V_{zener} = 12V$, $\pm V_{HI} = \pm 56V$, and
        a. $e_{in} = +7V_{DC}$.
        b. $e_{in} = -5V_{DC}$.

5-29   For the circuit in Problem 5-28,
        a. what is the maximum V_{CE} required of the two transistors?
        b. select two transistors to replace the 2N3904 and 2N3906.

5-30   For the composite amplifier in Figure 5-29, with possible gain variation in the suspended supply amp of 3.4 to 488, $\pm V = \pm 9V$, and $\pm V_{HI} = \pm 56V$, calculate the maximum possible output voltage from the overall amplifier.

5-31   Calculate all indicated voltages and the currents through each resistor in the circuit in Figure 5-36. Assume that the system is stable (that the integrator's capacitor is neither charging nor discharging). Also assume that U2's input offset voltage is negligible, but that U1's input offset voltage is 5mV. You may ignore all bias currents.

Hints: Do the calculations in the order indicated on the schematic, V_A, V_B, … . Also, be sure to indicate the direction each current is flowing. Carry four significant digits in your calculations. The input offset voltage of U1 makes its noninverting input pin 5mV **more negative** than its inverting pin.

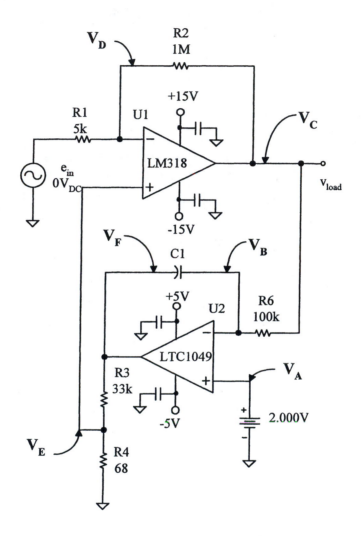

Figure 5-36 Schematic for Problem 5-31

Op Amp DC Characteristic Measurements Lab Exercise

Problem 1

Measure the following characteristics of a 741C op amp: V_{ios}, I_B^+, I_B^-. Calculate $I_{B\ average}$, I_{os} (measured directly, not calculated from I_B^+ and I_B^-).

Procedures for Problem 1

1. Specific procedures and schematics will _not_ be given. You are expected to develop these yourself.

2. Draw an accurate, detailed, complete schematic of the circuit used to measure each of the parameters. Measure and adjust all resistors accurately.

3. Record the procedures used **as you perform them**. Do not make the measurements and try to remember later what you did.

4. Tabulate the raw data. Show your data reduction calculations. Tabulate the measurements of the parameters above along with the manufacturer's worst case specifications.

Problem 2

Apply techniques for compensating for bias currents and offset current.

Procedures for Problem 2

1. Build an inverting amplifier with a 741C, with R_i = 470kΩ, R_f = 4.7MΩ, and ±15V supplies.

2. Measure and record the voltage at the output.

3. Discuss sources of the offset and possible compensation techniques.

4. Change the circuit to compensate for **offset current**.

5. Record the effect of the compensation.

6. Change the circuit to compensate for **bias currrents**. Record the effect of the compensation.

7. Place a hot soldering iron near the op amp and record the effect on the output voltage.

8. Summarize the conclusions you draw from these compensation techniques.

Op Amp Speed Lab Exercise

A. Closed Loop Frequency Response
1. Build the circuit in Figure 5-37.

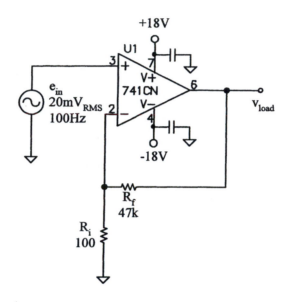

Figure 5-37 Noninverting amplifier

2. Apply DC power; then turn the function generator ON. Set the loaded generator to the proper values.

3. Display the function generator's output on channel A of the oscilloscope and on the digital multimeter, and display the amplifier's output on channel B.

4. Verify the **input's** DC, RMS, and frequency values with the digital meter.

5. Verify the **output's** DC and RMS values with the digital meter. Also assure that the output's phase is correct. Record the theory, actual and error. (Because of the high gain, it is not unusual to have hundreds of millivolts of DC offset at the output.)

6. Complete the second and third columns of a table similar to Table 5-2 by:

a. changing the input frequency to that indicated in each row as measured with the digital meter.
b. verifying that the input amplitude is $20mV_{RMS}$ as measured with the digital meter.
c. verifying that the output is **not** distorted.
d. measuring the output amplitude as indicated by the digital meter and then calculating and recording the amplifier's gain.
e. calculating the gain bandwidth product of the upper frequencies.

Table 5-2 Closed loop frequency responses

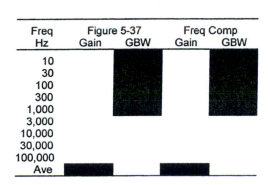

Freq Hz	Figure 5-37 Gain	GBW	Freq Comp Gain	GBW
10				
30				
100				
300				
1,000				
3,000				
10,000				
30,000				
100,000				
Ave				

7. Calculate and enter the average gain bandwidth product of the upper frequencies.

B. Frequency Compensation
1. Place a 10nF capacitor in parallel with R_f in Figure 5-37.

2. Repeat all of the steps of the previous section. Since the gain bandwidth product has been changed by the feedback capacitor, do your initial check at 10Hz.

C. Slew Rate

1. Build the circuit in Figure 5-38.

2. Apply the input.

3. Measure the output's rise time.

4. Calculate the op amp's slew rate. Remember, the rise time is measured from the 10% to the 90% points. So the change in voltage is only 80% of the output's peak-to-peak voltage.

5. Compare the slew rate of your op amp to the manufacturer's specification.

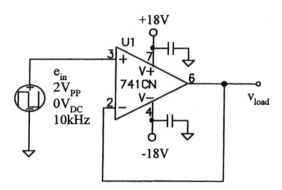

Figure 5-38 Voltage follower for slew rate measurement

D. Analysis and Conclusions
  1. Closed loop frequency response
   a. Create a plot of frequency versus gain from Table 5-2. Two curves are appropriate, one for the simple 741 and the second for the amplifier with the feedback capacitor.
   b. Discuss the results of these plots:
      1) Compare the simple 741 measured gain bandwidth product to the specifications.
      2) Compare the gain bandwidth product with C_f to theory.
   c. Explain why C_f produces a more robust circuit.

  2. Slew rate
   a. Make a table that compares the slew rate you measured and the manufacturer's specification for slew rate.
   b. What conclusions do you draw?

6

Layout of Analog Circuits

Where you place the components and how you choose to interconnect them have as much effect on the success or failure of your circuit as which parts you choose. These mechanical details, though perhaps annoying, are central to the performance of your design. This is equally true in prototype fabrication as it is when developing a circuit for production. Although the techniques for proper layout are most obvious when designing a printed circuit board, they apply every time you build a circuit: on a protoboard with solderless connections, on a vector board with point-to-point soldering, or on a wire-wrap board; and to surface mount parts or to parts with leads. Proper layout cannot make a poor design function properly, but poor layout will most certainly make a good design fail. Worse, the failures may not occur for a month or two. So your design has gone into production and has been sold and distributed widely. Then, with thousands in the hands of your customers, the circuits begin to malfunction.

During this chapter, you will learn the techniques necessary to assure that your circuits are designed to work, not doomed to fail. You must consider mechanical details right at the beginning. You then partition the circuit, separating key functional blocks. Each block is laid out individually. Critical path identification allows you to locate and place the important parts first. Supportive and other parts are then added. Spacing rules must be followed. Finally the traces are run: along the critical path first, then to supportive parts, next to the other parts, and finally to the supplies and ground. Trace size and placement are discussed. Connection to pads, ground planes, and other traces has its own rules you must follow. Step-by-step, through the chapter, the examples take you through the layout of a simple power amplifier, from initial mechanical definition to final ground trace placement.

Objectives

After studying this chapter, you should be able to do the following:
1. Partition a given circuit into low level analog, power, rf, and digital elements.
2. Define the mechanical requirements of the circuit.
3. Identify the critical analog path.
4. Properly place components.
5. Define proper spacing and orientation.
6. Select trace sizes.
7. Interconnect components, pads, and traces.
8. Discuss concerns.

6-1 Partition the Circuit

It is unusual today to find a circuit that includes only one type of circuit. Low level analog circuits may receive their input from a high frequency (rf) circuit. They are controlled by a digital circuit (microprocessor) and send the signal on to a power amplifier. But to assure that the large output is indeed correct, a sample is fed back to the low level analog circuit for automatic gain correction.

Though often used together, each of these circuits has its own requirements, and each may contaminate the signal. Low level analog (the major thrust of most of this text) amplifies **everything** that it sees. This requires particularly clean power and ground lines and minimum opportunity for interference. In rf circuits, lead inductance is critical, as are impedances (input, output, and even trace). Radio frequency circuits radiate contamination (unwanted signals) everywhere, right back into the low level analog circuit. Every time that the digital circuit changes state, it creates rf interference in the middle to high MHz range. Both rf and low level analog circuits must be protected from these transitions. Each level change also places spikes on the power and ground lines, so these must be kept separate from other circuits. Power circuits often handle currents in the 0.1A to 10A range. These currents create significant magnetic fields, which may couple current into nearby low level, analog, high impedance inputs, completely overwhelming the actual signal. In the ground return line, 2A flowing through 0.1Ω of trace and contact resistance produces 200mV. To the load, this may be insignificant. But 200mV at the ground pin of a low level analog amplifier with a gain of 30 may cause a several volt change at its output. This changes the current out of the power amplifier, which changes the ground signal to the amplifier, which changes its output, which changes the current out of the power amplifier, and so on.

To operate properly, low level analog circuits, radio frequency signals, digital subsystems, and power stages must all be kept separate. Your first task is to examine the circuit schematic and identify each of these blocks. Take four different color markers, and partition the system: low level analog, rf, digital, and power. You may need to redraw the schematic a few times, grouping components into these four types.

Each subsystem must be laid out separately. As much as possible, each should have its own power and ground. The grounds, then, are starred together at a single point, usually at the source of the raw power.

6-2 Mechanical Details

It is tempting to go directly to arranging the electronics and running traces. But the mechanical specifications seriously limit how you configure the electronics. So these hardware details must be attended to next.

At this point in the design, you must determine the following:

1. How big, and what shape, can the board be? How firm is this specification?

    This defines the playing field. All of the circuitry and its interconnections must fit inside this boundary. As the layout evolves, it may be convenient to be able to alter the size. ("Gee, it sure would make things work better if I had another 1/4" over here to run these traces.") Since this determines how your board fits into the overall design, you may have to get this information from someone else or from a team of people. So get as complete and detailed requirements as you can. These folks may be hard to get back together again, and this information will reduce the number of changes you are forced to make later on.

    Also, be sure that you know what else is around your board. A near-by motor generates considerable electrical interference. So you should consider moving the sensitive elements of your circuit as far away as possible, and perhaps including a shield. The locations of openings to the outside influence the placement of controls and heat producing elements.

2. How is the board to be mounted?

    Where and how large are the mounting holes and other hardware? Your board must be firmly supported. It will not float. There must be bolts or brackets or card guides. Plan for these at the beginning of the design, so that they do not interfere with the placement of parts and traces. Many circuit board failures are mechanical, caused by vibration and shock, so do not skimp on the mounting brackets and shock absorbers.

    Do **not** use the connectors to support the board. The connectors are designed to pass signals, not to provide structural integrity. Asking an edge connector to support your board is designing in a failure point, within the critical signal path. Also, even if the connector does not break, the stress and strain on conductors change their resistance, altering the signal that flows through them. Provide the strength with brackets and nuts and bolts. Let the connectors carry the signals, not the weight of the board.

3. How much vertical clearance is there?

    It is **very** discouraging to build, populate, test, troubleshot, and modify a board and then find that when you try to install it into the system, it does not fit. For example, the mounting holes all line up, but the heat sink from your output switch runs into the speaker on the board next to yours. It would have been easy to move that heat sink over 1/2" when you first placed it, if you had just known.

4. What type of input and output connectors are to be used and where are they located?

    Rarely is it acceptable to bring signals onto and off of a board by soldering wires directly to the board. These hard-wired connections make removing the board difficult and time-consuming. In a production board, it is usually important to be able to slip one board out and a replacement in. In a prototype, connectors are even more important. You may add and remove the board many times as the project develops. It is often important to be able to run the system with your board connected by some form of extenders lying on the bench. Hard-wired connections make all of this much more difficult. **Use connectors.**

Where must these connectors be placed? Their placement seriously affects the placement of components and the routing of traces. Find out early in the design. Try hard to keep the decision of which signals flow in which pins of the connectors. As you begin the layout, it is critical that certain signals be separated and that others be placed between them as shields. Where the signals are located within a connector may seriously alter their interaction (interference), and therefore the performance of the board. So it is good if you (rather than some systems designer) can make those decisions, and best if the precise pin assignment can be determined later in the circuit board layout.

5. What controls, jacks, and indicators (switches, potentiometers, LEDs, etc.) are required, and where and how are they to be mounted?

It is particularly convenient if the front panel controls and indicators can be mounted directly onto the board. Then the board is located so that these controls extend through the front panel. If you do this, then there is no need for front panel wiring and connectors to pass the signals to the controls. This greatly simplifies your hardware. You must plan, at this point in the circuit's layout, where each of these items is to be located.

Even if there are no front panel controls, commonly some adjustments must be made after the board is installed. Determine now what these are and where the adjustable components are to be located so that they can be accessed easily (perhaps even from outside the chassis). You also need to know whether the adjustment screw is top mounted or accessed from the side.

Even if there are no controls to be accessed, there may be parts that have special mounting brackets. Be sure to account for these needs. At this point it is a good idea to get all of the parts and lay them out on a cardboard cutout of the board. This helps visualize all of the placement and access requirements. It also reminds you that you are dealing with a three-dimensional object and that the height of the parts must also be considered.

6. How is the board to be tested?

Board testing, either automated or manual, is a key part of the fabrication of your circuit. Functional testing is done through the connectors. So you must find out now what signals are needed. You may be able to provide these test signals in the input and output connectors, or you may have to add separate connector(s) to handle these functional test signals. If the testing is to be automated, the placement of the test signal connectors may be fixed, and beyond your control. So find out now.

In-circuit testing is done with a set of spring loaded probes driven onto the board by an automated tester. The signals for the in-circuit tester probes must be available on test pads at defined locations on the board. You must account for these requirements now so that as you place components and run traces, the test points are conveniently located in the required places. Also, these test points must not invite electromagnetic interference.

Even if the testing is all to be manual, you should plan for test points. Probing a component may damage the part or its solder joint. Plan ahead. Provide pads adjacent to the components you plan to test. Make these test points easy to get at.

7. How many layers is the board to be?

A single layer, single sided board is certainly the simplest. The components are on one side, and the traces are on the other. If you can build a double sided board (with traces on both sides), then running traces is simpler. You may be able to place surface mount components on the back (trace side) of the board. This allows you to put the decoupling capacitors and input resistors **directly** below the ICs that they service. Lead lengths drop, reducing interference and lead inductance. Performance improves.

A multilayer board allows you to move the ground plane and the power supplies into the interior of the board. This frees up more space for signal runs on the outer surfaces. It also provides several shielding layers. But the manufacture of multilayer boards is much more complicated and expensive.

8. Are there any daughter boards to be connected to, or mounted onto, this board? If so, where do they go, and how are they mechanically supported?

This is another place where having all of the parts to move around on a cardboard mock-up of the board is a big help. When you are working with two-dimensional PCB layout software, it is very easy to overlook the height of the parts. Also, do not overlook the mechanical support needed for the daughter board.

9. What is the thermal environment? Where should heat sinks go, and how much room is there?

There are several issues here. First, certain components within your circuit may be particularly sensitive to heat. They need to be placed on the board away from areas of wide temperature variation. Knowing how the heat is going to flow around your board helps you find the best location for these parts.

Other parts on the board contribute to the system's heat. This must be taken into account as the total thermal management of the system is considered, so you may have to consult with the mechanical engineers on the project. At a minimum, be sure to move these heat producers away from the sensitive parts. Also, think about how air is going to flow across your board. You may even want to put the power parts on one of the board's edges. Then their heat sinks can be attached to the case or exposed to the outside.

Do not forget about the three-dimensional size of these heat producing parts. Again, the physical mock-up is a big help. It can cripple a project if you are forced to use a smaller heat sink, because the one you need won't fit where you put it on the circuit board.

Example 6-1

Begin the layout for the circuit in Figure 6-1. This board is developed, step-by-step, throughout the rest of this chapter. Begin the circuit board design by determining all of the mechanical details specified above. Start the drawing on paper at a 2-to-1 scale. You may also want to make a 1-to-1 cardboard mock-up. Find appropriate parts and determine mechanical constraints.

1. The board is to be a 2" by 2" square.

2. There are to be four mounting holes, each 1/8" in diameter. The holes are to be 1/4" from the edge of the board.

3. There is 3" of vertical clearance above the board.

4. There is to be one connector. It is a SIP5, with five pins. The pins are 0.1" apart. It is to be located on the center of one edge, 1/4" from the edge. Signals may be placed anywhere within the connector, regardless of what is indicated on the schematic.

5. There are no front panel controls.

6. There should be manual test points. The location is **not** critical.

7. The board is to be **single** sided.

8. There are no daughter boards.

9. There are no other heat producing elements to consider. Locate the heat sink for Q1 near the edge opposite the connector.

Solution

The mechanical layout is shown in Figure 6-2.

6-3 Component Placement

Now it is time to move the parts onto the board. Where you put the parts has a significant impact on the performance of the circuit. Particular care is needed for a circuit with small signals. These are easily contaminated by interference.

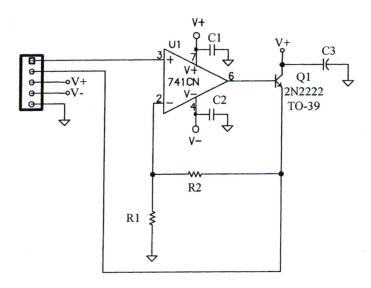

Figure 6-1 Schematic for Example 6-1

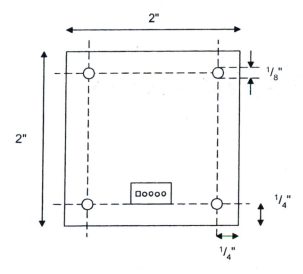

Figure 6-2 Mechanical layout for Example 6-1

High impedance parts convert even the smallest of induced noise currents into rather large noise voltages. High current lines create lots of electromagnetic flux and drop hundreds of millivolts across a few tenths of an ohm in the traces or the connector. High frequency signals broadcast like little radio stations all over your board. Any signal that changes abruptly creates harmonics in the MHz range on every edge, regardless of how rarely those changes occur. Where you put the parts, how far apart you put them, and how they are oriented all affect how much interference these parts receive, or generate, and how easily that interference is coupled into other circuit elements.

Begin with the schematic. Study the circuit. Determine the single, **critical** path. This is the route that the signal follows from input to output. For simple circuits, such as Figure 6-1, this is easy. For more involved boards, this may take some effort and some consultation with the designers. This is important. Put in the effort and get it right.

Now is also the time to begin to **think** about where this signal will flow and what pins in the connector to use. It is necessary to keep the input and the output as far apart as possible. Otherwise, part of the large output can couple back into the small input. In the worse case, the board becomes an oscillator. In general, small and large signals must be kept apart.

The critical path and the connector pin definition are shown in Figure 6-3.

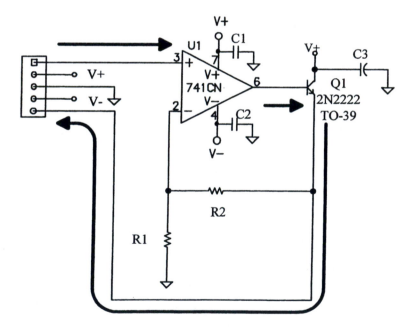

Figure 6-3 Critical signal identification and pin assignment

Pin 1 of the connector carries the input, and pin 5 has the output. Power and ground have been assigned to the pins between. This provides as good a separation as can be obtained with one connector. It would be electrically better if the system specifications allowed the output to be placed in a separate connector at the edge opposite to the input. Placing the supplies and ground between the input and output signals in the connector helps shield the input from the output.

The parts **in** the critical path are laid down first. For Figure 6-3, these are the input (pin 1), the op amp (input on pin 3, output on pin 6), the transistor (base to emitter), and the connector again (pin 5 this time). The smaller the signal, the closer together the parts should be placed. Parts tied to high impedance nodes (the noninverting input of the op amp) should also have as little trace tied to them as possible. Electromagnetic flux, intersecting these connections, induces current. This current, flowing into high impedance, produces a large noise voltage. So keep these lines as short as possible to reduce this noise reception. As the signal is amplified, at the output of the op amp, you may lengthen the connection. The lower the driving impedance, the easier it is for the circuit to override noise. Any long distances should be between points on a node driven from a low impedance (such as the emitter of the transistor).

How close is close enough? Following the suggestions above, it is tempting to place the pads of parts carrying low level signals side-by-side, touching. This is **not** a good idea. That much copper in one place causes the solder to cool too quickly, producing a cold solder joint. The proper minimum spacing between pads of diameter D is a distance of the same D as illustrated in Figure 6-4.

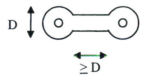

Figure 6-4 Proper minimum part spacing

Whenever possible, avoid placing components side-by-side. This looks neat and may make automated assembly easy. But like any parallel conductors, side-by-side placement enables current in one part to electromagnetically induce current into the adjacent component. Rotate one of the parts 90°, so that they are at right angles. If you must have the parts aligned, at least stagger them, so that they are not side-by-side.

Example 6-2

Place the components in the critical path for the circuit in Figure 6-3. Keep in mind that the transistor has a heat sink. Also remember that R1 and R2 must eventually be connected to the op amp, even though they are not in the critical path. Do **not** run any traces yet.

Solution

The initial placement is shown in Figure 6-5. This is **initial**. You will probably have to nudge these parts as the design developments. But this establishes the structure of the layout and assures that the parts in the critical path get the best positions on the board.

Many good printed circuit board layout software packages are available. Properly used, they may be helpful. But, none are used in these examples. The techniques being discussed apply to all types of fabrication, not just printed circuit boards. Also, the software's auto-place, auto-route, and rats nest display lead the novice into many poor decisions. The intention here is to illustrate good layout technique, not to demonstrate the "click here, drag there" that is unique to a particular software.

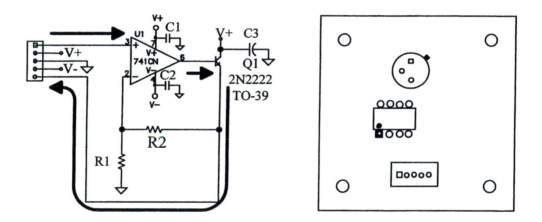

Figure 6-5 Placement of the parts in the critical path

Now that the critical path is established and the structure of the board is in place, identify all of those parts that connect to that critical path. Place these onto the board next. Remember that the smaller the signal or the higher the impedance, the closer the parts should be. It is not necessary to keep all of the parts attached to the critical path very close to those traces. The distance between components is set by the signal size and node driving impedance. So, as you place the parts, pull them back toward the input.

Example 6-3

Identify and place the other components attached to the critical path from Figure 6-5. Do **not** run any traces yet.

Solution

In this simple problem, R2 is the only part that is hanging onto the critical path, but is not in the critical path. It is placed as close to the op amp's pin 2 as practical, since the signal level at that end may be tiny. The resistor runs out to the right because the critical path comes from the transistor's emitter to pin 5 on the connector.

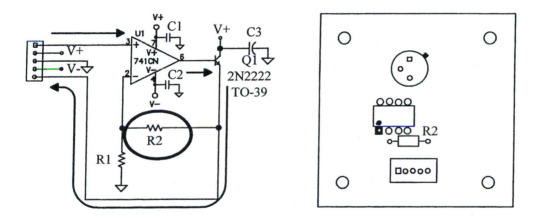

Figure 6-6 Other critical path parts

Now it's time to move the rest of the parts onto the board. Remember to keep parts that carry small signals as close together as possible. Similarly, decoupling capacitors should be close to the supply pin of the part being decoupled. For automated fabrication, jiggle the parts to keep them on a standard grid. Compactness counts, especially where the signal is small. Remember to avoid placing parts side-by-side to reduce interference. It's a good idea to visualize where the main traces will run, so you can avoid having to make major revisions when you begin to route the board. But do not run the traces yet. At this point, they get in the way, making part placement more difficult. Once all of the parts are on the board, 80% of the work is done.

Example 6-4

Place the remainder of the parts on the board from Figure 6-6.

Solution

The complete part placement is shown in Figure 6-7.

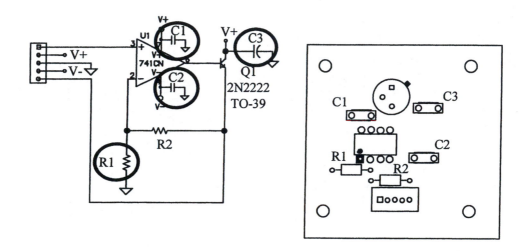

Figure 6-7 Complete part placement

Notice that R2 has been shifted slightly. When C2 was placed on the board, that area became rather crowded. Running traces there may be a problem. But by moving R2 down, you open this area without jeopardizing connections to the op amp's pin 2 (since R1 also has to tie to that pin).

6-4 Traces

Now it's finally time to begin to connect the parts. But, before you start drawing lines, you need to know several guiding principles.

Trace width is set by the amount of current that the trace has to carry, the thickness of the copper on the board, and the temperature rise that you are willing to tolerate. MIL-STD-273 provides a series of graphs that indicates the relationships of these parameters. For analog signals carrying less than 750mA, set the trace at 0.025" wide (0.025", or 25mils). There may be a few times when you have to reduce the width to 0.015" (15mils), but wider is better.

When traces join component pads, you must be sure that there is not too much metal. If there are too many traces joining a pad, the solder cools too quickly, forming a cold solder joint, which eventually fails. This is the same problem that you prevented by keeping two pads at least a diameter apart. So how many traces can you connect to a pad? The sum of the widths of all of the traces entering a pad should be less than 60% of the pad's diameter, as shown in Figure 6-8 and as given in the following equation.

$$W1 + W2 + \ldots < 0.6D$$

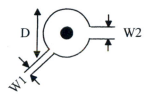

Figure 6-8 Trace width to pad diameter relationship to prevent cold solder joints

To bring two traces into a 0.065" (65mil) diameter pad (typical for an IC pin or 1/4W resistor), each trace must be 0.020" (20mils) wide or less. So if you are going to use 25mil traces, you have to drop their width as they enter the pad, or create a larger pad.

Another trace-related failure happens where traces meet. Applying flux and solder to the board during assembly leaves pockets of acid behind. The cleaning process is unable to remove this acid from junctions formed by traces joining at acute angles. Within a month of production, this residue can eat through the trace. The solution is to assure that no traces meet at angles less than 90°. Since it is not possible to make a perfect 90° angle, when you try to join two traces at a right angle, one side will be slightly greater than 90°, leaving the other slightly less than 90°. The solution is to form a trapezoid where traces meet. This problem and its solution are shown in Figure 6-9.

Finally, points tend to radiate electromagnetic energy, just like little radio transmitter antennas all over your board. So, when a trace needs to change direction, change the direction gently in an arc or in two 45° bends rather than one abrupt right angle. This also eliminates the possibility of an interior angle of less than 90°, for acid build-up. Depending on your artistic talent or your software package, you may even be able to convert the angles to smooth arcs. Look at Figure 6-10.

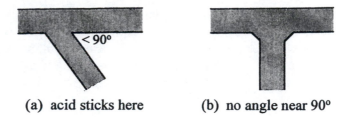

(a) acid sticks here   (b) no angle near 90°

Figure 6-9 Where traces meet

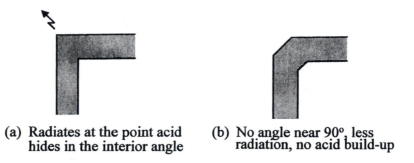

(a) Radiates at the point acid   (b) No angle near 90°, less
    hides in the interior angle   radiation, no acid build-up

Figure 6-10 Changing direction gently reduces radio frequency emissions

Now it's time to run the traces for the critical path. Since they are being placed first, they should all fit onto the back (trace side) of the board. Even if you are building a two-sided or multilayered board, since the critical path goes down first, it gets the best location. It should not have to jump back and forth between layers. Signal integrity is jeopardized each time it changes sides. As this trace is run, you may need to slide components slightly to provide adequate clearance.

Example 6-5

Run the traces for the critical path from Figure 6-7. Follow the guidelines explained in the previous part of this section.

Solution

The critical path traces are shown in Figure 6-11.

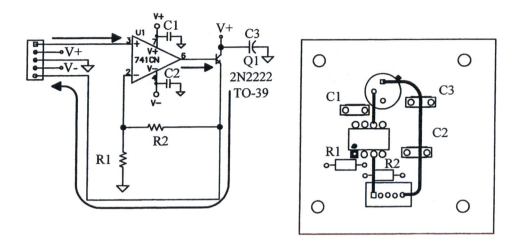

Figure 6-11 Trace placement along the critical path

Run traces to the other components associated with the signal. Power and ground come later. Place the traces in the following order:

1. From the critical path to any components that are connected to it. Do not worry yet about the other end of these components. For Figure 6-11, this means that you should next run a trace to the right end of R2.

2. Once all of the components touching the critical path are tied to it, then connect the other ends of those components, **if** those ends carry signals. Power and ground connections come later. For Figure 6-11, the left end of R2 is tied to the right end of R1 and to pin 2 of the op amp.

3. Make any other **signal** connections. For Figure 6-11, there are no other signal connections.

The result of these steps is shown in Figure 6-12. Notice that the trace from R2 to R1 to the op amp has been narrowed to keep W1 + W2 < 0.6D. Since the signal here is very low level, narrowing the trace is not a problem. Also, notice the trapezoid where R2 ties to the critical path. Resistor R2 has also been moved slightly to the left, so that its pad aligns with R1 and the op amp.

In making these connections, keep the runs as direct and simple as you can. Remember, these traces contribute to the signal. So, be very sensitive to sources of interference and the possibility of degrading the connection. There is little coupling between the current running through a component and the trace that passed under that component at right angles. But avoid long parallel lines or traces that run alongside a component. These two form a transformer, and current is coupled back and forth.

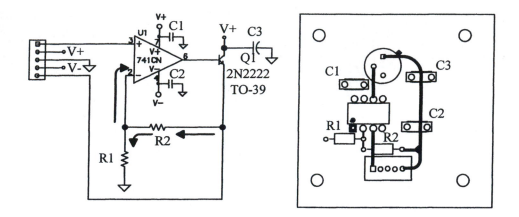

Figure 6-12 Other signal traces run

Vias carry a trace between the two sides of the board. Because of thermal expansion and contraction, and board vibration and twisting, there is a much greater risk of a via failing than a simple trace. Every time you send the signal (or some circuit that contributes to that signal) through a via, you increase the board's vulnerability to thermal and mechanical hazards. Keep the critical path, and as much of the support connections as possible, on one side.

With all of the signals run, it's time to connect power. Run ±V first and ground last. When you decouple ±V properly, these runs are at signal (ac) ground. So they can be snaked in and out around the board without fear of creating interference. In fact, they make reasonably good shields, isolating low level inputs from the high level output. If you have to put traces on the component side, then make them ±V. Often, a ground plane filling the unused areas of both sides is a good idea.

When running power to a component, be sure that the trace goes to the decoupling capacitor first, then to the power pin of the transistor or IC. This gives the capacitor a chance to short out any noise on the bus before it reaches the IC. Similarly, having the decoupling capacitor between the IC and the rest of the power distribution allows the capacitor to short out any glitches or oscillations when the IC suddenly needs charge. The charge can come from the capacitor, and it never has to reach out to the bus. Later, the capacitor will replenish its charge from the bus.

It is also critical to keep the leads and traces between the IC power pin and the capacitor (then to ground) as short as practical. Remember, all conductors have inductance. At higher frequencies, it does not take very long leads for the capacitor to look more like an inductor than a capacitor. At 6MHz, an inch long trace creates more reactance than a $0.01\mu F$ capacitor. The decoupling capacitor disappears, and its leads take over, defeating the purpose of the decoupler entirely.

Example 6-6

Run the ±V traces. Be sure to go to the decoupling capacitors first. Also remember the restriction of trace width and pad diameter. Be sure that you curve the traces. If you have to T a trace, make it at 90°, and add the trapezoid at the joint.

Solution

The solution is shown in Figure 6-13. Running V⁻ to C2 and then to the op amp's pin 4 was simple. But it proved to be a good idea to move R2 down toward the connector when R1 and R2 were placed. This gave enough room to bring the V⁻ trace under R2 and then angle it to C2 before running it over to the IC.

Getting V⁺ to the op amp's pin 7 was a little trickier. Leaving some room between the connector and the edge of the board allowed the V⁺ trace to loop around from the connector's pin 2. Otherwise, it may have been necessary to go to a double-sided board. You need to watch trace clearances, between traces and other pads. How much clearance you need depends on how large the voltage and how well the board can be manufactured. MIL-STD-275E specifies acceptable clearances. For voltages up to 100V, 0.005" (5mils) clearance is all right. Professional board shops can certainly work below that tolerance. But if you are etching the board yourself, add some extra room. The angled trace leading under and then to C1 bears a little watching when the first batch of boards is made.

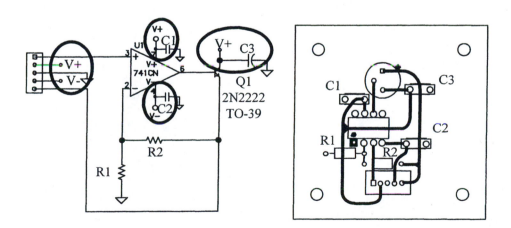

Figure 6-13 Power supply connections

Running a signal under an IC is just asking for interference. But the trace under the op amp is V^+. Well-decoupled, it is an ac ground. There should be no problem running V^+ under the op amp to get to C3. Do not forget to angle your bends and use trapezoids on the T junctions.

Test points were discussed in an earlier section. If their location is critical, then you must place them as components, during the component placement phase, and run traces to these points as you are running the critical path traces. However, given some flexibility in where you place the test point, it makes sense to get the rest of the board laid out first. Generally, it is easier to find and use the test points if they are along the edges of the board.

Example 6-7

Place test points on V^+, V^-, ground, the input, an intermediate point, and the output.

Solution

Test point placement is shown in Figure 6-14. It was convenient to line up the ground, V^+, the intermediate base voltage, and the output signal test points along the top edge. Because of the tightness of the critical path, the input and V^- test points could not be sent to the edge. So they were placed symmetrically on the two corners of the connector. They should be easy to find and access there.

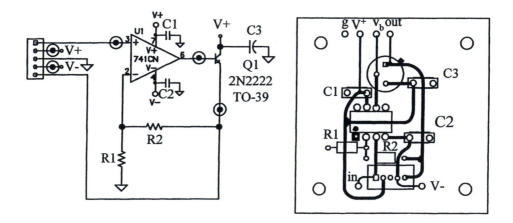

Figure 6-14 Test point placement

At last, it's time to run the ground traces. For single-sided boards, make the ground trace wide. It's OK to snake it in between and around and about. After all, it's a shield. Since ground has to run to several areas of the board, configure it as a **star**. This minimizes possible interaction of different parts of the circuit. This also allows you to separate grounds, putting the low level, ultraclean input circuits on one branch, and the higher current, output section on a different branch. But be careful not to create a ground loop.

If you have the time and facilities, it is often a good idea to spread ground into all of the unused areas of the board. This further shields and isolates the elements of your circuit. Be careful using area fills in a software package. Clearances may become tight. It is also a problem if an island of copper ends up filling part of the board but is not connected to ground. Be sure to remove ground from below inductors and transformers. That much conductor so near lowers those components' inductances.

While you are at it, if you have copper on both sides of your board, you may want to add a ground plane to the component side. This can also solve routing problems. There are several precautions when using ground planes. First, do not try to solder directly to the ground plane. There is just too much copper. The result is usually a cold solder joint. Instead, lay down a pad, isolated from the ground plane. Then run a trace from the pad to the ground plane. Solder to the pad. If you have several different types of circuits on your board (low level analog, rf, digital, power, etc.), give each its own ground plane and a separate return back to the system ground.

Example 6-8
Complete the board by running the ground traces.

Solution
The completed board is shown in Figure 6-15.

Two ground branches were run, one to the left and one to the right. They star together near the input connector. In order to have wide traces and not violate the W1 + W2 < 0.6D rule, the traces to R1 and C2 were placed beyond the parts, and then a drop was run to the ground end of each part.

Once the parts and traces were all in, the board was reviewed again. A few components and their associated traces were repositioned slightly to provide better clearance and cleaner connections.

Do not be misled by how straightforward this layout appears. In reality, the project, as small as it is, has been designed and changed several times. What may have seemed to be insight in placing a particular component is really the result of having done and redone it several times. That's OK. It is common to come to the end, and in looking over your work, to see a better way. This also occurs when you spend time looking at how others have laid out similar boards. It's normal to go back several times to modify the layout. That's part of gaining experience.

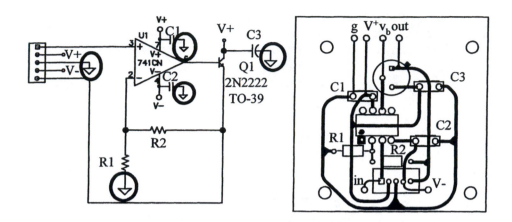

Figure 6-15 Complete example layout

6-5 Other Common Mistakes

In addition to the procedures and warnings you have already seen, there are several other common errors that lead to circuit layout failures. The first happens when placing the parts. It is tempting to use a large expanse of copper on the board as a heat sink. Most boards are made from fiberglass. This material expands quite a bit when heated. The potential failure occurs because the board expands vertically (from top to bottom) five times more than it does along the plane of the surfaces. This places stress on the through-hole solder joints and on the vias. Time after time, as the board heats and cools, expands and contracts, these connections are flexed. Eventually, they pop. Worse, the break often occurs within the board, where it cannot be seen. Keep heat away from the board. Be sure to properly size, or even oversize, the wattage of any part that dissipates heat. Apply heat sinks correctly and keep these away from the board. Allow as much clearance as possible between the board and parts that produce heat.

When running traces, especially as the layout gets crowded, you may be tempted to run a trace between the pins of a part. Most PCB software allows you to narrow the trace and lower the clearances enough to actually sneak the trace through. This is a **bad** idea. To accomplish this, the trace has to be very thin. This raises its resistance and increases the chance that it will not be properly produced. There is also a strong possibility that during the soldering steps, one or both of the adjacent pins will be shorted to the trace. Even if you are able to produce the layout reliably, running a signal that close to component pins is begging for interference, coupled from the trace to the pins or from the pins to the trace.

In both the part placement and the trace routing steps, the critical path is laid down first, and its supporting components and connections second. This is done to assure that the signal gets the best real estate with a minimum of potential interference. There is little reason to run any of these signals through a via, from the bottom to the top of the board. Vias are a big potential troublespot. Insertion, soldering, thermal stress, mechanical vibration, and twisting all increase the possibility that the via will eventually open. If this open is in the critical path, you have just lost the board. So do **not** use vias in the critical path or its support circuitry. For power and ground lines, redundancy can be built in. If one of these vias fails, there are often several others along the bus that can take over.

It is tempting to place a pad right in the middle of a trace, and solder the component to the trace. This allows for a tighter layout, but this violates the $W1 + W2 < 0.6D$ rule for traces and pads. When you try to solder to the trace, it wicks away the heat. The solder freezes on the top of the trace. It never gets hot enough to penetrate the surfaces of the trace and component. This penetration is necessary for a good metallurgical bond. Without it, cold solder joints occur. Eventually, this connection begins to break apart, and its resistance goes up. Even with very good inspection, cold solder joints are hard to detect and may not begin to fail for months after the board is placed into service. Instead of soldering to a trace, drop a run down from the trace to a pad, and connect to that pad. This is how the ground ends of R1 and C2 are connected in Figure 6-15.

Summary

If a circuit is to perform as designed, it must be built properly. Where the parts are placed and how they are interconnected may have as much to do with final circuit performance as do the components you select for the circuit. This is equally true regardless of the fabrication technique (protoboard, point-to-point, wire-wrap, or printed circuit board).

First, review the circuit. If there is a mix of technologies, then partition the circuit into subsystems: low level analog, digital, radio frequency, and power. Each of these should be laid out separately.

The mechanical details are determined next. How big is the board to be? How is it to be mounted? Where and how are you going to bring signals onto and off of the board? Are there controls that must be accessed? What is the thermal environment? Do not forget that you are dealing with a three-dimensional item, even though manual or software aided layout appears two-dimensional. Be sure to account for the vertical size of the components, connectors, controls, and heat sinks. All of these mechanical details define the playing field.

Using the schematic, identify the route that the signal takes through the circuit. With a little work, you should be able to find a single, critical path. Move the components in that critical path onto the board first. Be sure to place the parts so that the input and output signals are well separated. The smaller the signal, and the higher the node impedance, the closer together the parts should be. At their closest, though, the distance between parts must be no less than the diameter of their pads. This reduces the possibility of cold solder joints. As the signal grows, you may lengthen the distance that it travels. Finally, avoid placing parts side-by-side. Though this looks neat, allows for a compact arrangement, and may simplify automated assembly, this parallel arrangement encourages coupling between the components, like the primary and secondary of a transformer. Do **not** run any traces yet.

Next move the parts that touch the critical path onto the board. As before, keep the components connected to small signal, high impedance nodes close together. Observe the minimum spacing rule, and avoid placing parts side-by-side. Do **not** run any traces yet.

Now that the signal has the best place on the board, move the remaining parts onto the board. Remember that decoupling capacitors must be as close to the transistor or IC power pins as possible, not just hanging off the bus somewhere.

Finally, it's time to begin running traces. Wider is better, but default to 0.025" (25mils). When you are running traces into a pad, the sum of the widths of the traces must be less than 60% of the pad's diameter: $W1 + W2 + W3 + \ldots < 0.6D$. This prevents cold solder joints. Assure that traces do not meet in angles less than 90°. Otherwise, acid from the solder and clean steps may be caught in the forks and may eventually eat through the traces. To avoid problems with manufacturing tolerances, convert right angle trace joints to trapezoids. Curve or step traces around corners. Do not use 90° direction changes. They radiate.

With these rules in mind, first place the traces that makeup the critical path. Then run the traces from that path to the components that hang onto (but are not in) the main signal flow. Run any remaining signal traces. Run power to the decoupling capacitors first, then to the IC or transistor it services. If you have to put traces on both sides of the board, leave the signal traces on one side, and move power and ground to the other.

Test points allow easy postproduction performance verification. Probing pads or components directly with instruments may damage the board. Locate test points conveniently near the edge of the board. There may be specifications defining where they go.

The last traces to be run are for ground. A star configuration usually works well, but be sure that you do not create a ground loop. In fact, it may be a good idea to fill the remaining space with a ground plane on both sides of the board. If you do, remove the plane from under inductors and transformers. Check closely to see that there are no isolated islands. Do **not** try to solder directly to the ground plane or to traces. The expanse of copper wicks the heat away too quickly. A cold solder joint results.

Heat causes a fiberglass board to expand and contract much more vertically than in the plane of the board. This breaks through-hole solder joints and vias. So make every effort to keep heat up and away from the board. Do **not** use a plane as a heat sink.

Running signals between the pins of an IC invites interference. Vias are the weak link in the signal chain. Keep the critical path traces on one side of the board. Use vias only when absolutely necessary, and then for power and grounds.

Problem

Complete the layout for the circuit in Figure 6-16.

1. The board is to be 3" by 3".

2. Provide 3/16"-diameter mounting holes, 3/8" from each edge, in all four corners.

3. Place the connector that carries the input and the power onto the board, centered on one edge.

4. Provide a two-pin output connector (to go to an off-board speaker) on the opposite edge. Select an appropriate, commercially available, two-pin audio connector.

5. Keep the load ground and the signal ground separate, but include both in the input connector.

6. There are no vertical constraints.

7. The component sizes are shown in the figure.

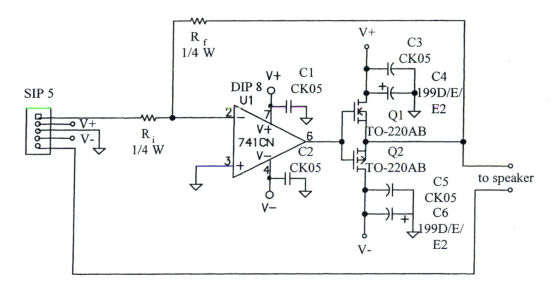

Figure 6-16 Layout problem schematic

7

Single Supply Amplifiers

The operational amplifier circuits you have seen so far have been supplied with bipolar (±V) power. This is done to allow the output voltage to swing both positive and negative.

However, a very large portion of electronics today are powered from a single DC supply and process signals that never go negative. Many industrial measurement and control applications are unipolar and are based around the $+5V_{DC}$ supply used by the digital portion of the system. A second class of single supply amplifiers deals strictly with AC signals. For these applications the precise DC level of the AC signal is not important. Audio amplifiers (e.g., entertainment systems, telephones, intercoms), active filters, oscillators, rf communications, and AC instrumentation all traditionally operate from a single power supply, often two 1.5V batteries. The DC level for these applications is blocked while the signal is passed between amplifier stages using RC coupling.

In this chapter you will learn how to use op amps for DC and AC signal processing while powered from a single supply voltage. Using these techniques, you will be able to process unipolar signals, both DC and AC, without the necessity of providing a negative power source.

Objectives

After studying this chapter, you should be able to do the following:
1. Explain why a traditional op amp, such as the 741, does not work well when powered from a single supply.
2. Compare bipolar and CMOS op amps optimized for single supply operation to the 741.
3. Explain why the output of an op amp must be offset from ground when powered from a single supply in an AC application.
4. Draw the schematic for an inverting amplifier circuit and a noninverting amplifier circuit using a single supply op amp.
5. Calculate component values necessary to bias the circuits above for class A operation.
6. Properly apply a discrete and an IC based supply splitter.

7-1 Operating the Op Amp from a Single Supply

No earth ground connection is required on an op amp. It simply requires a certain minimum difference in potential between the V^+ and V^- supply pins. This could be a difference of 24V, with all signals and instruments referenced at half this difference (i.e., ±12V). However, the op amp performs equally well with the same 24V supply if circuit common is defined as the most negative potential. That is, connect +24V to the V^+ pin and ground the V^- pin. This allows you to build a single regulated 24V power supply, rather than two (±12V) supplies, or to operate from a single, appropriately sized battery.

Circuits which combine both analog and digital electronics are most conveniently powered from a single supply, typically +5V, but increasingly +3V or +3.3V. Many battery operated circuits are powered from two or four 1.5V batteries. In both cases any op amps used must be able to operate from the single polarity power supply, which has a rather limited range. For traditional op amps, such as the 741, two problems occur.

Saturation voltages cause the first problem. The op amp's transistors require "headroom," a difference in potential between their collectors and emitters. Generally this is about 2V. This means that the output voltage can be driven no higher than 2V below the more positive supply, and it must be at least 2V above the more negative supply.

$$V^- + 2V < V_{load} < V^+ - 2V$$

For a 741 operating from a single +5V power supply, V^- is ground. This means that

$$2V < V_{load} < 3V$$

The output can vary only between 2V and 3V. In most circuits this is not acceptable.

The second problem is the range of voltages over which the **input** can vary, while causing appropriate changes in the output voltage. Usually this input range is smaller than the power supply range. If an input voltage equal to V^- is applied, the op amp's output is driven into saturation and may latch up. However, ground is the negative supply voltage when operating from a single supply, and 0V is usually a valid, minimum input. Traditional op amps would not output the proper voltage with an input of 0V and a V^- of ground.

The LM324 is an IC optimized for single supply operation. There are four op amps within the package. Its pinout is given in Figure 7-1. It can be powered with a single polarity power supply as low as 3V. When the op amp is powered from 5V and drives its output into a load resistance of 10kΩ or more, the output swings from a minimum of 20mV to over 4V. Its input voltage range includes the negative supply, so 0V input does not cause the op amp to latch up, or even saturate. It can also be powered from the traditional power supplies, up to ±16V.

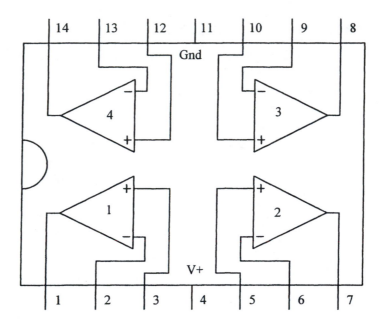

Figure 7-1 Pin diagram of the LM324 and LMC660
(*Courtesy of National Semiconductor*)

The LM324 is made with bipolar transistors. If your application requires exceptionally high input impedance, low bias currents, or an output voltage that swings to within 100mV of the supply voltages, then the CMOS LMC660 is available. The LMC660's output swing and bias currents are indeed notably better than those of the LM324. However, the LMC660 suffers from some of the problems common to CMOS op amps. Its maximum supply voltage is +16V, not ±16V. In fact, the LMC660 is optimized for single supply operation, not performing well from dual (±V) supplies. So, though it is pin compatible, you cannot just drop an LMC660 into the place of other quad op amps. Table 7-1 compares the key specifications of the 741C, the LM324, and the LMC660.

Example 7-1
A temperature sensor outputs 10mV/°F. It is to be used as the input in a microprocessor based system. The A-D converter has a full scale of 5V. Design the amplifier that is necessary to convert the voltage from the sensor at 150°F to the full scale of the A-D converter. Only a +5V power supply is available.

Table 7-1 Op amp typical specification comparison

	741C	LM324	LMC660	
input offset voltage	2.0	2.0	1.0	mV
bias current	80nA	45nA	2pA	
offset current	20nA	5nA	1pA	
gain bandwidth	1.5	1.0	1.4	MHz
slew rate	0.7	0.5	1.1	V/µs
power supply	±22	±16	16	V
V_{load} (+5V supply)	2-3	0.02-4	0.1-4.8	V
supply current	1.7	0.7	1.5	mA
I_{load}	25.0	40	22	mA

Solution

At 150°F, the temperature sensor outputs

$$V_{sensor} = \frac{10mV}{°F} \times 150°F = 1.5V$$

This is the input of the amplifier. The output of the amplifier should be 5V. The LMC660 outputs about 4.8V maximum. Assuming that the A-D converter's range cannot be altered to provide this needed headroom, some of the converter's range must be sacrificed. Setting $V_{out\,max} = 4.5V$, the amplifier's gain is

$$gain = \frac{V_{out\,max}}{E_{in}} = \frac{4.5V}{1.5V} = 3$$

$$gain = 1 + \frac{R_f}{R_i}$$

or $$\frac{R_f}{R_i} = 2$$

Selecting $R_i = 1.1k\Omega$ gives $R_f = 2.2k\Omega$. The schematic is in Figure 7-2.

7-2 AC Inverting Amplifier

When an AC signal is applied to an op amp powered from a single polarity voltage, an additional problem occurs. When the source inputs a signal that should cause the op amp's output to swing negative, the output can only go down to just a little above ground. At that point the op amp's output is in negative saturation. The negative portion of the output is clipped as illustrated in Figure 7-3.

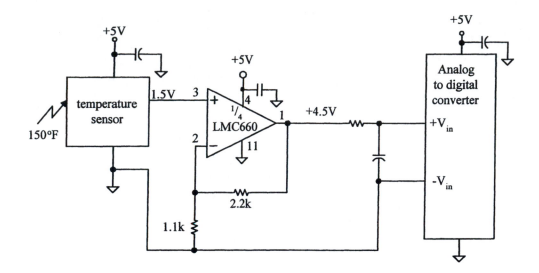

Figure 7-2 Schematic for Example 7-1

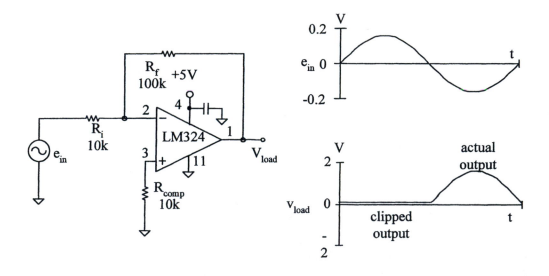

Figure 7-3 Clipped output from a single supply op amp

This problem can be solved by adding a DC level to the output. Offsetting the op amp's output by +2.5V (half of the supply voltage) removes the clipping, as shown in Figure 7-4.

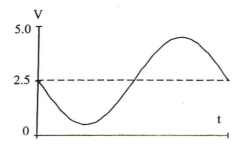

Figure 7-4 DC offset to remove output clipping

The simplest way to add this DC offset to the output of an inverting amplifier is shown in Figure 7-5. Resistors R_f and R_i form the traditional inverting amp configuration, giving a gain of $-R_f/R_i$ to the input signal, e_{in}.

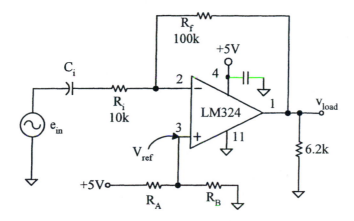

Figure 7-5 Biased inverting amplifier

The 6.2kΩ resistor from the output of the LM324 to ground is necessary to assure that the output transistors are properly biased. Without this DC path to ground, the signal may have crossover distortion.

Instead of connecting the noninverting input to ground, as is done in Figure 7-3, it is tied to a positive reference voltage divider. This V_{ref} is obtained from the positive power supply through the voltage divider, R_A and R_B. As long as R_A and R_B are small (100kΩ) compared to the input impedance of the op amp, the reference voltage is stable and predictable.

The key to the DC operation of the circuit in Figure 7-5 is the input DC blocking capacitor. To the bias voltage on the noninverting input, V_{ref}, the capacitor appears as an open circuit. The resulting circuit is shown in Figure 7-6. For the DC voltage, the circuit is actually a voltage follower. The voltage at the inverting input is driven to that at the noninverting input, V_{ref}. With negligible I_B^-, there is no voltage drop across R_f. The output DC, then, equals the voltage at the inverting input (which is the voltage at the noninverting input).

$$V_{load\ DC} = V_{ref}$$

Remember, this relationship holds because the input capacitor removes R_i from the DC circuit, dropping the DC circuit gain to 1 for the reference voltage. Be sure to select a capacitor does not leak DC current. Be careful with electrolytic capacitors.

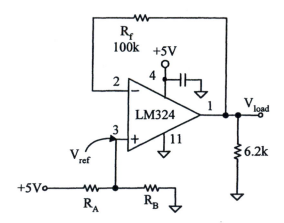

Figure 7-6 V_{ref} as the bias input to the noninverting amp

Example 7-2

Calculate the AC and the DC output voltages for the circuit in Figure 7-5 if:

$R_f = 220k\Omega$ $R_i = 33k\Omega$ $R_A = 100k\Omega$ $R_B = 68k\Omega$

$V+ = 5V$ $e_{in} = 200mV_P$

Solution

AC:

$$gain_{AC} = -\frac{R_f}{R_i} = -\frac{220k\Omega}{33k\Omega} = -6.67$$

$$v_{load} = -6.67 \times 200mV_P = 1.33V_P \; (180° \text{ phase shift})$$

DC:

$$V_{load} = V_{ref}$$

$$V_{ref} = \frac{R_B}{R_A + R_B} V^+$$

$$V_{ref} = \frac{68k\Omega}{68k\Omega + 100k\Omega} 5V = 2.0V$$

$$V_{ref} = 2.0V$$

The composite output is shown in Figure 7-7.

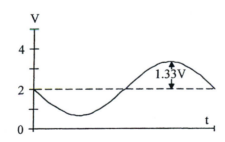

Figure 7-7 Composite output for Example 7-2

Example 7-3

What happens if the input to the circuit in Example 7-2 is increased to 350mV$_P$?

Solution

The schematic and resulting DC bias for simulation using *Electronics Workbench*™ are shown in Figure 7-8. The oscilloscope display is given in Figure 7-9. However, to duplicate these results, you must properly edit the model of the LM324. When you double click on the op amp, its Opamp Properties appear on the screen. The values used in the simulation are also shown in Figure 7-8.

Be aware that you must change these values in the Opamp Property boxes, regardless of the supply voltages connected to the IC in the schematic.

The DC level is correct. The peak amplitude should be

$$V_{load\,P} = 6.67 \times 350mV_P + 2V_{DC} = 4.33V_P$$

The oscilloscope display in Figure 7-9 shows the positive peak reaching 4.33V$_P$ and the negative peak being clipped at about ground. In reality, the LM324 clips the negative peak at about +0.2V.

The coupling capacitor in Figure 7-8 must be selected so that practically all of the source voltage is dropped across R$_i$ and little is dropped across the capacitor. Also, there must be little phase shift across that input capacitor. So the capacitor's reactance (X$_c$) must be much smaller than R$_i$.

$$X_C \ll R_i$$

Practically, make

$$X_C \leq 0.1R_i$$

The worst case is at the lowest frequency that the amplifier has to pass.

$$\frac{1}{2\pi f_{lowest}C} \leq 0.1R_i$$

$$C \geq \frac{10}{2\pi f_{lowest}R_i}$$

If you select a capacitor that is too small, part of the source voltage is dropped across the capacitor. The voltage that makes it to R$_i$ is smaller and shifted in phase. The result is an output that is also too small and not 180° out of phase with the input.

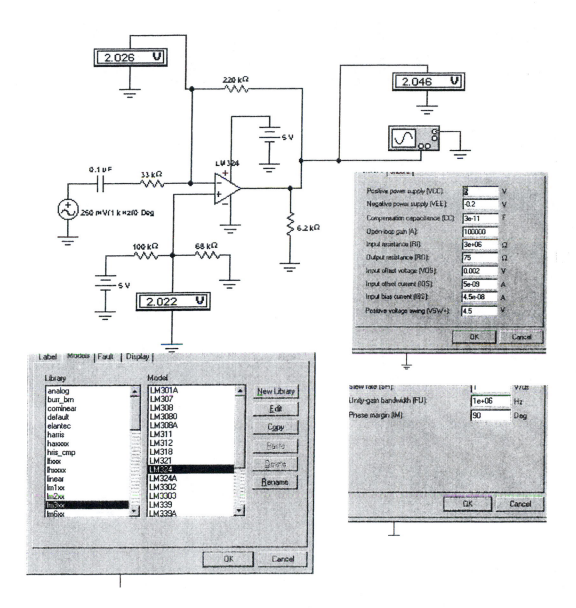

Figure 7-8 *Electronics Workbench^{TM}* simulation of Example 7-3

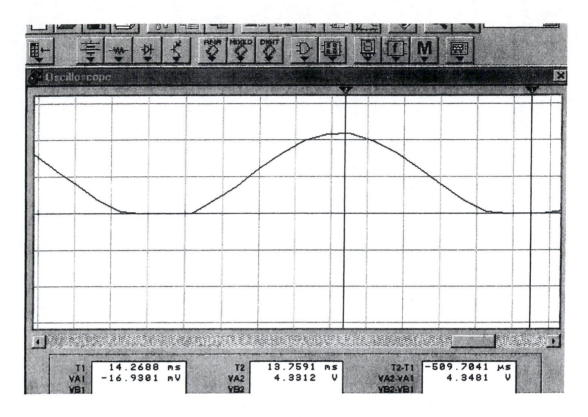

T1	14.2688 ms	T2	13.7591 ms	T2-T1	-509.7041 μs
VA1	-16.9301 mV	VA2	4.3312 V	VA2-VA1	4.3481 V
VB1		VB2		VB2-VB1	

Figure 7-9 Resulting waveform for Example 7-3,
produced with *Electronics Workbench*^{TM}

Example 7-4

   a. Select a properly sized input capacitor for the circuit from Example 7-3, assuming that the lowest frequency that the amplifier must pass is 1kHz.

   b. Illustrate the effect if the input capacitor is too small.

Solution

   a. The proper size capacitor is

$$C \geq \frac{10}{2\pi f_{lowest} R_i} = \frac{10}{2\pi(1kHz)(33k\Omega)} = 48nF$$

   Pick $C = 0.1\mu F$.

b. To investigate the effect of having an input capacitor which is too small, select

$$C = 4.7\text{nF}$$

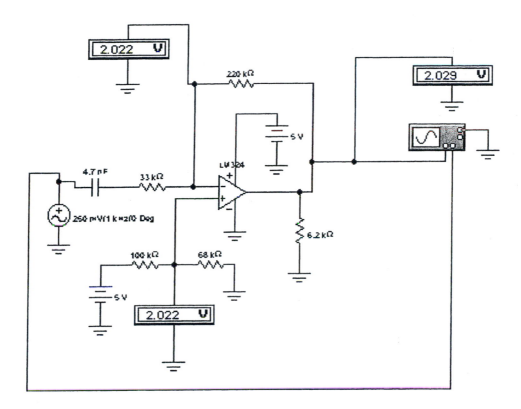

Figure 7-10 (a) *Electronics Workbench^{TM} simulation for Example 7-4*

There should be a 180° phase shift between the input and the output. The positive peak of the output should line up with the negative peak of the input. Instead, the output peaks 122µs too soon. That means that there is only a -136° phase shift across the amplifier. Also, you saw in Example 7-3 that the output amplitude should be 4.33V. But Figure 7-10 shows that the peak output amplitude is only 3.67V. The output DC is correct, so the capacitor is not passing all of the input into the amplifier.

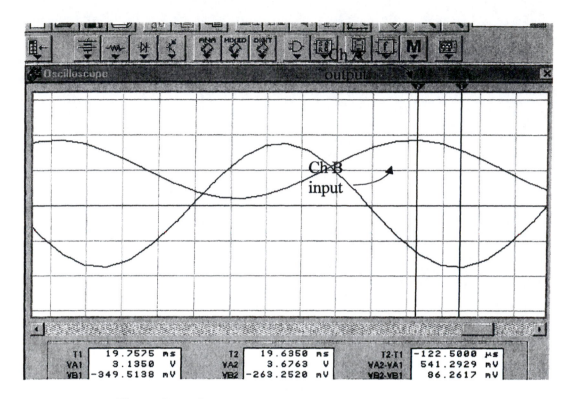

Figure 7-10 (b) Simulation waveform for Example 7-4
using *Electronics Workbench*™

7-3 AC Noninverting Amplifier

A noninverting amplifier can also be built from an op amp running from a single supply.
The schematic is given in Figure 7-11.

The basic noninverting amplifier has been modified by the addition of C_i, R_A, R_B,
and C_f.

$$v_{ACload} = \left(1 + \frac{R_f}{R_i}\right)e_{in}$$

$$V_{DCload} = \frac{R_B}{R_A + R_B}V^+$$

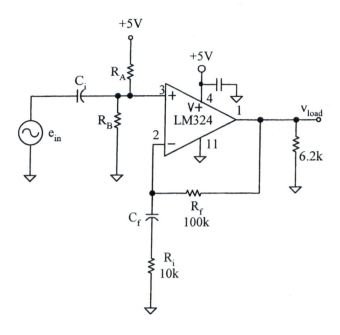

Figure 7-11 Full schematic of a noninverting single supply op amp

The DC circuit is shown in Figure 7-12. Resistors R_A and R_B form a voltage divider, placing V_{ref} on the noninverting input. The capacitor, C_f, appears as an open to the DC. This removes R_i from the circuit, converting it into a voltage follower similar to the inverting circuit of Figure 7-6. Consequently, the output DC is just the DC at the noninverting input (if it is assumed that there is no loss across R_f caused by I_B^-).

The input impedance is set by R_A and R_B. The +5V supply is an AC ground. This puts R_A and R_B in parallel.

$$Z_{in} = \frac{R_A R_B}{R_A + R_B}$$

The input capacitive reactance, X_c, must be small compared to this input impedance.

$$C_i \geq \frac{10}{2\pi f_{lowest} Z_i}$$

As with the inverting amplifier, X_{cf}, the capacitive reactance in series with R_i, must be smaller than R_i. That is,

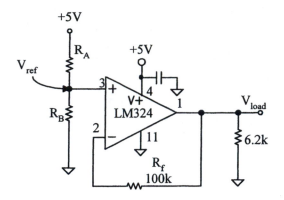

Figure 7-12 DC circuit of a noninverting single supply op amp

$$C_f \geq \frac{10}{2\pi f_{lowest} R_i}$$

Example 7-5

Design a noninverting amplifier, using a single +5V supply, with an AC gain of 15, a frequency range of 20Hz to 20kHz, and an input impedance greater than 100kΩ.

Solution

To properly set the DC bias, the output DC should equal half of the supply.

$$V_{DC\,load} = \frac{R_B}{R_A + R_B}(+5V) = 2.5V$$

This is done by setting $R_A = R_B$.

To make $\qquad Z_{in} = \dfrac{R_A R_B}{R_A + R_B} > 100k\Omega$

pick $\qquad\qquad R_A = R_B = 220k\Omega$

This makes $\qquad Z_{in} = 110k\Omega$

Now set the gain.

$$A_{AC} = 1 + \frac{R_f}{R_i} = 15$$

$$\frac{R_f}{R_i} = 14$$

$$R_i = \frac{R_f}{14}$$

Pick $\qquad R_f = 220k\Omega$. Then

$$R_i = \frac{220k\Omega}{14} = 15.7k\Omega$$

The lowest frequency of interest is 20Hz.

$$C_f \geq \frac{10}{2\pi f_{lowest} R_i} = \frac{10}{2\pi(20Hz)(15.7k\Omega)} = 5.06\mu F$$

Any capacitor larger than this will work, such as 10μF.

$$C_i \geq \frac{10}{2\pi f_{lowest} Z_{in}} = \frac{10}{2\pi(20Hz)(110k\Omega)} = 723nF$$

Select $\qquad C_i = 1\mu F$

7-4 Single Supply Splitter

If you need only one or two op amps biased from a single supply, the techniques of the previous sections are adequate. For more extensive circuits requiring many op amps, the separate stage capacitors and biasing resistors can be replaced with a single supply splitter circuit.

This splitter divides the V^+ in half, producing a common or $V_{ref} = \frac{1}{2} V^+$. You can then use this common to replace ground. By RC coupling into and out of this common referenced (rather than ground referenced) circuit, you have solved the biasing problems.

Figure 7-13 is the schematic of a four-stage amplifier biased from dual supplies. The same circuit is shown in Figure 7-14 using a single supply and individual stage biasing and coupling. Finally, Figure 7-15 is the same single supply amplifier circuit using a supply splitter and common referencing. Individual stage biasing (Figure 7-14) is certainly the most complex (and expensive). Use of a supply splitter to produce a circuit common gives a circuit (Figure 7-15) that is almost as simple as the dual supply circuit.

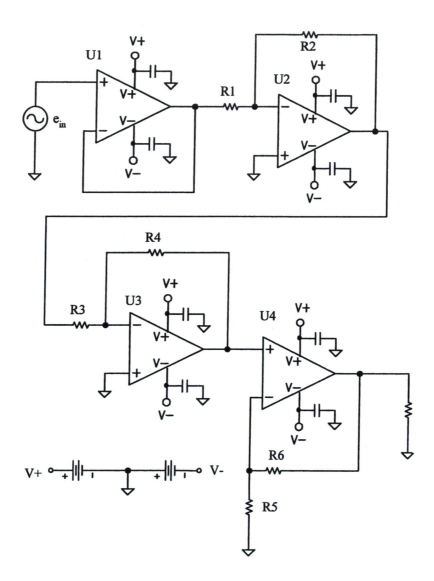

Figure 7-13 Four-stage amplifier biased from dual supplies

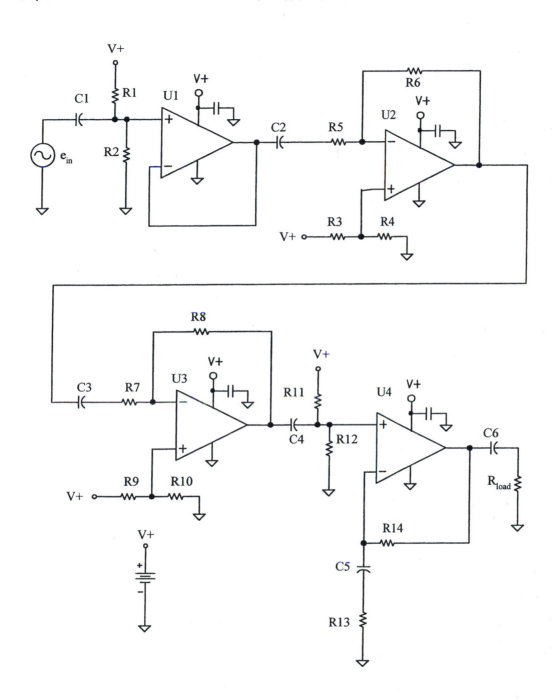

Figure 7-14 Amplifier biased from a single supply using individual biasing

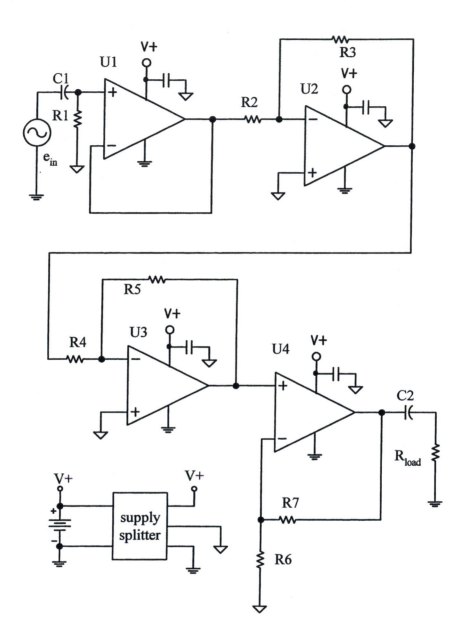

Figure 7-15 Amplifier biased from a single supply using a supply splitter

The simplest form of supply splitter is shown in Figure 7-16. It consists of a voltage divider with shunt capacitors. The 1kΩ resistors evenly split V⁺, establishing the common or bias reference potential. The two capacitors insure good AC grounding. If the amplifier is operating at the upper ranges of the audio frequency band or above, place the 10µF electrolytic capacitors in parallel with 0.1µF film capacitors. Most electrolytic capacitors have poor high frequency characteristics.

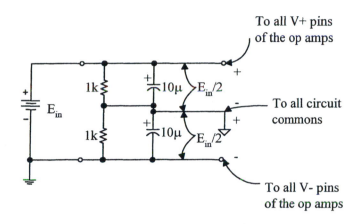

Figure 7-16 Simple supply splitter

If you draw only I_B^+ from the common connections, the circuit in Figure 7-16 works adequately. However, it consists of four components and eight pins. It draws current from the power supply and does not work well when currents at the common are above the tens of microamperes. Remember, the cost of a system is affected more by the number of components and pins than by the cost of those components.

The 7660 switched capacitor voltage converter that you saw in Chapter 4 can also be used to precisely split a voltage, providing hundreds of microamperes or more. However, there is switching noise associated with its output. A minimum of three components (the 7660 and two quality capacitors) with a total of 12 pins are needed.

The TLE2425 virtual ground generator is a significant improvement. It takes the +5V and ground, available in most digital systems, and precisely splits it to provide an analog common at half the supply, providing milliamperes of current. The schematic of a typical application is given in Figure 7-17. The TLE2425's package has only three leads. No external components are needed. So to split +5V, you only really need one component (the TLE2425) and three leads.

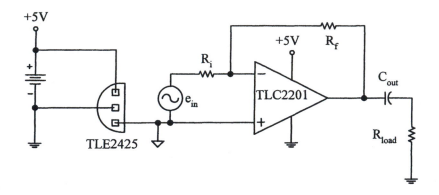

Figure 7-17 Supply splitter application using the TLE2425
(Courtesy of Texas Instruments)

Summary

Traditional op amps, such as the 741, require 2V of headroom. When the op amp is operated from a single +5V supply, the minimum output is +2V and the maximum is +3V. This is not acceptable when building analog processing circuits powered from the same +5V supply that the digital portions of the system use. Battery powered operations have similar problems.

The LM324 is a quad op amp, bipolar IC, optimized for single supply operation. Its performance specifications are equal to or better than those of the 741, but it requires only 1V of headroom below the positive supply, and its output can drop as low as 0.02V. In addition, ground (which equals its negative supply in single supply operation) is a legal input. The LMC660 is a CMOS op amp, also optimized for single supply operation. Its outputs can swing to within 100mV of both supply voltages. Its input currents are less than 1pA. However, the maximum power supply voltage is 16V, not ±16V.

Single supply AC amplifiers can be built with op amps. However, a DC voltage ($\sim\frac{1}{2} V^+$) must be induced at the output to allow the output signal to be driven both up and down by the input. For the inverting and the noninverting amplifiers, this is done by applying a reference voltage to the noninverting input. To prevent this reference from being amplified by the circuit's gain, be sure to place a capacitor in series with R_i. This opens that leg, converting the amplifier into a voltage follower for DC but keeping its gain for AC. Also be sure to RC couple into and out of the amplifier. Remember to select the values of the capacitors large enough so that at the signal's lowest frequency, the capacitive reactance is small compared to the resistance in series with the capacitor.

The circuit's complexity may be significantly reduced by using a supply splitter. This may be a pair of equal resistors, paralleled by capacitors for AC coupling. It divides the power supply voltage in half. This half supply voltage can then be used as the reference within the amplifier (everywhere that you usually use circuit common). The TLE2425 virtual ground generator can simplify generating this reference.

You have seen in these chapters how to build and power signal amplifiers. The op amp allows you to do this much more easily, more precisely, and more cheaply than with discrete components. These signals being fed into your IC amplifiers must be generated. In the following chapter, you apply the characteristics of op amps to produce triangle waves, square waves, and sine waves. Special purpose function generator, phase locked loop, and direct digital synthesizer ICs are also presented.

Problems

7-1   a.   Carefully draw the output from the circuit in Figure 7-18.
        b.   Explain why the output is shaped as it is.
        c.   Modify the schematic to improve the quality of the output signal.

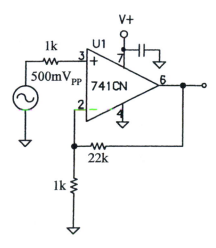

Figure 7-18 Schematic for Problem 7-1

7-2   For the circuit in Figure 7-19, calculate the following:
   a. Output AC signal peak-to-peak voltage.
   b. V_{ref}.
   c. $V_{out\,DC}$.
   d. Input impedance.
   e. Draw the composite output waveform.

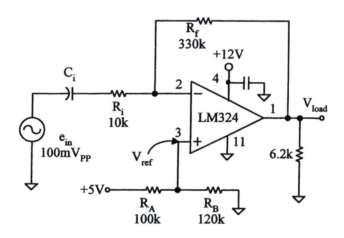

Figure 7-19 Schematic for Problem 7-2

7-3   Design an inverting amplifier, using an LM324 op amp operating from a single +5V supply. The amplifier must have an input impedance of at least 20kΩ and a gain of -15. The lowest signal frequency is 300Hz.

7-4   For the circuit in Figure 7-20, calculate the following:
   a. Output AC signal peak-to-peak voltage.
   b. V_{ref}.
   c. $V_{out\,DC}$.
   d. Input impedance seen by the voltage source, e_{in}.
   e. Draw the composite output waveform.

7-5   Design an inverting amplifier, using an LM324 op amp operating from a single +5V supply. The amplifier must have an input impedance of at least 20kΩ and a gain of 15. The lowest signal frequency is 300Hz.

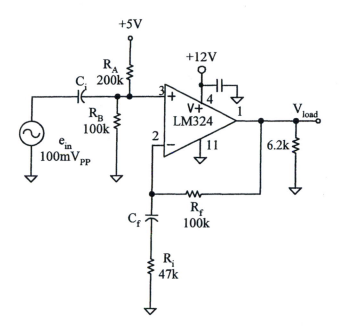

Figure 7-20 Schematic for Problem 7-4

Single Supply Op Amp Lab Exercise

A. Design

Design an amplifier using the following guidelines:

1. LM324 op amp

2. single +5V supply

3. $Z_{in} \geq 10k\Omega$

4. $A_v = -20$

5. $e_{in} - 50mV_{RMS}$ at 1kHz

6. V_{ref} set properly to produce a DC output equal to $2V_{DC}$

B. Circuit Evaluation

1. Build the circuit that you designed in Section A.

2. Set V_{ref} to ground (+ input of the op amp tied to ground).

      a.   Display the output on the oscilloscope using DC coupling.

      b.   Record the wave shape and its amplitude.

      c.   Explain the cause of the distortion.

3.   Set V_{ref} to the value calculated in step A6.

      a.   Measure the AC RMS output voltage, and calculate the actual AC gain. Compare this to that specified in step A4.

      b.   Measure the DC output voltage. Compare this to that specified in step A6.

      c.   Vary V_{ref} and observe the effect on the output signal using the oscilloscope on DC coupling. Record your observations.

      d.   Increase e_{in} until both the output upper and the lower peaks are clipped off. Record the upper clipped output level $(+V_{sat})$ and the lower output level $(-V_{sat})$.

      e.   Return the input level and reference to those specified in Section A. Confirm proper AC and DC operation.

4.   Replace the LM324 with an LMC660. Repeat steps 3a-d. Compare the results for the LMC660 with those for the LM324.

5.   Replace the LMC660 with an LM741C. Repeat steps 3a-d. Explain the results.

8

Waveform Generators

During the previous chapters you have seen how to condition signals with op amps. However, these signals must be produced, system control commands must be generated, and measurements and timing must be performed. Those three functions (signal generation, command generation, and measurement and timing) depend heavily on precision waveform generation.

In this chapter you will see how to produce square waves, ramps, and sine waves. These can come from circuits based around op amps or from special purpose integrated circuits. Circuits that produce all three wave shapes from stand-alone ICs and ICs that respond to digital commands will be covered.

Objectives

After studying this chapter, you should be able to do the items listed below for each of the following circuits:

Circuits

Astable multivibrator - 555   Function generator - 2206
Crystal controlled oscillator   Phase locked loop
Op amp ramp (triangle) generator   PLL with a 2206
Op amp sawtooth generator   Digital frequency synthesizer

Tasks

1. Draw the schematic.
2. Qualitatively and quantitatively describe circuit operation.
3. Calculate component values necessary to meet given frequency and amplitude specifications.
4. Discuss advantages, disadvantages, limitations, and precautions of the given circuit, and compare the circuit with others that produce the same wave shape.

8-1　Square Wave Generators

Square waves are the simplest waveforms to produce. They are used in system control (both analog and digital), tone generation, and instrumentation. They can also be used to produce both ramps and sine waves. In this section you will see square wave generators made with a special timer IC and with a crystal for a high level of accuracy and stability.

8-1.1　The 555 Timer

The 555 timer is an integrated circuit specifically designed to perform signal generation and timing functions. It allows a great deal of versatility. It operates from a wide range of power supplies, sinking or sourcing 200mA of load current. Proper selection of only a few external components allows timing intervals of several minutes or frequencies as high as several 100kHz. The block diagram of the 555 is given in Figure 8-1.

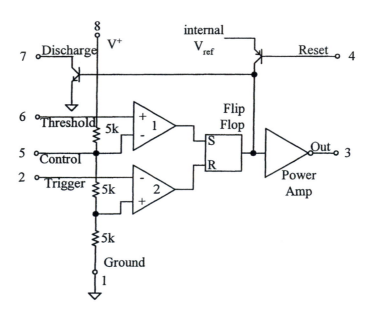

Figure 8-1 Block diagram of a 555 timer IC

The core of the 555 timer is a voltage divider and two comparators. The reference voltage to comparator 1 is $^2/_3V^+$, while the reference voltage to comparator 2 is $^1/_3V^+$. Whenever the voltage on the **threshold** input exceeds $^2/_3V^+$, comparator 1 forces the flip-flop to output a high. A high from the flip-flop saturates the discharge transistor and forces the output from the power amplifier to go low. Even if the voltage at the threshold input falls below $^2/_3V^+$, comparator 1 cannot cause the flip-flop to reset. Comparator 1 can only force the flip-flop's output high (set).

To reset the flip-flop, changing its output to a low, the voltage at the **trigger** input must fall below $^1/_3V^+$. When this occurs, comparator 2 resets the flip-flop, sending its output low. This low from the flip-flop turns the discharge transistor off and forces the power amplifier to output a high. Comparator 2 can only reset the flip-flop.

In summary, to force the output of the 555 timer low, the voltage on the **threshold** input must exceed $^2/_3V^+$. This also turns the discharge transistor **on**. To force the output of the timer high, the voltage on the **trigger** input must fall below $^1/_3V^+$. This also turns the discharge transistor **off**.

A voltage may be applied to the control input to override the level at which this switching occurs. When the timer is not in use, a 0.01µF capacitor should be connected between pin 5 and ground to prevent noise coupled onto this pin from causing false triggering.

Connecting the **reset** (pin 4) to a logic low places a high on the output of the flip-flop. The discharge transistor goes **on**, and the power amplifier outputs a low. This condition continues until **reset** is taken high. This allows synchronization or resetting of the circuit's operation. When the timer is not in use, **reset** should be tied to V^+.

An astable multivibrator can be produced by adding two resistors and a capacitor to the basic timer IC, as shown in Figure 8-2.

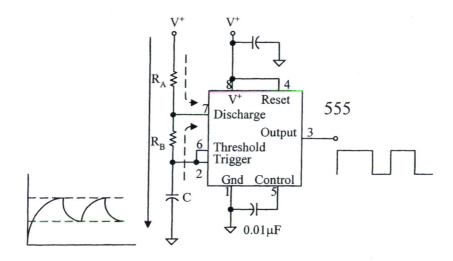

Figure 8-2 Basic square wave generator

When power is first applied, the capacitor holds no charge. This places the **trigger** below $\frac{1}{3}V^+$. The output goes high, and the discharge transistor goes off.

The capacitor charges through R_A and R_B (solid line) until its voltage exceeds $\frac{2}{3}V^+$. This voltage on the **threshold** input causes the flip-flop to be set, the output to drop to a low, and the discharge transistor to be turned on. The capacitor now discharges (dotted line) through R_B, and the discharge transistor (inside the IC) to ground. Current also flows from V^+ through R_A into the discharge transistor. Resistors R_A and R_B must be large enough to limit this current and prevent damage to the discharge transistor.

As the capacitor discharges, its voltage falls. When the voltage on the **trigger** input falls below $\frac{1}{3}V^+$, the output goes high again, and the discharge transistor turns off. This cycle repeats itself for as long as power is applied.

The length of time that the output remains high is the time for the capacitor to charge from $\frac{1}{3}V^+$ to $\frac{2}{3}V^+$.

$$v(t) = A\left(1 - e^{-t/RC}\right) \qquad \text{charge of an RC circuit}$$

For the time it takes the circuit to charge from 0 to $\frac{2}{3}V^+$,

$$\frac{2}{3}V^+ = V^+\left(1 - e^{-t/RC}\right)$$

$$e^{-t/RC} = \frac{1}{3}$$

$$-\frac{t}{RC} = \ln\left(\frac{1}{3}\right) = -1.099$$

$$t = 1.099RC$$

For the time it takes the circuit to charge from 0 to $\frac{1}{3}V^+$,

$$\frac{1}{3}V^+ = V^+\left(1 - e^{-t/RC}\right)$$

$$\frac{1}{3} = 1 - e^{-t/RC}$$

$$e^{-t/RC} = \frac{2}{3}$$

$$-\frac{t}{RC} = \ln\left(\frac{2}{3}\right) = -0.405$$

$$t = 0.405RC$$

So the time to charge from $^1\!/_3V^+$ to $^2\!/_3V^+$ is

$$t_{high} = 1.099RC - 0.405RC = 0.69RC$$

$$= 0.69(R_A + R_B)C$$

The output is low while the capacitor discharges from $^2\!/_3V^+$ to $^1\!/_3V^+$.

$$v(t) = Ae^{-t/RC}$$

$$\frac{1}{3}V^+ = \left(\frac{2}{3}V^+\right)e^{-t/RC}$$

$$\frac{1}{2} = e^{-t/RC}$$

$$-\frac{t}{RC} = \ln\!\left(\frac{1}{2}\right) = -0.69$$

$$t = 0.69RC$$

$$t_{low} = 0.69R_BC$$

Notice that both R_A and R_B are in the charge path, but only R_B is in the discharge path.

$$T = t_{high} + t_{low} = 0.69(R_A + R_B)C + 0.69R_BC$$

$$= 0.69(R_A + 2R_B)C$$

$$f = \frac{1}{T} = \frac{1.45}{(R_A + 2R_B)C}$$

With this circuit configuration, it is impossible to have a duty cycle of 50% or less.

Remember, $\qquad\qquad t_{high} = 0.69(R_A + R_B)C > t_{low} = 0.69R_BC$

However, the circuit in Figure 8-3 allows the duty cycle to be set at practically any value.

During the charge portion of the cycle, diode CR1, forward biases, shorting out R_B.

$$t_{high} = 0.69R_AC$$

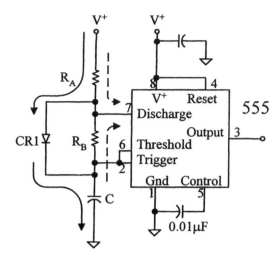

Figure 8-3 Adjustable duty cycle, rectangular wave generator

However, during the discharge portion of the cycle, the diode is reverse biased.

$$t_{low} = 0.69 R_B C$$

$$T = t_{high} + t_{low} = 0.69(R_A + R_B)C$$

$$f = \frac{1}{T} = \frac{1.45}{(R_A + R_B)C}$$

Resistors R_A and R_B could be variable or even digital potentiometers. This would allow you to adjust the frequency and pulse width (perhaps under program control). However, a series fixed resistance of at least 100Ω should be added in series with R_A and R_B if you are going to make them adjustable. This limits the peak current to the discharge transistor should the rheostats be adjusted to their minimum values.

Although the 555 timer has been used in a wide variety of often unique applications, it is very hard on its power supply lines, requiring quite a bit of current and injecting many noise transients. This noise often is coupled into adjacent ICs, falsely triggering them. The 7555 is a CMOS version of the 555. Its quiescent current requirements are considerably lower than those of the 555, and the 7555 does not contaminate the power supply lines. It is pin compatible with the 555. So this CMOS version of the 555 should be your first choice when a 555 timer is used.

8-1.2 Crystal Controlled Oscillators

The crystal controlled oscillator provides the most accurate, most stable frequency of all of the waveform generators commonly available.

At the heart of the oscillator is a piezoelectric crystal. Applying an electrical difference in potential across two of the faces of the crystal causes the crystal to deform along one of its mechanical axes. An electrical impulse, or step, causes the crystal to vibrate, mechanically oscillating back and forth. The frequency of oscillation is determined almost exclusively by how the crystal is cut. There is a very small tolerance (often on the order of one part per million) caused by variation in temperature.

So the piezoelectric crystal makes a very precise mechanical vibrator. However, just as the electrical disturbance causes a mechanical vibration, mechanical vibrations produce an electrical signal. As the crystal vibrates in response to the original electrical shock, the crystal in turn produces an electrical signal at the same stable frequency at which it is vibrating.

Of course, both the mechanical vibrations and the resulting electrical signal die out. However, if you electrically stimulate the crystal again and again, it continues to produce its electrical output. The key to keeping these oscillations going is to synchronize the stimulating pulses with the vibrations. It's just like pushing a swing.

Some form of electronics is necessary to use the signals from the crystal to trigger an output step in phase with these signals. This output step can then be fed back into the crystal to keep it oscillating. One of the simplest of these crystal oscillators is shown in Figure 8-4.

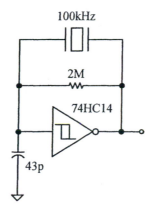

Figure 8-4 Schmitt trigger gate type crystal oscillator
(*Courtesy of Linear Technology*)

A CMOS Schmitt trigger is used. The gate must be CMOS so that its input does not load down the crystal. The 2MΩ feedback resistor provides a bias path and works with the 43pF capacitor as a low pass filter. This low pass filter sets the input DC voltage at the average of the output waveform. Its critical frequency should be much lower than the frequency of the oscillator.

$$\frac{1}{2\pi RC} << f_{crystal}$$

With a 50% duty cycle output, the input should be at $2.5V_{DC}$. This is in the middle of the Schmitt trigger's hysteresis region. When the signal through the crystal swings the input up above the Schmitt trigger's upper threshold, the output falls to a logic low. When the signal from the crystal drives the input below the Schmitt trigger's lower threshold, the output goes to a logic high.

"Although these types are popular, they are often associated with temperamental operation, spurious modes of operation or downright failure to operate. ... It is not uncommon in circuits of this type for gates from different manufacturers to produce markedly different circuit operation. In other cases, the circuit works, but is influenced by the status of other gates in the same package. Other circuits seem to prefer certain gate locations within the package. In consideration of these difficulties, gate oscillators are generally not the best choice in a production design; nevertheless, they offer low, discrete component count, are used in a variety of situations and bear mentioning."[1]

Replacing the Schmitt trigger digital gate with a comparator IC solves these problems but requires a few more components. The schematic is given in Figure 8-5. The 2kΩ resistors bias the noninverting input to half of the supply. The other 2kΩ resistor and the capacitor form a low pass filter, setting the DC voltage at the inverting input to the average of the output waveform. This is also half of the supply if the waveform has a 50% duty cycle.

The crystal provides positive feedback from the output of the comparator to its noninverting input. A transient at the output causes the crystal to vibrate at its resonant frequency. This, in turn, generates a signal of the same frequency at the noninverting input. The signal is riding above and below the DC level set by the two 2kΩ resistors. When the composite wave at the noninverting input rises above the average value held by the capacitor, the output of the comparator is driven up. This output step adds energy to the vibrating crystal, keeping it going. When the signal from the crystal drives the noninverting input below the average value held by the capacitor, the comparator's output swings to ground. Again, this step is properly synchronized to add energy to the vibrating crystal. This crystal is operating in **series** resonance. So be sure to order the correct type.

"Above 10MHz, AT-cut crystals operate in overtone mode. Because of this, oscillations can occur at multiples of the desired frequency."[2] A series RC network can be added to reduce the gain as frequency increases as shown in Figure 8-6.

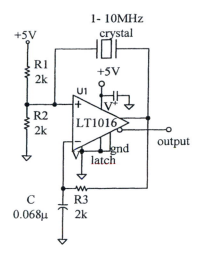 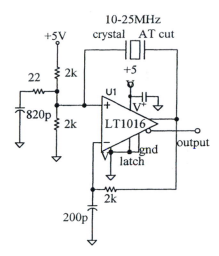

Figure 8-5 Comparator based crystal oscillator (*Courtesy of Linear Technology*)

Figure 8-6 10MHz to 25MHz crystal oscillator *(Courtesy of Linear Technology)*

8-2 Ramp Generators

Ramps, triangle waves, and sawtooth waves are used extensively in sweeping (i.e., causing a variable to change at a linear rate) in many industrial, measurement, and signal and data conversion circuits. You depend on a ramp generator every time that you use an oscilloscope to move the beam linearly across the screen. Sine waves can also be derived from triangle waves. Op amps can be used to produce these important timing and driving waveforms.

8-2.1 Triangle Wave Generator

The voltage across a capacitor, when charging toward a fixed **voltage**, is exponential. However, the output from a triangle wave generator varies **linearly** with time. These two wave shapes are compared in Figure 8-7. Clearly, charging the capacitor with a fixed voltage does not produce a triangle wave.

The key relationship for current and voltage across a capacitor is

$$i = C\frac{dv}{dt}$$

Rearrange. $$dv = \frac{i}{C}dt$$

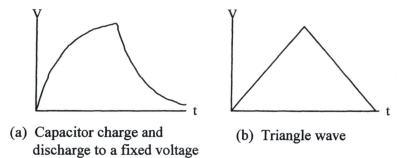

(a) Capacitor charge and
    discharge to a fixed voltage

(b) Triangle wave

Figure 8-7 Capacitor charging waveform

Take the integral.
$$\int dv = \frac{1}{C} \int idt$$

$$v = \frac{1}{C} \int idt$$

If you charge the capacitor with a constant current, I, then

$$v = \frac{I}{C} \int dt = \frac{I}{C} t$$

Charged with a constant current, the voltage across the capacitor charges **linearly** with time, exactly as shown in Figure 8-7(b).

The circuit in Figure 8-8 drives a constant current into the capacitor. This causes the output voltage (which is also the voltage across the capacitor) to vary linearly with time.

The current charging the capacitor is provided by the source, E, through R. Since the op amp's inverting input is at virtual ground,

$$I = \frac{E}{R}$$

Ideally, none of I flows into the op amp. All of it goes to charge the capacitor. When E is applied, the voltage across the capacitor, and therefore at that output, ramps linearly downward as the capacitor charges. This continues at a **constant rate** until the output reaches $-V_{sat}$.

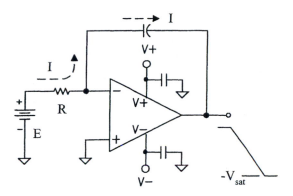

Figure 8-8 Charging a capacitor with a constant current

The **rate** at which this ramp drops is the important parameter.

$$V = \frac{I}{C}t$$

$$\text{rate} = \frac{\text{volts}}{\text{second}} = \frac{V}{t}$$

$$\text{rate} = \frac{I}{C}$$

For the op amp,
$$I = \frac{E}{R}$$

$$\text{rate} = \frac{E}{RC}$$

The rate or slope of the ramp is directly proportional to the charging voltage and inversely proportional to the time constant, RC.

The RC network, as implemented in Figure 8-8, allows you to produce a linear ramp. To build a ramp generator, you only need to add a two-level, **noninverting** comparator as illustrated in Figure 8-9.

A positive potential at e_{in} drives the noninverting input high, sending the output to $+V_{sat}$. The output stays at $+V_{sat}$, even when e_{in} falls and then goes negative, because R_f and R_i form a voltage divider between e_{in} and V_{sat}. The large, positive $+V_{sat}$ keeps the noninverting input positive even for small **negative** values of e_{in}.

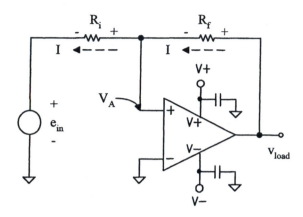

Figure 8-9 Two-level, noninverting comparator

The output does not switch to $-V_{sat}$ until e_{in} has gone negative enough to pull the noninverting input to just below ground. The input voltage necessary to do that is the lower trigger point. You can derive that voltage as follows. The current in the outer loop, I, is

$$I = \frac{V_{sat} - e_{in}}{R_f + R_i}$$

$$V_A = e_{in} + IR_i$$

$$V_A = e_{in} + R_i \left(\frac{V_{sat} - e_{in}}{R_f + R_i} \right)$$

$$V_A = \frac{e_{in}R_i + e_{in}R_f + V_{sat}R_i - e_{in}R_i}{R_f + R_i}$$

$$V_A(R_f + R_i) = e_{in}R_f + V_{sat}R_i$$

The output switches when V_A goes to ground. Here, e_{in} is the lower trigger point, V_{LT}.

$$0 = V_{LT}R_f + V_{sat}R_i$$

$$V_{LT} = -\frac{R_i}{R_f} V_{sat}$$

When the input falls below this lower trigger, the output goes to -V_{sat}. This large negative on the end of R_f holds the noninverting input of the op amp negative, keeping the output at -V_{sat}. Even when e_{in} goes a little positive, -V_{sat} through the voltage divider (R_f and R_i), keeps the noninverting input of the op amp negative. This, in turn, forces the output to stay at -V_{sat}. When e_{in} goes positive enough to drive the noninverting op amp input above ground, the output switches to +V_{sat}. That is the upper trigger point and can be derived just as the lower trigger point was derived.

$$V_{UT} = \frac{R_i}{R_f} V_{sat}$$

The input/output relationship of the noninverting comparator is illustrated in Figure 8-10.

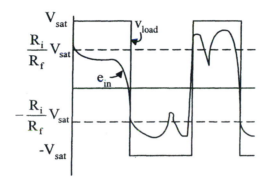

Figure 8-10 Input-output relationship of a two-level comparator

For applications above the audio range, use a comparator IC such as the LM311 instead of an op amp. The comparator is much faster than a comparably priced op amp. This means much sharper edges out of the comparator and crisper points on the triangle, as well as a higher possible output frequency. Be sure, however, that you configure the comparator to swing its output both positive and negative. Otherwise, the capacitor will be able to charge, ramping down, but it can never discharge.

The linear charge circuit from Figure 8-8 can be combined with the noninverting comparator of Figure 8-9 to produce a triangle wave generator. Its schematic is shown in Figure 8-11. The output of the linear charge circuit (U1) is a ramp. This ramp is the input to the comparator (U2). The square wave output of the comparator provides the charge voltage for the linear charge circuit.

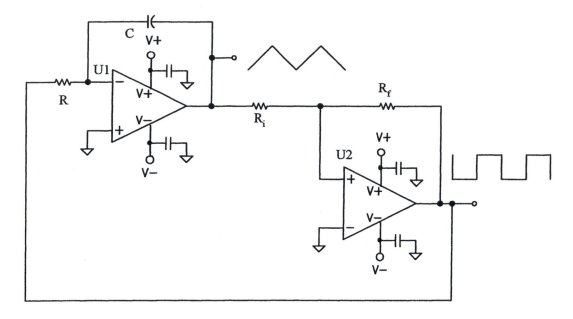

Figure 8-11 Triangle wave generator

Assume that the output of U2 is initially low (-V_{sat}). Current from the output of U1 flows through its feedback capacitor and through R. The capacitor charges up at a linear rate of

$$\frac{V_{sat}}{RC}$$

The output of U1 ramps up. When it reaches the upper trigger voltage of the comparator,

$$\frac{R_i}{R_f} V_{sat}$$

the comparator's output switches to +V_{sat}.

With +V_{sat} as the charge voltage for the linear charge circuit, the current through the capacitor reverses. It now flows from the output of U2, through R, through C, and into U1. The capacitor charges down at the same rate as it charged up. The charge on the capacitor and the output of U1 go to zero and then negative. When the output of U1 goes negative enough to cross the lower trigger point of the comparator, the comparator's output switches to -V_{sat}. This reverses the charge direction of U1, and the cycle repeats. The result is a square wave out of the comparator U2 and a triangle wave at the output of U1.

The rate of rise and fall of the triangle wave and the trigger points of the comparator set the amplitude and the frequency. The rate of charge of the capacitor is

$$\text{rate} = \frac{V_{sat}}{RC}$$

The triangle charges between V_{LT} and V_{UT}, so the distance that the triangle wave has to travel is

$$\text{distance} = V_{UT} - V_{LT} = 2\frac{R_i}{R_f}V_{sat}$$

The time it takes to travel a given distance is

$$\text{time} = \frac{\text{distance}}{\text{rate}}$$

So for the triangle wave,

$$t = \text{pulse width} = \frac{2\dfrac{R_i}{R_f}V_{sat}}{\dfrac{V_{sat}}{RC}} = 2\frac{R_i}{R_f}RC$$

But this is only the time for one half cycle. The full period is

$$T = 2t = 4\frac{R_i}{R_f}RC$$

and the frequency is

$$f = \frac{1}{T} = \frac{R_f}{4R_iRC}$$

Varying **any** of the components affects the frequency.

The amplitude of the square wave is determined by the saturation voltage of U2. You can set this more precisely with zener diodes.

The peak-to-peak amplitude of the triangle wave is determined by the trigger levels of the comparator. The triangle wave must go from the lower trigger point to the upper trigger point (and vice versa) to cause the comparator to switch.

$$v_{PP} = V_{UT} - V_{LT} = 2\frac{R_i}{R_f}V_{sat}$$

The amplitudes of the square wave and of the triangle wave are directly proportional to the output level (either V_{sat} or V_{zener}) of U2. In addition, R_i and R_f, which set the comparator trigger points, help set the triangle amplitude but (along with R and C) also alter the frequency.

Example 8-1

For the circuit in Figure 8-11, let:

$R = 10k\Omega$ $C = 0.1\mu F$ $R_i = 100k\Omega$ $R_f = 820k\Omega$ $\pm V = \pm 12V$

a. Calculate the amplitude of the triangle and the square waves.

b. Calculate the signals' frequency.

c. Verify your calculations with computer simulation.

Solution

a. Square wave amplitude

$$v_{PP} = V_{sat} - (-V_{sat}) \approx 10V - (-10V)$$

$$v_{pp} \approx 20V_{PP}$$

Triangle wave amplitude

$$v_{PP} = 2\frac{R_i}{R_f}V_{sat}$$

$$= 2\frac{100k\Omega}{820k\Omega}10V$$

$$v_{PP} = 2.44V_{PP}$$

b. $$f = \frac{R_f}{4R_iRC}$$

$$= \frac{820k\Omega}{4 \times 100k\Omega \times 10k\Omega \times 0.1\mu F}$$

$$f = 2.05kHz$$

c. The circuit simulation results are shown in Figure 8-12.

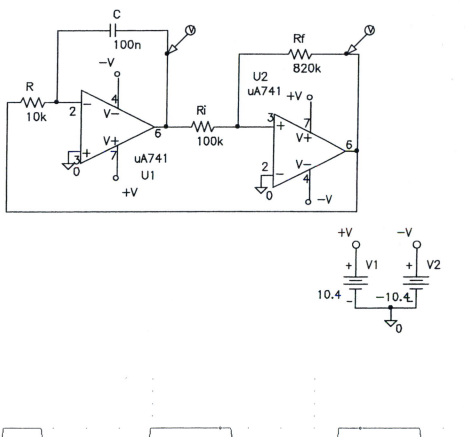

Figure 8-12 Simulation results for Example 8-1(a) and (b)

The output of U2 is a square wave. By altering the supply voltages to ± 10.4V, the output of the comparator is set to ± 10V. This accounts for the fact that the model of the 741 has only a 0.4V saturation, rather than 2V which is typical for the hardware.

The triangle output has an amplitude of

$$v_{PP} = 1.3643V - (-1.3486V) = 2.71V$$

There is also a slope to the edges of the square wave. This has altered both the amplitude of the triangle wave and the frequency.

$$f = \frac{1}{T} = \frac{1}{1.0627\text{ms} - 475.3\mu\text{s}} = 1.7\text{kHz}$$

In Figure 8-13 the op amp used for U2 has been replaced by an LM111 comparator. The resulting triangle amplitude is

$$v_{PP} = 1.2296V - (-1.2406V) = 2.47V$$

This is remarkably close to the 2.44V_{PP} predicted by theory.

$$f = \frac{1}{T} = \frac{1}{5.0987\text{ms} - 4.6120\text{s}} = 2.05\text{kHz}$$

The frequency also more closely matches that obtained by theory. Notice that it is necessary to run the simulation for much longer with the LM111 than with the UA741, in order to give the oscillator time to start.

8-2.2 Sawtooth Generator

You may find it necessary to provide a relatively slow linear ramp with a rapid drop (or rise in the case of a negative ramp) at the end. This is a sawtooth wave. Also, in applications such as time base generators and power control circuits, the sawtooth must be triggered by some control signal. The circuits of this section allow you to meet these requirements.

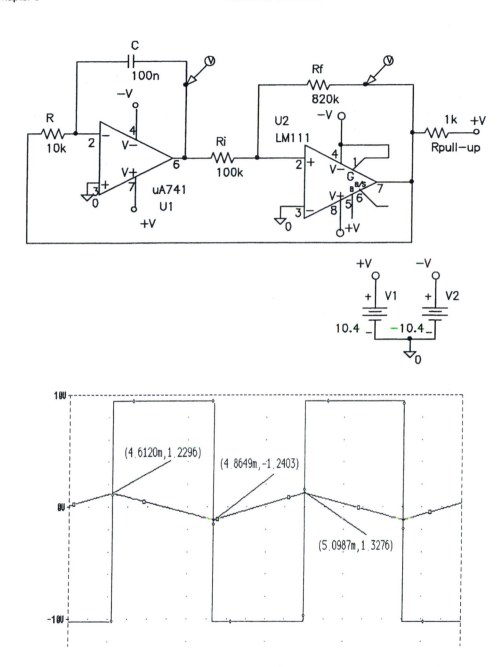

Figure 8-13 Example 8-1 with a comparator IC

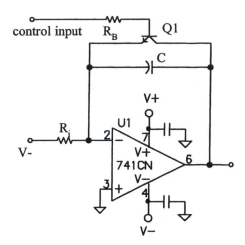

Figure 8-14 BJT controlled ramp generator

The circuit in Figure 8-14 gives you the ability to control ramp generation with an external signal. An NPN bipolar junction transistor is placed across the charging capacitor. Notice that the emitter of the transistor is tied to the inverting input of the op amp, which is at virtual ground. Resistor R_B limits the base current to protect the transistor and whatever is producing the control input. However, R_B should be kept small enough to assure that the transistor is driven into saturation.

With a zero or negative control input voltage, Q1 is off. The capacitor charges up from the op amp's output, through the capacitor, through R_i and to V^-. The charge rate is

$$\text{rate} = \frac{V^-}{R_i C}$$

So the slope of the output voltage can be controlled by changing either R_i or V^-. Resistor R_i could be a potentiometer, either mechanical (adjustable with a screwdriver) or digital. Changing the V^- used to charge the capacitor also changes the charge rate. But do not change the negative supply to the op amp. This charging voltage could come from a potentiometer, from a voltage out, digital to analog converter, or from an analog circuit.

When you apply a positive control input voltage, Q1 turns on. If this voltage is large enough to force Q1 into saturation, you have effectively placed a short around the capacitor. The capacitor rapidly discharges. The output voltage falls to zero (actually ~0.2V) and stays there as long as your positive control voltage keeps Q1 saturated. Examples of the waveforms you can expect are given in Figure 8-15.

A programmable unijunction transistor (PUT) can be used to replace the BJT across the capacitor in Figure 8-14. The PUT is a thyristor with an anode, a cathode, and an anode gate. Normally the PUT is off, presenting a large resistance between the anode and the cathode. However, when the voltage on the anode (with respect to a grounded cathode) exceeds the voltage on the gate by approximately 0.7V, the PUT goes on. Like other thyristors, once fired, the anode-to-cathode resistance falls to a short. The gate loses control. The PUT remains on until the anode current falls below the holding current. Then the PUT goes off again. The anode voltage must be taken above the gate voltage to fire the PUT again.

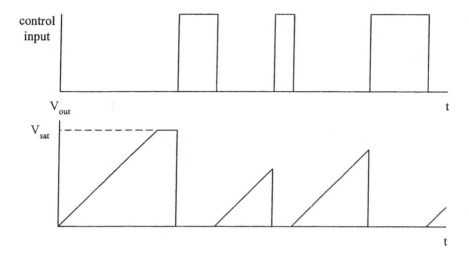

Figure 8-15 Control input signal's effect on the ramp generator's output

A PUT controlled sawtooth generator is shown in Figure 8-16. When power is first applied, the PUT is off. The capacitor begins to charge up, and the output rises. This continues until the output voltage (which is also the anode voltage of the PUT) is ~0.7V above the control input (the gate voltage). When that level is reached, the PUT goes on. The capacitor is shorted. It immediately discharges through the PUT. The output voltage (voltage across the capacitor) falls. When the current through the PUT falls below its holding current, its goes off, and the cycle repeats. There is about 1V left across the PUT when it goes off. The output waveform is shown in Figure 8-17.

The period of the PUT sawtooth generator depends on the charge rate (V/RC) and on the control voltage as shown in Figure 8-17.

$$\text{rate} = \frac{\text{distance}}{\text{time}}$$

$$T = \frac{(V_{control} + 0.7V) - 1V}{\dfrac{|V^-|}{RC}}$$

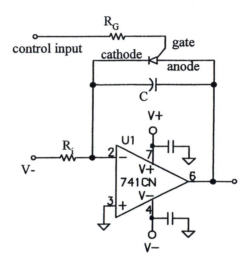

Figure 8-16 PUT controlled sawtooth generator

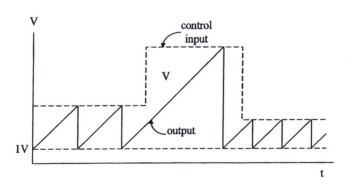

Figure 8-17 Output from a PUT controlled sawtooth generator

$$T \approx \frac{V_{control}RC}{|V^-|}$$

$$f \approx \frac{|V^-|}{V_{control}RC}$$

The PUT controlled sawtooth generator can be used as a voltage-to-frequency generator.

You must be aware that for the PUT to turn off, the current through it must fall below its holding current, I_H (specified by the manufacturer). When the PUT is on, not only does the capacitor discharge current flow through the PUT, but also a current equal to that used to charge the capacitor. This current flows through R and to V^-. To get the PUT off, be sure that

$$I = \frac{V^-}{R} < I_H$$

Otherwise, once the PUT goes on, it is held on by this charge current, even when the capacitor has fully discharged. You can lower this charge current by increasing R or decreasing the negative voltage. However, both affect the charge rate and therefore the frequency. So you have to balance changes in either R or V^- with an appropriate change in C.

8-3 Function Generator

The circuits that you have studied so far generate either a square wave or a triangle wave. A function generator simultaneously produces these two signals as well as a sine wave. In addition, it is easy to vary the frequency from 0.01Hz to over 1MHz without significant distortion of any of the signals. Often frequency and amplitude can be controlled with a voltage. This diversity and flexibility often make the function generator the most popular of waveform generators.

There are several monolithic integrated circuits available that allow you to build a function generator with the addition of only a few external components. The XR-2206 is widely applied and quite flexible, while still being easy to use. It produces a triangle wave, a square wave, and a sine wave. The output amplitudes can be adjusted with external potentiometers (mechanical or digital) and with an external voltage. The frequency can be adjusted through a range extending from below 0.01Hz to over 1MHz. This can be done with external resistors, a capacitor, a logic level, and an analog voltage level.

The block diagram of the XR-2206 is given in Figure 8-18. Its heart is the current controlled oscillator. A constant current is set with either external resistor, R_{t1}, or

R_{t2}. Which resistor sets the constant current is determined by the logic level applied to the FSK (frequency shift keying) pin.

As you saw in the previous section, charging and discharging a capacitor with a constant current produces a ramp. That capacitor is connected to the C_t pins. This ramp is compared to two levels. When the ramp exceeds the upper level, a comparator within the current controlled oscillator block changes state, causing the constant current to reverse direction. The ramp now heads down at the same rate. When the ramp drops below the comparator's lower threshold, the current reverses direction again. The capacitor charges up again.

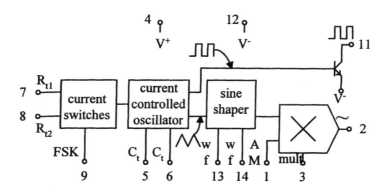

Figure 8-18 XR-2206 block diagram

Power is applied between the V^+ pin (pin 4) and the V^- pin (pin 12). This difference in potential must be

$$10V \leq V^+ - V^- \leq 26V$$

Power may be single supply, usually with V^- (pin 12) grounded. Or you may provide a split supply.

$$\pm 5V \leq V^+ - V^- \leq \pm 13V$$

The control functions are referenced to V^-, so running the IC single supply greatly simplifies generating external control signals.

The frequency is controlled by three sets of pins: C_t (connected between pins 5 and 6), the FSK logic level (pin 9), and the current through external timing resistors (at pins 7 and 8).

$$f = 0.32 \frac{I}{C_t}$$

where

$$1nF \le C_t \le 100\mu F$$

and

$$1\mu A \le I \le 3mA$$

This current must flow **out** of the IC and is determined by the external resistors, R_{t1} and R_{t2}. Which time resistor is used is set by the FSK signal at pin 9 with respect to V^-. Assuming that you are running the XR-2206 single supply, with pin 12 grounded,

$$V_{FSK} > 2V \rightarrow R_{t1} \text{ (pin 7)}$$

$$V_{FSK} < 1V \rightarrow R_{t2} \text{ (pin 8)}$$

The XR-2206 regulates the voltage at both R_t pins, keeping it 3V above V^-. So, in the simplest case, you connect both the timing resistor, or digital potentiometer, and V^- to ground. This then sets

$$I_t = \frac{3V}{R_t}$$

To produce a 50% duty cycle square wave, all that is left is to add an external pull-up resistor. Look back at the XR-2206's block diagram (Figure 8-18). The square wave pin is driven by an open collector transistor. When the output should go high, the transistor is open. To produce an output voltage, connect a resistor from whatever high level voltage you want, to pin 11. When the output should go low, the internal transistor shorts to V^- (not necessarily ground). Be sure to take this into consideration when selecting the power for the IC. Usually, it is simplest to connect V^+ to +12V (or more) and V^- to ground. Pick the pull-up resistor to limit current into the open collector transistor to 2mA or less.

$$R_{pull-up} > \frac{V_{logic} - V^-}{2mA}$$

Figure 8-19 shows the XR-2206 as a simple square wave generator. Normally, the XR-2206's output duty cycle is 50%. This is accurately maintained within a few percent across the entire frequency range.

The circuit in Figure 8-20 allows you to produce a variable duty cycle. Remember, when the voltage at the FSK pin is greater than 2V above V^-, R_{t1} on pin 7 is used to set the current that charges the timing capacitor. When the FSK pin voltage is within 1V of V^-, R_{t2}, at pin 8, is used to generate the capacitor charge (or discharge) current.

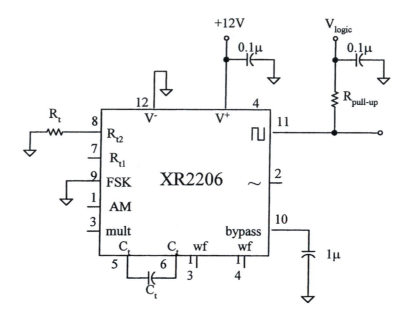

Figure 8-19 Simple square wave generator

To provide a variable duty cycle, the FSK pin is connected to the square wave output pin. When the oscillator drives this pin high, timing resistor R_{t1}, at pin 7, sets the capacitor current and therefore the ramp-down time.

$$t_{high} = \frac{C}{0.64 I_{Rt1}}$$

where

$$I_{Rt1} = \frac{3V}{R_{t1}}$$

When the oscillator drives the square wave and therefore the FSK pin low, timing resistor R_{t2} on pin 8 is enabled. The low time is now set by

$$t_{low} = \frac{C}{0.64 I_{Rt2}}$$

where

$$I_{Rt2} = \frac{3V}{R_{t2}}$$

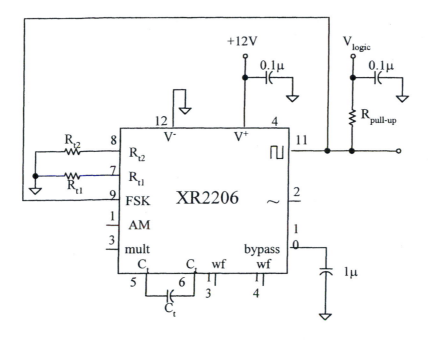

Figure 8-20 Variable duty cycle

Be sure to set the minimum values of these two resistors to at least 1kΩ, to insure that the timing current does not exceed its maximum, 3mA. So if you plan to use potentiometers (either mechanical or digital), put a 1kΩ fixed resistor in series with each.

    A triangle or sine wave is also generated at pin 2. Leaving the waveform pins (pins 13 and 14) open produces a triangle wave. Placing a 180Ω resistor between these pins rounds the peaks of the triangle into a sine wave. You can tweak the value of this waveform resistor to adjust the sine wave's distortion.

    The maximum amplitude of the triangle wave and the sine wave are controlled by the AC resistance between the multiplier pin (pin 3) and ground.

$$v_{P\ sine} = \frac{60\mu V}{\Omega} \times R_{mult}$$

$$v_{P\ triangle} = \frac{160\mu V}{\Omega} \times R_{mult}$$

For simple operation, be sure to tie the AM pin (pin 1) to either V⁻ or V⁺.

The DC level of the triangle or sine wave is set by the DC level on the mult pin. So, if you are powering the XR-2206 from dual supplies and want your sine wave to ride above and below ground, tie R_{mult} to ground as shown in Figure 8-21.

Choose the power supplies carefully. Remember that

$$10V \le V^+ - V^- \le 26V$$

So running the IC from ±5V to ±12V is all right. But, a single +5V supply is too small, and ±15V is too large.

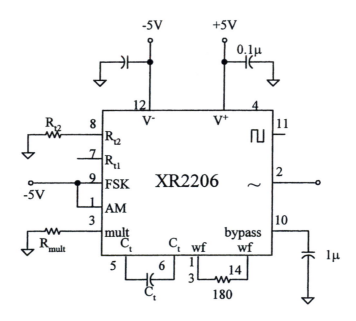

Figure 8-21 Simple sine wave generator

External control is more easily done if the XR-2206 has V^- set to ground. If you do this, the triangle and sine waves' DC level must be offset to about half of the single power supply voltage. This can be done by adding a DC voltage to the mult pin with a voltage divider. Since these resistors appear to be in parallel to AC, keep their resistance small compared to R_{mult} and be sure to take their value into account when setting the sine and triangle wave magnitude. Figure 8-22 shows the single supply operation with a sine or triangle wave output.

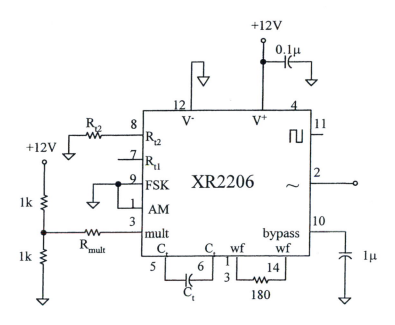

Figure 8-22 Single supply sine wave generator

Finally, remember that R_{mult} can be a mechanical potentiometer for manual adjustment, or a digital potentiometer if you want a microprocessor to adjust the output amplitude.

The resistance R_{mult} sets the **maximum** output voltage. This amplitude can be reduced by applying a voltage to the AM (amplitude modulation) pin, pin1. The relationship between the AM pin voltage and the output amplitude is shown in Figure 8-23. When the voltage on the AM pin is at either V^- or V^+, the output at pin 2 is the maximum set by R_{mult}. Setting V_{AM} halfway between the positive and negative supply voltages drives the output amplitude to zero. If you are powering the XR-2206 from dual supplies, grounding the AM pin causes the output to fall to zero. If a single supply (e.g., 12V) is used, the AM pin must be set to ½ V^+ (e.g., 6V) to force the output to zero.

As V_{AM} increases or decreases from this null point, the amplitude grows. It reaches a maximum when V_{AM} is 4V away from the midpoint. If the IC is powered from dual supplies, any voltage more negative than -4V causes a maximum output amplitude. As V_{AM} becomes less and less negative, the amplitude falls, reaching zero when the AM pin is grounded. Setting this pin positive causes the amplitude to increase linearly with increasing voltage, until V_{AM} is 4V. Beyond that V_{AM}, the amplitude is a maximum.

Single supply operation is similar, with the null point at $\frac{1}{2}V^+$ and the amplitude increasing as V_{AM} moves away from that point.

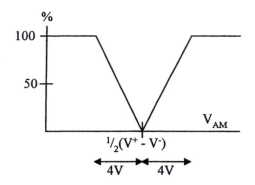

Figure 8-23 AM voltage control of amplitude
(Courtesy of Exar Corporation.)

Example 8-2
>Given that $R_{mult} = 68k\Omega$ and the XR-2206 is powered from a single +15V supply, calculate the output RMS sine wave amplitudes for V_{AM} that equals:
>
>**a.** ground   **b.** 9.5V   **c.** 5V

Solution
>**a.** With the AM pin tied to the negative supply (ground in the case of single supply operation), the output amplitude is maximum.
>
>$$v_{P\,sine} = \frac{60\mu V}{\Omega} \times 68k\Omega = 4.08V_P$$
>
>But the problem asks for the RMS voltage.
>
>$$v_{RMS} = \frac{4.08V_P}{\sqrt{2}} = 2.88V_{RMS}$$
>
>**b.** Look again at Figure 8-23. The midpoint is $\frac{1}{2}V^+ = 7.5V$. As the voltage increases from that point, the amplitude increases at a rate of
>
>$$\frac{100\%}{4V} = 25\frac{\%}{V}$$

At $V_{AM} = 9.5V$, this voltage is 2V above the midpoint, so

$$v_{RMS} = 25\frac{\%}{V} \times (9.5V - 7.5V) \times 2.88V_{RMS} = 1.44V_{RMS}$$

c. Below the midpoint, the slope of the line in Figure 8-23 is negative.

$$v_{RMS} = -25\frac{\%}{V} \times (5V - 7.5V) \times 2.88V_{RMS} = 1.8V_{RMS}$$

The AM function allows you to control the amplitude of the sine or triangle wave with an external voltage. This voltage could come from a microprocessor and a voltage out DAC, such as the MAX512 you saw in Chapter 3. Or the voltage could be generated by additional analog electronics. The integrator in the Composite Amplifier section of Chapter 5 could be combined with an ideal rectifier or true RMS converter IC (Chapter 10) to provide automatic level control. In either case, be sure to scale and offset the voltage to fit **entirely** on one edge of the curve in Figure 8-23. It must not cross the null point and head up the other side Also, be sure that you pick the edge that will provide stable, negative system feedback.

"As the bias level approaches $V^+/2$, the phase of the output signal is reversed; and the amplitude goes through zero. This property is suitable for phase-shift keying and suppressed-carrier AM generation. Total dynamic range of amplitude modulation is approximately 55dB." [3] This means that the XR-2206 can be used to produce the signal for AM radio transmission. The audio information is coupled directly to the AM pin. For 100% modulation, this signal should be $4V_p$ riding evenly above and below ground. Power the XR-2206 from dual supplies. Set the frequency from the XR-2206 to the carrier frequency. The output from the IC is the carrier signal whose amplitude varies with the audio information. Of course, be certain to adhere to all FCC regulations when you couple this rf signal to a power amplifier and antenna.

Applying a voltage to the AM pin allows you to control the amplitude. You can also alter the frequency with an externally generated voltage. This is done by placing a voltage on the left end of R_t. Look at Figure 8-24.

The current through R_t is now set by

$$I_t = \frac{3V - V_{op\,amp}}{R_t}$$

Varying the voltage out of the op amp changes the current drawn from the XR-2206, which in turn changes the frequency.

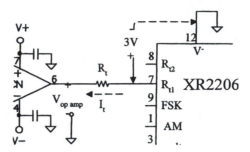

Figure 8-24 Controlling frequency with voltage

$$f = \frac{0.32 I_t}{C_t}$$

An op amp is used to set this voltage for two reasons. First, the op amp's output impedance is much smaller than R_t, so only the voltage at its output need be considered. If you were setting this voltage with a potentiometer (mechanical or digital), you would have to calculate I_t based on the resistances that the potentiometer presents at each of its settings.

The second reason that an op amp is used to drive R_t is its ability to produce an inversion and an offset. To force I_t (and therefore the frequency) to zero, $V_{op\ amp}$ must be set to 3V. This puts the same voltage on each side of R_t, leaving no voltage **across** the resistor. For voltage control of frequency, it is appropriate that 0V into the circuit should produce 0Hz. So the op amp should be configured as an inverting summer, with a 3V offset. Zero volts to that summer produces 3V to the left end of R_t, which drives the frequency to zero.

The op amp inverting summer also corrects the last problem. To increase the frequency, you must increase I_t. But to increase I_t, $V_{op\ amp}$ must be set **down**, below 3V. The more negative $V_{op\ amp}$ becomes, the larger the difference in potential across R_t, the larger I_t becomes, and therefore the higher the frequency. This relationship is opposite to what is usually required for voltage control of frequency. Configuring the op amp as an inverting summer accomplishes this.

Example 8-3

For the circuit in Figure 8-25, calculate all indicated voltages and currents and the output frequency and amplitude for:

a. $E_{in} = 2V$
b. $E_{in} = 4V$

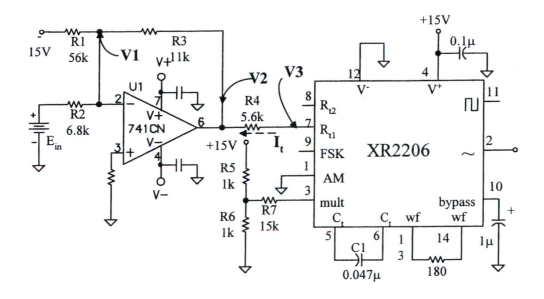

Figure 8-25 Voltage control of frequency for Example 8-3

Solution

a. Because there is negative feedback, assuming that the op amp is not saturated, V1 is at virtual ground.

$$V1 = 0V$$

$$V2 = -\frac{11k\Omega}{56k\Omega}(-15V) - \frac{11k\Omega}{6.8k\Omega}(2V) = -0.29V$$

$$V3 = 3V$$

$$I = \frac{3V - (-0.29V)}{5.6k\Omega} = 588\mu A$$

$$f = \frac{0.32 \times 588\mu A}{0.047\mu F} = 4kHz$$

b.

$$V1 = 0V$$

$$V2 = -\frac{11k\Omega}{56k\Omega}(-15V) - \frac{11k\Omega}{6.8k\Omega}(4V) = -3.52V$$

$$V3 = 3V$$

$$I = \frac{3V - (-3.52V)}{5.6k\Omega} = 1.16mA$$

$$f = \frac{0.32 \times 1.16mA}{0.047\mu F} = 7.9kHz$$

Using an inverting summer to provide a 3V offset and to invert the frequency control input voltage has provided a linear relationship between input voltage and output frequency. Doubling the input voltage (2V to 4V) has doubled the frequency (4kHz to 8kHz).

There is one final consideration when dealing with the frequency control pins. The voltage at the R_t pins is 3V above V^-, and current must flow **out** of those pins. So the voltage on the other end of the timing resistor must be more negative than 3V above V^-.

$$V_{op\ amp} < V^- + 3V$$

If you run the XR-2206 from a single positive supply, then V^- is ground. Powering the op amps from a dual supply then gives a wide range of voltage for $V_{op\ amp}$. So the op amp can swing the XR-2206's frequency through a wide range.

However, if you choose to run the XR-2206 from a dual supply, the voltage at the R_t pin is only 3V above the negative supply. Since most op amps saturate when their output comes within 2V of the negative supply, this leaves only 1V output range for $V_{op\ amp}$, greatly restricting its effect on frequency.

A reasonable compromise is to power the XR-2206 from ±5V. This puts the R_t pins at −2V. Then power the op amp from ±15V. The output of the op amp can then swing down to about -13V, giving an 11V range through which $V_{op\ amp}$ can move to set the frequency. Finally, remember that if you choose to power the XR-2206 from dual supplies, then both the FSK and the AM functions respond to voltages with respect to the voltage on V^-. So having −5V on V^- may complicate the FSK and AM signal conditioning.

8-4 Digitally Controlled Frequency Synthesizers

The ramp and sinusoidal signal generators that you have seen so far have all employed analog techniques to produce their output signals. These circuits are relatively straightforward. However, they suffer from the same limitations that plague most other analog electronics. To a large degree, the circuits' performances are dependent on the values of resistors and capacitors within the circuit. For precise results, low tolerance components must be used. This translates into increased initial cost. Variation in temperature, power supply voltage, output signal frequency, and level all alter the values of the resistors and capacitors in the circuits. So the output drifts around. The analog solution is to purchase components whose values do not change significantly with changes in the analog environment, again at a higher cost.

In addition to the inability to produce an inexpensive, precise, stable signal using only analog techniques, there is the problem of control. Certainly it may often be desirable to set the frequency (or amplitude or offset) of the waveform generator with an embedded microprocessor. A digital potentiometer or voltage out DAC can accomplish these tasks. But this still puts the control back into the analog domain where it falls victim to the problems already mentioned. So, in addition to having the microprocessor set the signal generator's output parameters, it must then monitor the resulting waveform, and change its commands as the signal drifts. This technique of digital control requires not only the microprocessor and the digital potentiometer or DAC, but also signal rectification and filtering, an analog to digital converter, and counters to find the frequency. Oh, and don't forget the **real time** control algorithms you need to write, the increased RAM and ROM, ... Although this way of using a microprocessor to manage a waveform generator has worked, there are simpler, set-it-and-forget-it circuits. Once commanded, they are self-regulating, stable, and reliable.

In this section of Chapter 8, two techniques for producing stable, precise, easily digitally adjusted sinusoids are presented. The phase locked loop wraps digital electronics around the analog voltage controlled waveform generator, digitally adjusting the frequency for inaccuracies and variations in the analog circuit's performance. A purely digital technique, the digital synthesizer, is a considerably more complex technique. But since it is now fully integrated into a single IC, precise sine waves are a simple matter of streaming in the correct serial word.

8-4.1 Phase Locked Loop Fundamentals

The phase locked loop (PLL) plays the same role in the frequency (and phase) world as the op amp does in the voltage world. The op amp has two voltage inputs, noninverting and inverting. The inverting input is normally used for feedback from the output. Similarly, the PLL has two inputs. Digital frequencies are applied. The PLL's feedback is connected to the circuit's output. The op amp changes its output voltage to whatever value is necessary to drive the difference in **voltage** between the two inputs to zero. The PLL changes its output phase and frequency to whatever **frequency** or **phase** is necessary to make the two input frequencies and phases track. Placing a voltage divider in the feedback path of an op amp causes the output voltage to be increased by the amount of the feedback voltage division (resulting in amplification). Placing a frequency divider in the feedback loop of a PLL causes the output frequency to be increased by the amount of the feedback divider. A firm grasp on the similarities of the PLL and the op amp simplifies your analysis and design of circuits with PLLs.

The phase locked loop contains three blocks: the phase detector, the low pass filter, and the voltage controlled oscillator (VCO). These are shown in Figure 8-26, connected in a "frequency follower" configuration. The phase detector receives two digital signals, one from the input and the other feedback from the output. The loop is locked when these two signals are at the same frequency and have a fixed phase difference. (A locked PLL is analogous to an op amp not being saturated.) The output of the phase detector is a digital signal whose **average** value is proportional to the difference in phase between the input and the feedback signals. The low pass filter removes the digital variations from its input, leaving only the DC voltage. This varies as the difference in phase between the input and the feedback signals varies. Finally, the voltage controlled oscillator (VCO) outputs a digital signal whose frequency and phase are controlled by the DC from the filter. Some VCOs also output a sine wave. Since stable generation of a sinusoid is a major concern of this chapter, that is an important option.

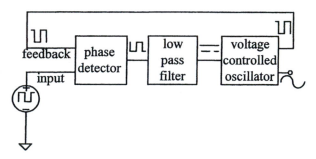

Figure 8-26 Phase locked loop block diagram as a frequency follower

Variations in input frequency cause similar variations in the output frequency. A small increase in frequency at the input is initially interpreted as an increase in the difference in phase between the input and the feedback signals. The pulse width of the output from the phase detector goes up, causing a rise in the DC out of the low pass filter. An increase in the voltage into the VCO causes its phase, and then frequency, to increase. So increasing the input frequency causes the output frequency to increase. As long as the input frequency does not change so much to cause the loop to lose its lock, the output's frequency tracks the input's. Of course, to produce a higher frequency, the VCO must have a larger input DC. This requires a larger difference in phase between the input and the feedback signals. But the frequencies stay locked, though their phase relationship may change. Since a microprocessor can easily alter an output digital frequency, this phase locked loop frequency follower allows the control of a sinusoidal frequency by an embedded (microprocessor) controller.

The big problem with analog signal generators is variation in frequency caused by component value shifts, because of either initial tolerance or drift. Such a variation in either the low pass filter or VCO components may cause the output frequency to go up. This is initially seen as an increase in phase at the output and also at the feedback input. But increasing the feedback phase reduces the difference in phase between the input and the feedback. The average value output from the phase detector goes down. This lowers the DC from the low pass filter, lowering the frequency from the VCO. Since the initial problem was an increase in frequency from the VCO, the loop has corrected itself. True, the phase difference at the input of the phase detector may be a bit changed. But, the output frequency is kept locked with the input's frequency, even though component values vary from the nominal.

It is now time to look more carefully at the detailed function of each of the blocks. The phase detector outputs a digital wave whose **average** value is proportional to the difference in phase between the input and the feedback digital signals. Best performance is often obtained if these signals have a 50% duty cycle.

The simplest phase detector is an exclusive OR gate. Its truth table is given in Table 8-1. When the two inputs are the same (0,0 or 1,1), the output is low. Whenever the inputs are at different logic levels (0,1 or 1,0), the output goes high. Figure 8-27 illustrates how this relationship implements phase detection. In Figure 8-27(a) the input leads the feedback by 60°. When both signals are low, the output is low. When the input goes high, but the feedback stays low, the output goes high. After 60°, the feedback goes high too, driving the output low. A complementary effect occurs on the trailing edges of the signals. The result is an output signal whose frequency is twice that of the input, with a duty cycle of 33%.

Table 8-1 Exclusive OR truth table

Input	Feedback	Output
0	0	0
0	1	1
1	0	1
1	1	0

When the feedback signal lags by 90°, as shown in Figure 8-27(b), the unmatched portion increases, increasing the duty cycle of the output to 50%. A reduction of the phase difference between the input and the feedback signal is shown in

Figure 8-27(c). The mismatched portion of the signals is reduced, lowering the duty cycle of the output. In this case, the duty cycle falls to 17%.

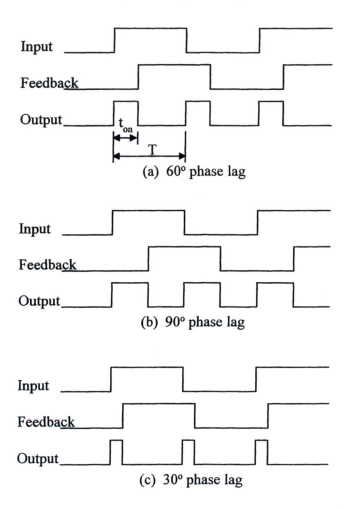

Figure 8-27   Exclusive OR waveforms produce phase detection

There is a linear relationship between phase angle difference and the duty cycle of the output from the phase detector.

$$D = \text{duty cycle} = \frac{\theta}{180°}$$

The average value of a rectangular signal is

$$V_{DC} = D \times V_P$$

$$V_{DC} = D \frac{\theta}{180°}$$

So, from the exclusive OR gate type of phase detector, there is a linear relationship between the phase angle difference of its inputs and the average value of its output signal. The output is a maximum of V_{III} when the two inputs are 180° out of phase. It is reasonable to assume that the feedback lags the input.

Of course, you could choose to use any logic IC exclusive OR gate. However, the CD4046 contains all of the active electronics needed to produce a phase locked loop within its single package, including the exclusive OR phase detector. The schematic symbol is shown in Figure 8-28. The IC is CMOS, allowing you to power it from +5V to +15V. You can match input levels with traditional 5V logic or power the IC from the positive side of your analog supplies. The V_{DD} pin (pin 16) is for V^+, while V_{SS} (pin 8) is usually grounded. The exclusive OR phase detector's output is pin 2.

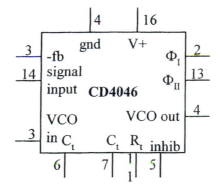

Figure 8-28 CD4046 phase locked loop schematic symbol

If there is a significant difference in frequency between the signal input and the feedback signal, the PLL may not be able to obtain a lock. An extreme case is shown in Figure 8-29. There are two examples. In Figure 8-29(a), the input frequency has been increased to twice that being fed back from the VCO. As shown, V_{output} = 50% V^+. When the input frequency goes up to three times the feedback signal (from the VCO), the result is as shown in Figure 8-29(b). Increasing the input frequency even more should **increase** the output from the phase detector to drive the VCO frequency even higher.

However, the exclusive OR type of phase detector actually outputs only $33\%V^+$ when the input frequency is three times too high. So increasing the input frequency has actually caused the phase detector's output to **decrease**. The VCO's frequency drops; it does not increase. Lock has been lost. It can be reestablished only if the input frequency is close to the VCO's frequency. Once the PLL is locked, changes in the input or feedback frequency must be in small steps.

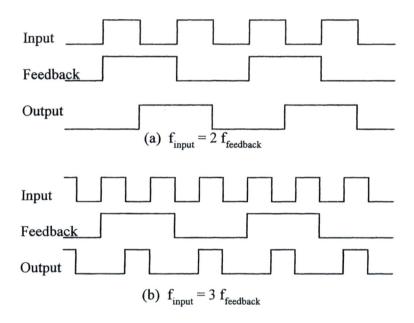

Figure 8-29 Exclusive OR loses lock for significant frequency differences

This tendency to lose lock is a major weakness of the exclusive OR type of phase detector. Within the CD4046 is a second phase detector. It shares the input and feedback signals with the Φ_I (exclusive OR) detector, but it has its own, Φ_{II} output (pin 13). It consists of digital logic and flip-flops for memory and a three-state output driver. When the input frequency is above the feedback frequency, the output is at V^+ most of each cycle. During the remainder of each cycle, the output is open. This V^+ output allows the filter capacitor to charge, increasing the voltage to the VCO, increasing the frequency from the VCO, and increasing the feedback frequency. When the Φ_{II} output is open, the filter capacitor holds its charge. If the input frequency is less than the feedback frequency, the Φ_{II} output is ground most of each cycle and open when it is not at ground.

Ground to the filter allows the capacitor to discharge, lowering the voltage to the VCO, lowering the frequency out of the VCO, and lowering the feedback frequency.

Once the input and feedback frequencies match, the Φ_{II} output is high for that part of each cycle that the input signal leads the feedback signal. The rest of the period, the output is open, so the capacitor in the filter slowly charges a bit more, increasing the VCO output. If the feedback signal's phase leads the input signal's phase, Φ_{II} is ground for that part of the cycle when feedback leads the input, and open the rest of the period. This allows the filter capacitor to bleed its voltage down, tweaking the VCO's output phase back into alignment with the input. When both frequency and phase match, the Φ_{II} output is open. In summary, the Φ_{II} phase detector of the CD4046 PLL drives the filter to whatever voltage is necessary to align the feedback frequency and phase with the input frequency and phase.

The second major element in the phase locked loop is the low pass filter. Its job is to extract the average value of the digital output of the signal from the phase detector. It should present a steady DC level to the input of the VCO which is equal to the average value of the phase detector's output. A simple, passive, first order, low pass filter is shown in Figure 8-30. The capacitor charges and discharges through the resistor in response to the highs and lows from the phase detector's output.

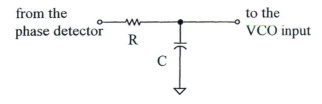

Figure 8-30 Simple, passive, first order, low pass filter

Conflicting requirements are placed on the filter. Any variation of the filter's output produces a change in the VCO's frequency, so ripple from the filter should be small. This is done by selecting a large RC time constant. Each cycle, the capacitor can only charge or discharge a small amount, providing a low output ripple and a steady VCO output frequency. However, the larger RC, the slower the filter is in responding to changes, slowing the overall loop's response to a change in input frequency.

The value of RC used must be chosen based on the lowest frequency, longest period output from the phase detector. A peak-to-peak ripple out of the filter of 10% of the V_{DC} is certainly the largest variation of practical consideration. This level of ripple would allow a 10% variation in the frequency from cycle to cycle. For any filtering at or more severe than this, the charge and discharge cycle of the capacitor is practically linear. This suggests the following guideline in selecting the filter's time constant:

$$\tau = R \times C = \frac{50\% T_{longest}}{\% \, ripple}$$

(The derivation of this equation is a straightforward application of the exponential charge and discharge of a capacitor, but it is omitted here because it is not central to understanding the performance of the PLL.)

Example 8-4
    a. Select filter components for a PLL whose lowest frequency from the phase detector is 5Hz. A 0.5% variation in frequency is allowable.
    b. How long will it take the output from the filter to stabilize?

Solution
    a. For a frequency of 5Hz, T = 0.2s.

$$\tau = \frac{50\% \times 0.2s}{0.5\%} = 20s$$

$$\tau = R \times C$$

Select C = 100µF as a practical upper limit.

$$R = \frac{\tau}{C} = \frac{20s}{100\mu F} = 200k\Omega$$

    b. It takes 5τ for an RC circuit to settle. This means that it takes 100s before changes in the capacitor's voltage are insignificant.

For simple or high frequency applications, this 5τ delay for the system to stabilize is acceptable. However, many of the more advanced applications require better settling times, more flexible (programmable) filtering, and an overall system approach to the performance of the entire loop. In these applications which require optimal performance, Laplace transforms for each block of the PLL must be used. These are then combined with the negative feedback to predict the loop's performance. Both steady state and response to changes to any of the parameters or inputs of the system can be examined. Given a full Laplace domain characterization of the loop, the constants, including R and C, can be chosen to provide the best trade-off in steady state versus transient performance.

The last block of the phase locked loop is the voltage controlled oscillator (VCO). It outputs a logic compatible square wave. The frequency of the square wave is directly proportional to its input DC voltage. Normally, the precise relationship is set by an external resistor and capacitor.

There is a VCO within the CD4046. Look at Figure 8-28. A capacitor between 50pF and 1μF is connected to the C_t pins (pins 6 and 7). The timing resistor is connected from the R_t pin (pin 1) to ground. It should be between 5kΩ and 1MΩ. The precise frequency is also a function of supply voltage. Use the graph in Figure 8-31 to pick an R_t and a C_t for a specific frequency.

There are several points from Figure 8-31 worth highlighting. The frequency on the vertical axis is the **center** frequency. It is the frequency when the VCO input DC is at $\frac{1}{2}V^+$. You can expect a linear variation in frequency as the input voltage changes. But the minimum input DC to produce an output square wave is about 1V. So, the maximum frequency is considerably above twice the center frequency. Secondly, the voltage to frequency relationship is sensitive to supply voltage. So select a higher supply if an output frequency of 1MHz or more is needed. But be aware that V_{hi} of the output square wave is almost V^+. Choosing $V^+ = 15V$ to get a higher frequency causes the output to no longer be compatible with 5V logic.

Finally, notice that there may be as much as $\pm 50\%$ variation in frequency from unit to unit. Initially, this is rather disappointing. You design the circuit for a frequency of 100kHz but end up with some models running at 150kHz and others at 50kHz. However, used in a PLL, this variation is eliminated. If a particular VCO is running too slowly, the feedback frequency is too low. The phase detector outputs a higher voltage, driving the VCO up to the correct frequency.

8-4.2 PLL Synthesizers

With all of the effort and details of the previous section, the phase locked loop succeeded in producing an output signal whose frequency is the same as the input's frequency. Wow! A piece of wire does the same thing. If you use the XR-2206 for the VCO, the output is a sine wave. That's something, but not worth the big build-up the PLL received. So, what's the payoff?

The real advantage comes when a frequency divider (digital counter IC) is placed in the PLL's feedback loop. Look at Figure 8-32. A highly stable, accurate clock is divided to 10Hz and used as the input signal. Assuming that the input loop is locked, then the feedback signal must also be 10Hz, accurate and stable. The programmable divider divides its input frequency by n, the digital code input on its parallel input lines. With its output at 10Hz, it must have an input frequency of $n \times 10Hz$, which is the frequency output from the VCO.

The 74169 is configured as a down counter. It decrements the data held in its registers on each clock pulse from the VCO. When it reaches 0, its output goes high. On the next clock pulse from the VCO, it reloads its registers from the data presented on D_0-D_3. At the next clock pulse from the VCO, it begins to decrement again.

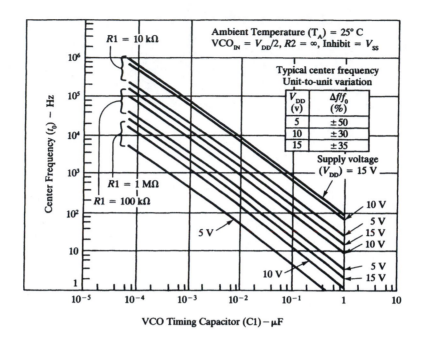

Figure 8-31 CD4046 VCO frequency *(Courtesy of Intersil Corporation, reprinted with their permission)*

So, if you were to set the data lines to 0010_2 (2), on the first pulse, the data would be loaded, setting the registers to 0010_2. The second pulse from the VCO would send the data to 0001_2. The third pulse causes the registers to go to 0, and the output pulse to be sent. In reality, it takes one additional pulse to load the data into the 74169. So the output frequency is

$$f_{VCO} = (n + 1) \times f_{in}$$

To alter the output frequency, you simply change the code input to the feedback divider. Once settled, the output frequency is accurately (n + 1) times the input frequency. Each step in the input code sends the output frequency up by an amount equal to the input frequency (10Hz in Figure 8-32). The input oscillator sets the size of the frequency steps at the output, and the data determines how many steps are taken. It is as accurate as the crystal oscillator from which the input is derived. Should the analog characteristics of the filter or the VCO change, the output frequency shifts accordingly, changing the feedback frequency, changing the output of the phase detector, changing the DC voltage into the VCO, and changing the frequency back to where it is supposed to be.

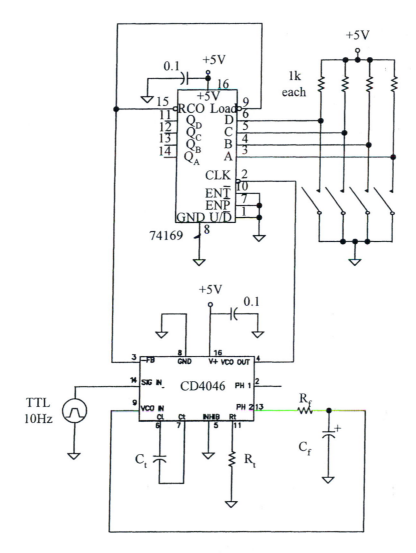

Figure 8-32 PLL synthesizer

You most certainly can cascade the digital dividers or use a PLD or microprocessor as the frequency divider. However, be aware that the frequency from the CD4046's VCO can sweep over only about a factor of 100 (e.g., from 200Hz to about 2kHz). So you need to provide several timing capacitors, each selectable from an analog switch or multiplexer. When the digital controller sends the code to the dividers, it must also choose the correct timing capacitor.

A phase locked loop frequency synthesizer schematic is shown in Figure 8-33. IC U1 is a commercially packaged 100kHz crystal oscillator. Its output is a TTL compatible 50% duty cycle square wave. Following the crystal oscillator is a counter block. It divides by 10^4. You can implement this with four 7490s, an LSI counter IC, or a PLD. The result is that the frequency out of this block is an accurate, stable 10Hz which becomes the input signal to the phase locked loop.

The feedback divider also is not specified. It may be four 741169s, cascaded properly; a single LSI device; a PLD (or part of the PLD used in the input divider); or a microprocessor peripheral. It may even be a single chip microcomputer IC. Whichever implementation you choose, it must have the ability to divide its input frequency by any value between 2 and 65,535. Therefore, the input to the 16-bit counter may be at any frequency between 20Hz and 1.3MHz (in 10Hz increments).

The VCO must generate this frequency. It consists of the XR-2206, the inverting summer, and associated components. You have seen this arrangement before in Figure 8-25. Please look back at that section if you are a little rusty on this circuit's performance. Switches S1-S3 have been added to allow the XR-2206 to sweep across the entire frequency range, from 20Hz to 1.3MHz. Practically, a sweep in frequency of 100:1 is as much as can be expected from the XR-2206 and a single capacitor. So three ranges have been provided. The maximum negative that the inverting summer can provide is -13V. This occurs at the maximum frequency of each range.

$$I_t = \frac{3V - (-13V)}{5.6k\Omega} = 2.9mA$$

This is just below the XR-2206's maximum I_t.

With S1 closed,
$$f_{max} = \frac{0.32 \times 2.9mA}{1\mu F} = 928Hz$$

This is the top end of the low frequency range. Similarly, for S2 closed, f_{max} = 42kHz, which is the midrange. The upper frequencies are reached with S3 closed, giving 1.97MHz, more than enough to satisfy the feedback signal requirements. There are many ways to implement these three switches. If the feedback dividers are set with external thumb wheel switches, then a rotary switch works well. If you are using a microcomputer or other digital logic to control this signal generator, then three analog switches or a four-channel multiplexer could be used.

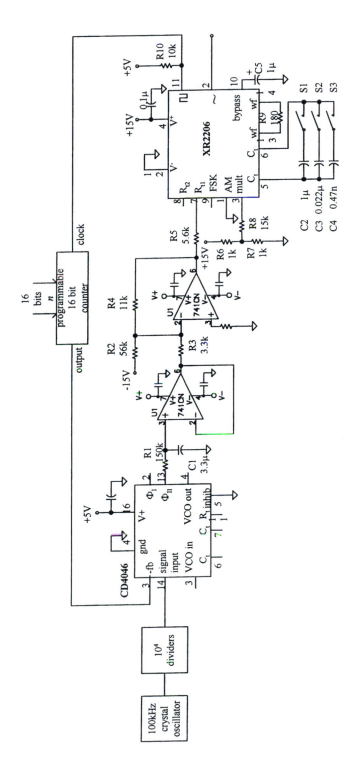

Figure 8-33 PLL synthesizer with the XR-2206 as the VCO

367

The offset of the inverting summer is set to 3V.

$$V_{offset} = -\frac{11k\Omega}{56k\Omega}(-15V) - \frac{11k\Omega}{3.3k\Omega}(0V) = 2.95V$$

This assures that a zero input from the filter sets the inverting summer output to 3V, which means there is no voltage across R5 and no current from the XR-2206.

The gain of the inverting summer is set so that a 5V input from the filter swings the inverting summer's output from 3V to -13V, producing a maximum current through R5 and a maximum frequency.

$$V_{out\,max} = -\frac{11k\Omega}{56k\Omega}(-15V) - \frac{11k\Omega}{3.3k\Omega}(5V) = -13.7V$$

The voltage follower prevents R3 from discharging the filter capacitor. The filter smooths out the signal from the phase detector. That CMOS circuit is powered from +5V to insure that it is compatible with its TTL input and feedback signals.

Assuming a 5% ripple, $\qquad \tau = RC \approx \dfrac{50\%\,T_{longest}}{\%\ ripple} = \dfrac{50\% \times 50ms}{5} = 0.5s$

$$\tau = RC = 150k\Omega \times 3.3\mu F = 0.495s$$

Example 8-5

Calculate all quantities for the circuit in Figure 8-34 with the loop locked.

Solution

Assuming that the loop is locked, the frequency at the feedback input of the CD4046 must be the same as at its input.

$$f_A = 2Hz$$

The input of the divide-by-n counter is at a frequency n times above its output.

$$f_B = 2Hz \times 386 = 772Hz$$

The speed of the encoder's shaft determines the frequency out of the encoder, at the rate of 0.5Hz/rev/min.

$$speed = \frac{f_B}{\dfrac{0.5Hz}{rev/min}} = \frac{772Hz}{\dfrac{0.5Hz}{rev/min}} = 1544 \text{ rev/min}$$

The motor turns in response to voltage applied to its armature. One volt produces 150rev/min, so

$$V_D = \frac{1V}{150 \text{rev/min}} \times 1544 \text{rev/min} = 10.3V$$

The transistor, Q1, serves as a current amplifier. There is a 0.7V base-emitter potential difference between V_D and V_E.

$$V_E = V_D + 0.7V = 10.3V + 0.7V = 11.0V$$

Resistors R_f and R_i form a voltage divider.

$$V_F = \frac{R_i}{R_f + R_i} V_D = \frac{10k\Omega}{22k\Omega + 10k\Omega} \times 10.3V = 3.22V$$

Since there can be no significant voltage difference between the two inputs of a unsaturated op amp,

$$V_G = V_F = 3.22V$$

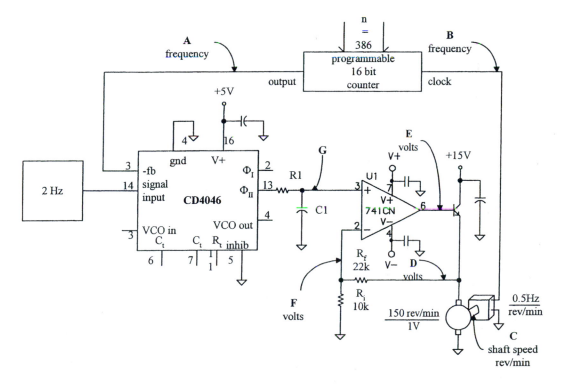

Figure 8-34 Phase locked loop motor speed control

8-4.3 Totally Digital Synthesizer

The PLL synthesizer of the previous section solves several of the problems with purely analog function generators. It is accurate and easily controlled digitally, and it exhibits good long-term stability. However, it still suffers from limitations of its analog components. The loop must remained locked for reliable performance. You must wait for the loop to stabilize when changing frequencies. At low frequencies, this wait may be significant. Frequency is still determined by an analog voltage or current. Any noise injected into this point in the circuit causes jitter.

The block diagram of a purely digital frequency synthesizer is shown in Figure 8-35. The only analog components in the circuit are the output half of the digital to analog converter (DAC), a passive low pass filter, and the output amplifier. None of these blocks affect the signal's frequency. The frequency is determined totally by the digital elements of the circuit.

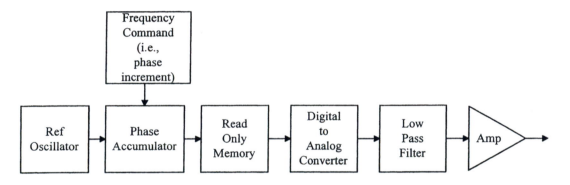

Figure 8-35 Digital frequency synthesizer

The values of the sinusoid are stored in the ROM. These values, addressed sequentially, are converted into a voltage by the DAC. The ROM is addressed by the phase accumulator. Each time the phase accumulator is clocked by the reference oscillator, a different address is presented to the ROM. The ROM outputs the data at the address, and the DAC converts that data into a voltage.

The phase increment indicates to the phase accumulator how much to add to the current address when the reference oscillator clocks the accumulator again. If the phase increment is one, each time the reference oscillator clocks the phase accumulator, the address advances one place. The very next data point is output, and the DAC may make a small change in its output. The system must step through every single address in the ROM, filling the accumulator, one step at a time, before a complete sine wave cycle is observed at the output. This generates the lowest frequency that the synthesizer can output.

$$T_{longest} = T_{ref\,oscillator} \times 2^n$$

where $T_{longest}$ = the period of the lowest output frequency
$T_{ref\,oscillator}$ = the period of clock
n = the number of bits in the phase accumulator

$$f_{lowest} = \frac{f_{ref\,oscillator}}{2^n}$$

If you change the value of the phase increment to two, each time that the reference oscillator clocks the phase accumulator, the address steps two. Now, every other memory location is output. It takes precisely half as many steps to run the accumulator through its full range, outputting a complete sine wave cycle. Taking two steps at a time, it takes exactly half as long to create a full sine. So the output period is half of its longest value, and the frequency has precisely doubled.

Similarly, if the value of the phase accumulator is three, it takes only one-third the time to get all the way to the end of the cycle (since you now are going three steps at a time). The period is one-third of its longest value, and the frequency has tripled.

So

$$\text{frequency resolution} = f_{lowest} = \frac{f_{ref\,oscillator}}{2^n}$$

The larger the phase increment, the fewer steps it takes to complete a full cycle, and therefore the higher the frequency. You must take at least two steps, one on the positive cycle of the sine wave and one on the negative. Actually, there must be something more than two steps per cycle. At exactly two steps per cycle, the phase increment is $180°$. The first output to the DAC is at $0°$, which is 0. The next output is at $180°$, which is also 0. The next step is at $360°$, also 0. So, in practice, to create a sine wave that is not too distorted, set

$$f_{out\,max} \le \frac{f_{ref\,oscillator}}{32}$$

Example 8-6

A direct digital synthesizer has a 21-bit accumulator. Determine the reference oscillator's frequency needed to allow 1Hz increments in the output. Also, calculate the highest practical frequency.

Solution

$$\text{frequency resolution} = f_{\text{lowest}} = \frac{f_{\text{ref oscillator}}}{2^n}$$

$$f_{\text{ref oscillator}} = 2^n \times f_{\text{lowest}} = 2^{21} \times 1\text{Hz} = 2.0972\text{MHz}$$

$$f_{\text{out max}} \leq \frac{f_{\text{ref oscillator}}}{32} = \frac{2.0972\text{MHz}}{32} = 65.5\text{kHz}$$

The reference oscillator clocks the accumulator. Inputs to the accumulator include the reference oscillator and the data bits that indicate how large each step should be (the phase increment that sets the frequency). The accumulator consists of a summer and a latch. The summer adds the data at the phase increment inputs to the data currently being held in the latch. When the reference oscillator clock occurs, this sum is stored in the latch. The adder then sums this new data with the phase increment and presents it to the latch to be stored by the next clock.

The size of the ROM needed depends on the number of data bits into the DAC. This connection is more easily seen if the data stored in the ROM are for a triangle wave. To produce a triangle wave, the data advance one step each time the address is advanced. An 8-bit DAC has 256 (i.e., 2^8) steps. So the ROM must have 256 locations for the ramp up and 256 locations for the steps down. A 12-bit DAC would require $2 \times 4096 = 8192$ locations.

When you are storing sinusoidal values in the ROM, the calculation of ROM size becomes a little more complicated. The ROM must have enough locations so that the steepest rise requires the DAC to change its output by only one step. The equation for the data in the ROM is

$$\text{value} = \left(\frac{1}{2} \times 2^n\right)\sin\theta + \left(\frac{1}{2} \times 2^n\right)$$

For an 8-bit DAC this becomes

$$\text{value} = 128\sin\theta + 128$$

So, as the phase (from the phase accumulator) goes from $0°$ through $90°$, $180°$, $270°$, and finally to $360°$, the values output from the ROM go from 128 to 255, back to 128, and then to 0, and end up back at 128 again.

How small a step is necessary to cause a step in value of one at the steepest part of the curve? The sine is steepest at $\theta = 0°$. The size angle step needed to produce a step of one in amplitude can be calculated.

$$128 + 1 = 128 \sin \theta + 128$$

$$1 = 128 \sin \theta$$

$$\theta = \arcsin\left(\frac{1}{128}\right) = 0.447°$$

For an 8-bit DAC, there must be enough addresses so that the angle is advanced only $0.447°$ from address to address. There are $360°$ worth of data in the ROM. At $0.447°$ per address, there must be

$$\text{\# of addresses} = \frac{360°}{\dfrac{0.447°}{\text{address}}} = 805 \text{ addresses}$$

Since ROMs come in 2^n sizes, select a ROM with 1024 memory locations, where each location contains an 8-bit data word. You need 10 bits ($2^{10} = 1024$) from the phase accumulator as address bits for the ROM. Use the upper 10 bits. Leave the lower output bits from the phase accumulator unconnected.

A similar calculation for a 10-bit DAC indicates that it requires a ROM with 4096 locations, using 12 bits from the phase accumulator. Similarly, a 12-bit DAC requires a ROM with 16,384 memory locations and 14 bits from the phase accumulator.

So the size of the DAC determines the size and complexity of the ROM and its addressing. But how big do you make the DAC? That depends on how much noise you are willing to tolerate at the output. Digital to analog converters typically have an error of $\pm\frac{1}{2}$ least significant bit. For an 8-bit converter, this means that there is 1/256 or 0.4% uncertainty in the output. A 12-bit DAC has an uncertainty of 1/4096 or 0.02%. This uncertainty in the output translates into a low level, broad band noise, producing a signal-to-noise ratio of -48dB for an 8-bit converter and -72dB for a 12-bit DAC.

Using the 12-bit DAC produces a much cleaner output signal. But remember that this then requires a 16k by 12-bit ROM. The 8-bit DAC, though providing more noise at the output, needs only a 1k by 8-bit ROM.

Once you have decided on the size of the DAC, several other features must be considered. The DAC must receive its input data in parallel. It must be able to complete its output change in tens of ns. Look carefully at the addressing and control bus structure of the DAC. Many DACs expect to be driven from a microprocessor bus and require chip select, data latch, and output enable pulses. The simpler the control structure, the better. Remember, you have to generate these signals. Finally, all DACs require a reference voltage. It must be at least as accurate as the DAC ($\pm\frac{1}{2}$LSB). The raw power supply voltage is **not** clean enough to be used as your reference!

The final block of the digital frequency synthesizer is the filter. It is the filter's responsibility to remove the sharp edges that happen when the clock steps the data,

sending a new value into the DAC. These steps occur at the reference oscillator's rate.
So, the filter is low pass, with its cut-off frequency set to

$$f_{\text{filter cut-off}} = 2 \times f_{\text{out max}}$$

This assures that even the highest frequency **signal** passes through the filter without
attenuation. But, since the reference oscillator is running 32 times faster than the highest
output signal,

$$f_{\text{filter cutoff}} = \frac{f_{\text{ref oscillator}}}{16}$$

The other information needed to design a filter is its order. Order tells of the
filter's complexity and how steeply it falls. For this synthesizer, a sixth-order filter is
fine. It has six energy storage elements, typically three inductors and three capacitors, or
six resistor-capacitor pairs.

The filter rolls off at a rate of -6dB/octave/order. An octave is a doubling of the
frequency. Remember, the clock's harmonics are 16 times above the filter's cut-off.
That is four octaves (2, 4, 8, 16). So, the noise introduced by the clock is attenuated by

$$\text{attenuation} = \frac{-\dfrac{6\text{dB}}{\text{octave}}}{\text{order}} \times 4\,\text{octaves} \times 6\,\text{order} = -144\text{dB}$$

This is more than enough to remove the effects of the clock from the output sine
wave. Remember that regardless of the sine wave's frequency, the noise is introduced at
the clock's frequency and is attenuated by -144dB. Both the noise frequency and the
filter are fixed. This greatly simplifies the filter.

Micro Linear produces a series of digital synthesizers they call Programmable
Sine Wave Generators. The ML2036 is shown in Figure 8-36.

Power is ±V ± 10%. Since this IC handles both analog and high frequency
digital signals, be sure to decouple these supplies well. The analog ground pin and the
digital ground pin should be connected together and then to a clean ground.

The reference oscillator is controlled by the **clk**$_{\text{in}}$. This **clk**$_{\text{in}}$ signal may be either an
externally generated, TTL compatible signal (50% duty cycle, f < 12MHz) or a parallel
resonant crystal between 3MHz and 12MHz. If you use a crystal, be sure to place it as
close to the IC as possible, and keep all traces as short as possible. This **clk**$_{\text{in}}$ signal is
divided by 2 and then by 2 again. These divided signals are available at **clk**$_{\text{out1}}$ and **clk**$_{\text{out2}}$,
to drive external logic. The reference oscillator's frequency is

$$f_{\text{ref oscillator}} = \frac{f_{\text{clk in}}}{4}$$

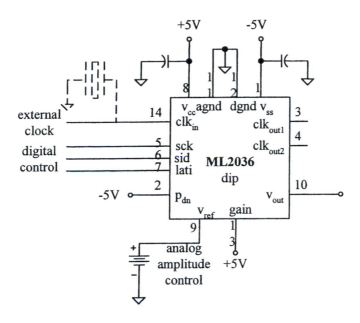

Figure 8-36 Digital synthesizer IC schematic

The ML2036 has a 21-bit accumulator. So combining the divide-by-4 logic with the relationship you saw earlier, the output sine wave's frequency is

$$f_{out} = f_{clk\,in}\frac{data}{2^{23}}$$

The data is sent into the ML2036 serially, using the **sck**, **sid**, and **lati** lines. The **data** word is 16-bits long. Begin by taking both the serial clock (**sck**) and the latch input (**lati**) signal low. Then place the data's least significant bit (LSB) on the **sid** first. Clock it into the internal shift register with a rising edge of the **sck** signal. Take the **sck** signal low again; then change the **sid** data bit. Clock this second bit in by sending the **sck** signal high again. Continue toggling the **sck** and the **sid** bits to latch in all of the 16 bits into the input shift register. When the last bit is entered and the **sck** has returned low again, take the latch input signal (**lati**) high for at least 50ns; then return it to a logic low. The data is transferred to the internal data register when **lati** falls.

The p_{dn} logic signal is used to inhibit the output or to turn the IC off. A logic low sets the output to 0V immediately. To enable normal operation, tie p_{dn} to -5V. Entering all 0s into the data, then, stops sine generation at the end of the current cycle. A logic high on p_{dn} turns the IC off, lowering the supply currents to less than 100μA.

The peak sine amplitude is set with the **gain** and v_{ref} signals.

$$V_{out\,P} = V_{ref} \qquad \text{when } \textbf{gain} = 1$$

$$V_{out\,P} = \tfrac{1}{2} V_{ref} \qquad \text{when } \textbf{gain} = 0$$

This output amplitude assumes a 1kΩ, 100pF load. It can swing to within 1.5V of the supplies.

$$V_{out\,P\,max} \leq V_{CC} - 1.5V$$

The output amplifier within the IC has a slew rate of 0.4V/μs. This also limits the output amplitude, depending on how fast you want to go. Applying the sine wave slew rate constraints from Chapter 5,

$$\text{slew rate} = 2\pi f_{max} v_P$$

$$v_{P\,max} = \frac{0.4\,\dfrac{V}{\mu s}}{2\pi f} = \frac{64\text{kHz}\,V_P}{f_{out}}$$

So you can drive v_{ref} from a DAC or analog control circuitry to automatically adjust the amplitude.

Example 8-7

The manufacturer indicates that the highest clock frequency should be 12MHz. Calculate the resulting highest sine wave output frequency and the largest V_P at that frequency.

Solution

Since you can enter a 16-bit data word,

$$f_{out} = f_{clk\,in}\frac{data}{2^{23}} = 12\text{MHz}\frac{2^{16}}{2^{23}} = 93.75\text{kHz}$$

$$v_{P\,max} = \frac{64\text{kHz}\,V_P}{f_{out}} = \frac{64\text{kHz}\,V_P}{93.75\text{kHz}} = 0.68V_P$$

This is a long way from the V_{CC} - 1.5V = 3.5V_P limit. However, at this high a frequency, the limiting factor is the amplifier's slew rate, not its saturation level. Applying more than 0.68V at the V_{ref} input will not produce a larger signal. If you need a larger signal, add an amplifier after the signal generator.

Summary

In this chapter you have studied square wave, triangle wave, and sine wave generators. The 555 timer is a special purpose IC, which is easily configured to produce a square wave. It senses the charge and discharge of an external RC network with two internal comparators. Internal control circuits either short out or allow the charge of the external timing capacitor between $\frac{1}{3}V^+$ and $\frac{2}{3}V^+$.

The crystal oscillator provides the most accurate and most stable square wave. A piezoelectric crystal provides feedback from the output to the input of a CMOS Schmitt trigger. A biasing resistor and filter capacitor to ground are also needed. More reliable performance is obtained by replacing the CMOS gate with a comparator IC. The crystal is placed between the output and the noninverting input, providing positive feedback. The noninverting input must also be biased to half of the supply. An RC low pass filter from the output to the inverting input assures a 50% duty cycle.

Ramps can be generated by charging a capacitor with a constant current. This is easily done by placing the capacitor in the negative feedback loop of an op amp. The charge rate is directly proportional to the applied voltage while being inversely proportional to the input resistance and capacitance. By driving a two-level noninverting comparator with the ramp generator and using the output of the comparator to drive the ramp, a triangle wave can be produced. The amplitude is determined by the trip points of the comparator. Frequency depends on this amplitude, the RC charge components, and the output levels of the comparator. One side benefit is the square wave generated by the comparator.

Sawtooth waveforms can be produced by placing a switch around the capacitor of the ramp generator. Using a bipolar transistor for this switch allows you to synchronize the sawtooth with some external signal. A programmable unijunction transistor (PUT) around the capacitor terminates the ramp when it reaches an externally set voltage.

Function generators produce a ramp by charging a capacitor with constant current. A two-level comparator monitors the ramp level and switches the charge directions (charging current) when the ramp crosses one of the levels. This results in a triangle wave and a square wave from the comparator. A sine wave is produced by rounding the tips of the triangle wave.

The XR-2206 implements all of these functions as well as control of frequency (FM), amplitude (AM), and DC level with external resistance and with DC voltages. Frequency shift keying (FSK) allows you to encode digital data for transmission by shifting the frequency between two levels depending on the logic signal applied to the FSK pin. This can also be used to produce a variable duty cycle square wave.

The phase locked loop consists of three elements. (1) The phase detector outputs a signal whose average value is equal to the difference in phase between the two input signals. (2) The low pass filter smooths its input into a steady DC. (3) The voltage controlled oscillator outputs a frequency set by its input DC. When locked, the PLL outputs whatever frequency is necessary to keep the input frequencies the same and to minimize their difference in phase.

Adding a digital divide-by-n counter in the feedback loop means that the output frequency generated by the VCO must be n times above the input frequency. Using the divider allows digital control of the output frequency, with the input signal setting the minimum frequency and the resolution. The XR-2206 may be used as the VCO. This allows you to input a crystal oscillator generated, accurate square wave and to output a sine wave whose frequency is accurately set by the digital divider in the feedback loop.

The direct digital synthesizer generates the signal by stepping through the locations in a ROM. Stored in the ROM are sinusoidal values. These values are output to a DAC to produce the sine wave. The sharp edges associated with digital steps are removed by a low pass, high order, passive filter. New addresses are fed to the ROM at a constant rate from a crystal reference oscillator. The size of the step is set by the external data. This determines the output sinusoidal frequency. Small steps take a long time to get through the data in the ROM, producing a low frequency sine wave. Take larger steps, and it takes less time to get to the end of the ROM. The frequency goes up. The resulting signal's frequency is just as accurate and as stable as that of the crystal oscillator driving it.

The Micro Linear ML2036 integrates all of these elements into a single package. You add a crystal and an external reference voltage. Data to set the frequency is clocked in serially.

Amplifiers with RC networks placed in their feedback path can be made to oscillate, generating a desired waveform. When you lower the gain of the amplifier and drive the frequency determining network with an external signal, you have converted the oscillator into an active filter.

Problems

8-1   Calculate the time high, the time low, and the frequency of the output from the circuit in Figure 8-37.

8-2   Explain the operation of the circuit in Figure 8-38 for:
a.   Enable = 2.5V.
b.   Enable = 0V.

8-3   Design a square wave generator using a 555 to produce a TTL compatible output. Provide two switch selectable frequencies: one at 0.3Hz, and the other at 10kHz.

8-4   Explain the advantages of a crystal oscillator over a generator that uses an RC pair to set the frequency.

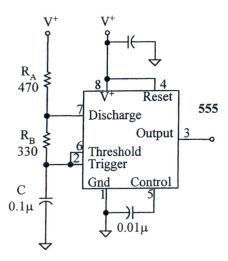

Figure 8-37 Schematic for Problem 8-1

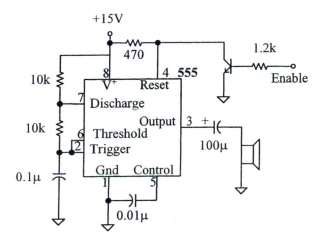

Figure 8-38 Schematic for Problem 8-2

8-5   Explain the advantages of the comparator-based crystal oscillator over the Schmitt trigger gate type of oscillator.

8-6   For the comparator-based crystal oscillator, explain the purpose of:
a. placing the crystal between the output and the positive input.
b. $R_{pull-up}$.
c. R1 and R2.
d. R3 and C.

8-7   Calculate the rate (slope) of the output from the circuit in Figure 8-39.

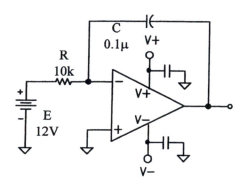

Figure 8-39 Schematic for Problems 8-7 and 8-8

8-8   For Problem 8-7, could you use a 741C op amp if E = 12V, R = 470Ω, and C = 0.001μF? Explain.

8-9   For the circuit in Figure 8-40, calculate the comparison levels (trigger points) if $\pm V = \pm 12V$ and:
a. $R_f = 10k\Omega$, $R_i = 4.7k\Omega$.
b. $R_f = 10k\Omega$, $R_i = 9k\Omega$.

8-10   Design a noninverting two-level comparator with a 2V difference between the positive and negative comparison levels when operated from $\pm 15V$.

8-11   Calculate the items listed below for the circuit in Figure 8-41:
a. frequency of the output.
b. rate of rise of the triangle wave (slope).
c. triangle wave amplitude.
d. square wave amplitude.

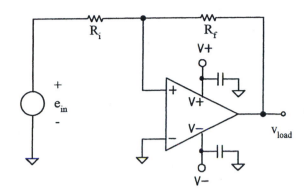

Figure 8-40 Schematic for Problem 8-9

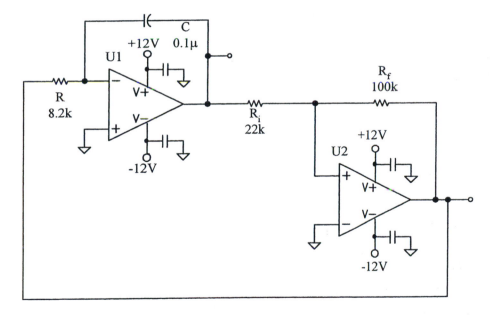

Figure 8-41 Schematic for Problem 8-11

8-12   **a.** Design a triangle wave/square wave generator with a triangle peak-to-peak amplitude of 10V. Provide two switch-selectable frequencies: one at 0.3Hz and the other at 10kHz. Use an LM311 for the comparator to provide fast rise time.

b. Design a clipper circuit for the output of the square wave to make it TTL compatible without changing the charge and discharge voltages fed back to the ramp generator.

8-13   For the circuit in Figure 8-42, draw the output and accurately label the voltage and time axes. The control input consists of positive pulses occurring at every zero crossing of the line voltage (i.e., a pulse every 8.3ms).

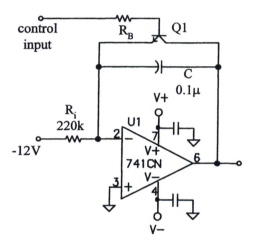

Figure 8-42 Schematic for Problem 8-13

8-14   **a.** Design a PUT controlled ramp generator that outputs a 4kHz signal when the control input is $4V_{DC}$ and $\pm V = \pm 12V$.
   b. What will the frequency of your circuit be when the control input voltage rises to $8V_{DC}$?

8-15   The circuit in Problem 8-14 is a voltage to frequency converter. However, an increase in the control voltage produces a **decrease** in the frequency. Design a signal conditioner using an op amp to drive its control input so that 4V into the signal conditioner produces an output frequency of 4kHz, and 8V into the signal conditioner produces 8kHz out of the sawtooth generator.

8-16   Calculate all indicated voltages, the sine wave output amplitude, DC level, and frequency for the circuit in Figure 8-43.

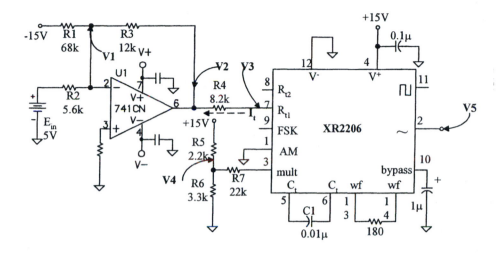

Figure 8-43 Schematic for Problem 8-16

8-17   Design a function generator that outputs a 1kHz sine wave with a $1V_{RMS}$ amplitude at $0V_{DC}$. Draw the schematic and determine all component values.

8-18   **a.** Given that the signals in Figure 8-44(a) are the input and feedback signals to an exclusive OR phase detector, draw the output.
   b. What is the average value of the output signal?
   c. Repeat parts a and b for the input signals in Figure 8-44(b).

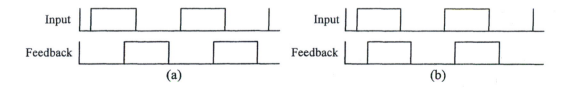

Figure 8-44 Exclusive OR phase detector inputs for Problem 8-18

8-19 Explain the difference between the exclusive OR phase detector and the CD4046 Φ_{II} phase detector. Which provides the wider lock range? Explain.

8-20 A phase locked loop operates from 20Hz to 200kHz.
   a. Select the value of the filter components for a 2% ripple.
   b. How long does it take the loop's output frequency to settle to within 5% of its final frequency?

8-21 Using the CD4046 for both the phase detector and the VCO, design a PLL that operates over the range of 20Hz to 200kHz. Draw the schematic, and determine all component values. Be sure to show how you plan to switch capacitors. Use a +5V power supply.

8-22 Alter your design for Problem 8-22 to produce a PLL synthesizer using a CD4046 for the phase detector and VCO. Set the upper frequency at 200kHz, and the frequency step size at 5Hz. Be sure to include the crystal oscillator design, the details of the input and the programmable feedback dividers, and the analog switch to alter the VCO range.

8-23 Replace the CD4046 of Problem 8-22 with an XR2206.

8-24 Calculate all indicated quantities for the circuit in Figure 8-45.

8-25 What is the resolution of the control circuit in Problem 8-24 (smallest change in motor speed)?

8-26 For a direct digital synthesizer, determine the frequency of the reference oscillator and the number of bits needed in the accumulator for a maximum output frequency of 1MHz with a resolution of 1Hz.

8-27 For the direct digital synthesizer of Problem 8-26, prove that a 10-bit DAC requires a 4k (2^{12}) ROM.

8-28 Find a reference that explains how to design a sixth-order Butterworth passive filter. Use it to design the sixth-order filter needed for the circuit in Problem 8-26.

8-29 Find a Micro Linear data book or their World Wide Web site. Retrieve a copy of the specifications of the ML2036. Answer the following questions:
   a. What is the range of acceptable crystal frequencies?
   b. What is the maximum output frequency?
   c. Prove with a calculation that this limit is **not** set by the digital synthesizer part of the IC.
   d. What causes this maximum frequency limit?

e. What crystal frequency would you choose for maximum output frequency resolution (smallest possible step in output frequency) ? (This is not given in the specs. Think about it!)

f. Calculate this smallest frequency step.

g. What code must be input into the IC given the crystal from parts e and f to provide a 15kHz output frequency?

h. Explain how the output amplitude can be controlled digitally to $1V_{RMS}$ in $10mV_{RMS}$ steps.

i. If you select $1V_{RMS}$ output, how accurate is the output amplitude?

j. If you select a $1V_{RMS}$ output, how much random noise do you expect at the output?

k. Complete a timing diagram for **sid**, **sck**, and **lati** which indicates how you would input the signals necessary to produce a 15kHz output sine wave.

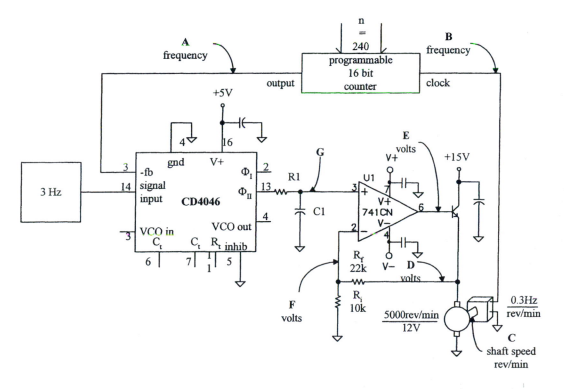

Figure 8-45 Schematic for Problem 8-24

XR-2206 Function Generator Lab Experiment

A. Frequency Adjustment

   1. Build the circuit in Figure 8-46.

   2. Calculate the size of R_t which limits the current from the R_t pin to less than its maximum value when there is a 15V drop across that resistor. Set the fixed value of R_t slightly greater than this value.

   3. Measure the value of C_t.

   4. Apply the power to the IC.

   5. Monitor the square wave output.

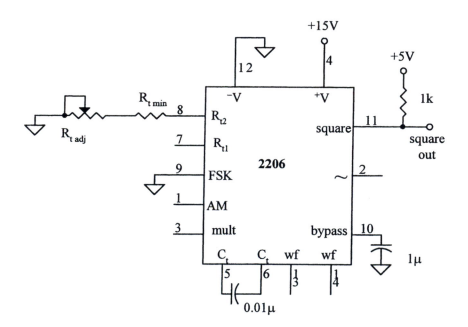

Figure 8-46 XR-2206 simple frequency configuration

   6. Adjust R_t to produce a 1kHz output frequency.

   7. Remove power, and measure the total value of R_t.

8. Compare this value with the theoretical value of R_t needed to provide a 1kHz frequency.

9. Alter R_t and C_t as necessary to produce a 1MHz frequency.

10. Measure C_t and R_t.

11. Using the $C_{t\,1MHz\,actual}$, calculate the theoretical value of $R_{t\,1MHz\,actual}$ needed to produce a 1MHz output.

12. Compare $R_{t\,1MHz\,theory}$ to $R_{t\,1MHz\,actual}$.

13. Compare the error in the timing resistor at 1MHz to the error in the timing resistor at 1kHz.

14. Return the timing resistor and capacitor to those values needed to produce a 1kHz output. Verify that the output signal is 1kHz.

B. DC Offset
1. Connect the two 1kΩ resistors to pin 3 as shown in Figure 8-47.

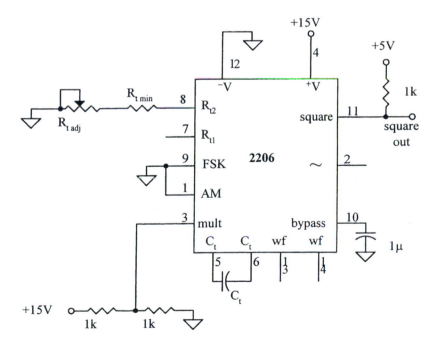

Figure 8-47 XR-2206 DC offset

2. Measure the DC voltage on pin 3 and on the sine/triangle pin (pin 2).

3. Compare the actual output DC with the theoretical output DC.

C. Amplitude Adjustment
   1. Add R_{mult}, as shown in Figure 8-48.

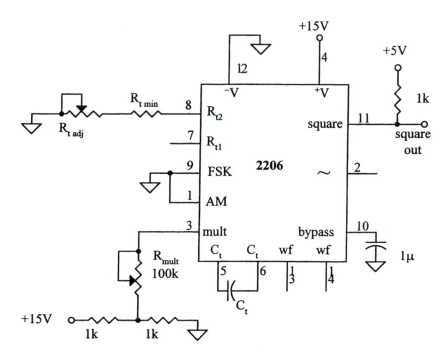

Figure 8-48 Amplitude adjustment

2. Monitor the output on the sine/triangle pin (pin 2) on the oscilloscope.

3. Adjust R_{mult} to produce a $1V_{peak}$ output.

4. Remove R_{mult} and measure its value.

5. Calculate the theoretical peak output.

6. Compare this theoretical value to the actual $1V_{peak}$ output.

7. Add the 180Ω resistor between the waveform pins (pins 13 and 14).

8. Monitor the output on the sine/triangle pin (pin 2) on the oscilloscope.

9. Adjust R_{mult} to produce a $1V_{peak}$ output.

10. Remove R_{mult} and measure its value.

11. Calculate the theoretical peak output.

12. Compare this theoretical value to the actual $1V_{peak}$ output.

D. VCO Operation

1. Add the inverting summer U1a as shown in Figure 8-49.

2. Adjust V_{freq} to set the output of U1a to 0V.

3. Confirm that the output amplitude, DC level, and frequency are correct.

4. Vary V_{freq} to 0V, 1V, 2V, 3V, 4V, 6V, 8V, and 10V. At each voltage, measure and record $V_{U1a\ out}$, $I_{frequency}$, $freq_{actual}$, $freq_{theory}$, and $error_{freq}$.

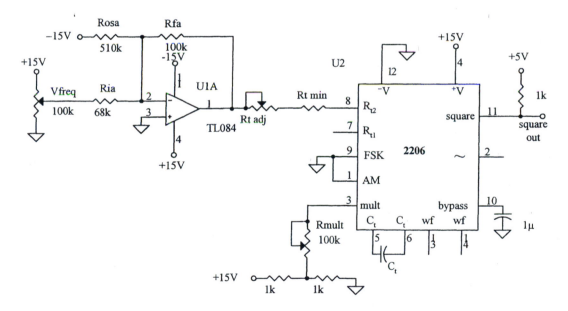

Figure 8-49 VCO control of the XR-2206

E. AM Operation

1. Add the inverting summer U1b as shown in Figure 8-50.

2. Adjust V_{am} to set the output of U1b to 0V.

3. Confirm that the output amplitude, DC level, and frequency are correct.

4. Vary V_{am} to 0V, 1V, 2V, 3V, 4V, and 5V. At each voltage, measure and record $V_{2206\,pin1}$, sine amplitude$_{measured}$, amplitude$_{theory}$, % error$_{amplitude}$.

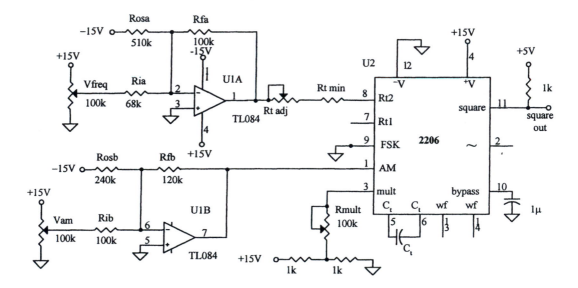

Figure 8-50 AM control of the XR-2206

F. Output Amplifier

1. Add the output amplifier as shown in Figure 8-51.

2. Monitor the output of U1C with both the oscilloscope and the digital multimeter.

3. Set V_{offset} to 5V, 0V, and -5V. Record the output DC and peak.

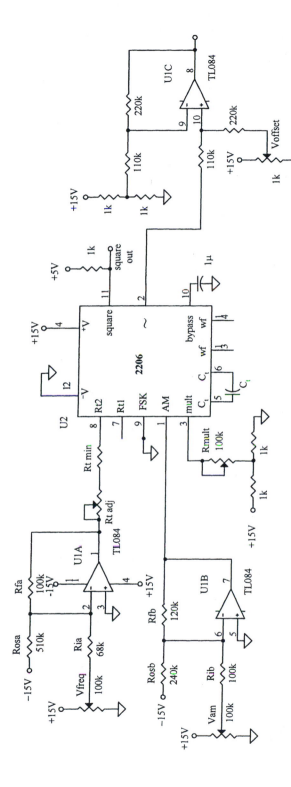

Figure 8-51 Final output amplifier

391

Phase Locked Loop Frequency Synthesizer Lab Exercise

A. Setup

    1. Build the circuit in Figure 8-52. Omit the 74169 and switches. Place a connection directly from the output of the XR-2206 to the -fb input of the CD4046.

    2. Place a short across the filter capacitor.

    3. Apply power to the circuit.

    4. Adjust the potentiometer to produce $3.00V_{DC}$ at the output of the op amp.

    5. Remove the short from across the filter capacitor.

    6. Connect channel 1 of the oscilloscope to the input signal generator, and channel 2 of the oscilloscope to the square wave output of the XR-2206.

    7. Connect the digital multimeter to the output of the filter.

B. Input Signal Tracking

    1. Set the input signal generator to produce a TTL compatible, 10kHz square wave.

    2. Are the input and the output signals tracking (i.e., phase locked)? If not, your circuit is not working correctly.

    3. Once the circuit is locked, move channel 2 of the oscilloscope to the XR-2206's sine wave output. Carefully record both channels of the oscilloscope.

    4. Record the following data: frequency, op amp output voltage, and filter output voltage.

    5. Increase the signal generator's frequency until the loop stops tracking. Record the following for the highest frequency for which the loop remains locked: frequency, op amp output voltage, and filter output voltage.

    6. Decrease the signal generator's frequency until the loop stops tracking. Record the following for the lowest frequency for which the loop remains locked: frequency, op amp output voltage, and filter output voltage.

    7. Review the data taken in steps B4-B6. Explain what causes the loop to lose lock at the top and the bottom.

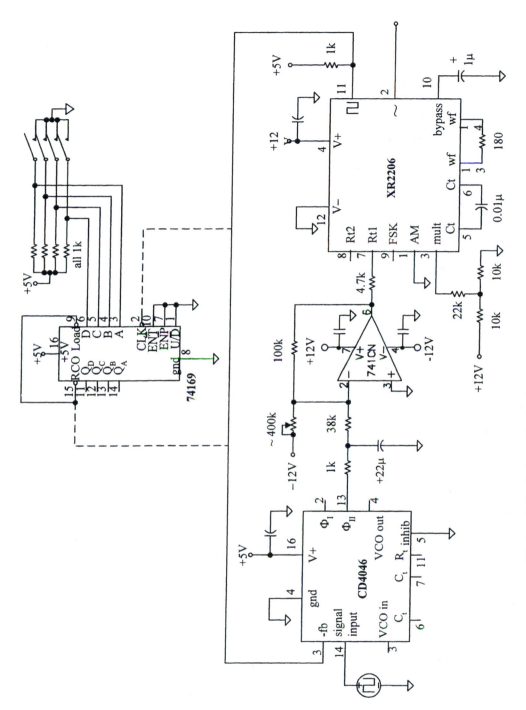

Figure 8-52 Schematic for the PLL synthesizer lab experiment

393

C. Frequency Multiplication

    1. Remove the connection from the output of the XR-2206 to the input of the CD4046. Connect the divider into the loop as shown by the dotted lines.

    2. Set the switches to 1000_2 (DCBA).

    3. Set the signal generator's frequency to 1kHz.

    4. Is the loop locked, and is the output frequency 9kHz? When it is, carefully record the oscilloscope's display.

    5. Alter the switches in binary steps from 0001_2 to 1111_2. At each switch setting, measure and record the frequency of the sine wave output.

References

1. Jim Williams, "Circuit Techniques for Clock Sources: Application Note 12," *1990 Linear Applications Handbook: A Guide to Linear Circuit Design*, Linear Technology, 1990, p. 12-1.

2. Williams, *Linear Applications Handbook,* p. 12-3.

3. Staff, "XR-2206 Monolithic Function Generator," *Exar Data Book*, Exar Corp., 1986, p. 6-15.

9

Active Filters

Electric filters are used in practically all circuits that require the separation of signals according to their frequencies. Applications include (but are certainly not limited to) noise rejection and signal separation in industrial and measurement circuits, feedback of phase and amplitude control in servoloops, smoothing of digitally generated analog (D-A) signals, audio signal shaping and sound reinforcement, channel separation, and signal enhancement in communications circuits.

Such filters can be built from passive RLC components, electromechanical devices, crystals, resistors, capacitors, and op amps (active filters). Active filters are applicable over a wide range of frequencies. They are also inexpensive and offer high input impedance, low output impedance, adjustable gain, and a variety of responses.

In this chapter, you will learn the characteristics, terminology, and mathematics of active filters. The analysis, design, and calculations behind several types of low and high pass filters, wide and narrow band pass filters, the notch filter, and the state variable filter are presented. Electronically controlled active filters allow adaptive and automatic filtering. You will see how to build filters that can be adjusted with an analog voltage, a data word, or a clock frequency.

Objectives

After studying this chapter, you should be able to do the following:

1. Describe in detail a frequency response plot.
2. Convert between ratio and dB gain.
3. Define the f_o, and f_{-3dB} and locate each on a frequency response plot.
4. Briefly state the purpose of Laplace transforms.
5. Analyze a wide variety of circuits to determine their transfer function, gain versus frequency response, and phase versus frequency response.
6. Draw the ideal and practical frequency response plots for a low pass filter, high pass filter, band pass filter, and notch filter.
7. Identify the stop band and the pass band of a given filter response.
8. Describe the roll-off rate, dB/decade, and dB/octave, and their relationship to filter order.

9. Given the response curve of a band pass or notch filter, determine the center frequency, low frequency cut-off, high frequency cut-off, bandwidth, Q, and notch depth.

10. Describe one application for the low pass, high pass, band pass, and notch filters.

11. List four disadvantages of passive filters and explain how active filters overcome them.

12. List four disadvantages of active filters.

13. Derive the transfer function and gain and phase equations for a first- and a second-order low pass, high pass, band pass, notch, voltage controlled, and state variable active filter.

14. Given filter specifications, determine type, order, and component values necessary to build the filter using a Sallen-Key, equal component, band pass, notch, state variable, voltage controlled or switched capacitor digitally controlled implementation.

15. Analyze a given active filter to determine f_o, f_{-3dB}, response shape, roll-off rate, and pass band gain.

16. List advantages, disadvantages, limitations, and precautions for designing and building each of these filters.

9-1 Introduction to Filtering

Filters are specified, analyzed, and designed somewhat differently than are the circuits discussed in the previous chapters. Performance is specified in terms of the frequency response, that is, how the gain and phase shift change with frequency. Analysis and design often use Laplace transforms and the circuit's transfer function. Angular frequency, f_o, f_{3dB}, bands, ripple, roll-off rate, center frequency, and Q are among the terms unique to filters. In this section, frequency response plots, basic transform math, and terminology common to most filters are presented. Later sections of the chapter apply these fundamentals to specific filter circuits.

9-1.1 Frequency Response

The gain and phase shift of a filter change as the frequency changes. Indeed, this is the purpose of a filter. Performance is often described with a graph of gain and phase versus frequency.

You have already seen such a frequency response when investigating the frequency response of an op amp, the gain bandwidth. The vertical axis is the gain, and the horizontal axis is the frequency. The frequency axis gives equal distance for each **decade** of frequency. The distance between 1Hz and 10Hz is the same as the distance between 100kHz and 1MHz. On a linear plot, the distance between 100kHz and 1MHz should be 100,000 larger. However, this would make it impossible to plot any large range of frequencies on a single graph, so the horizontal axis is scaled logarithmically.

A log scaled horizontal axis is shown in Figure 9-1. In Figure 9-1(a), each decade (factor of 10) increase or decrease moves you the same distance along the scale. There are two other key ideas. First, the starting point is not at zero. If you went farther left, each increment would lower the scale by 10 (to 0.1, 0.01, 0.001, ...), but you would never reach zero. So, start your log divisions at the lowest frequency of interest, not zero (DC). Second, the divisions between decades are not uniform (linear). This is easily seen in Figure 9-1(b), an expansion of a single decade. Five does not fall halfway between 1 and 10; 3¹/₃ does. Be very careful about this when interpolating between major scale divisions.

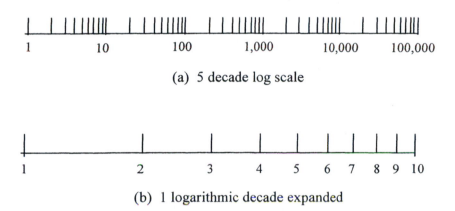

(a) 5 decade log scale

(b) 1 logarithmic decade expanded

Figure 9-1 Logarithmic scales

Gain, plotted on the vertical axis, is normally expressed in dB.

$$dB = 10 \ \log_{10} \frac{\text{power out}}{\text{power in}}$$

$$dB = 10 \ \log \frac{\dfrac{v_o^2}{R_{\text{load}}}}{\dfrac{e_{\text{in}}^2}{R_{\text{in}}}}$$

$$dB = 10 \ \log \left(\frac{v_o}{e_{\text{in}}}\right)^2 \left(\frac{R_{\text{in}}}{R_{\text{load}}}\right)$$

If you assume that

$$R_{in} = R_{load}$$

then

$$dB = 10 \ \log\left(\frac{V_o}{e_{in}}\right)^2$$

$$dB = 20 \ \log\frac{V_o}{e_{in}}$$

For this to be valid, the load resistance must equal the filter's input resistance. This is seldom the case. However, the definition is normally used anyway.

You can readily convert ratio gain (v_o/e_{in}) to dB gain with any scientific calculator. Notice that log base 10 is used, not the natural log (log base e or ln). It is handy to know several dB and ratio points. These are listed in Table 9-1. Decreasing the ratio gain by 10 subtracts 20dB; increasing it by a ratio of 10 adds 20dB. Doubling the gain adds 6dB; halving it subtracts 6dB. Cutting the gain by $\sqrt{2}$ gives -3dB. A ratio of 1 ($v_o = e_{in}$) is a gain of 0dB.

Table 9-1 Ratio and dB gain comparison

v_o/e_{in}	dB	v_o/e_{in}	dB
1000	60	0.707	-3
100	40	0.5	-6
10	20	0.1	-20
2	6	0.01	-40
1	0	0.001	-60

A typical low pass filter frequency response is given in Figure 9-2. The vertical axis is scaled in decibels. The highest gain is called the **pass band gain**, A_o. In Figure 9-2, A_o = +10dB.

The horizontal axis is scaled logarithmically (two decades in the figure). By plotting dB versus log frequency, you are actually plotting log gain versus log frequency. So any linear relationship is displayed as a straight line.

A parameter of major importance is the **-3dB frequency**, f_{-3dB}. At that frequency, the gain has fallen 3dB below A_o. In Figure 9-2, the pass band gain A_o=10dB. At f_{-3dB} the gain falls to 7dB.

$$A_o = 10dB$$

$$f_{-3dB} = 2500Hz$$

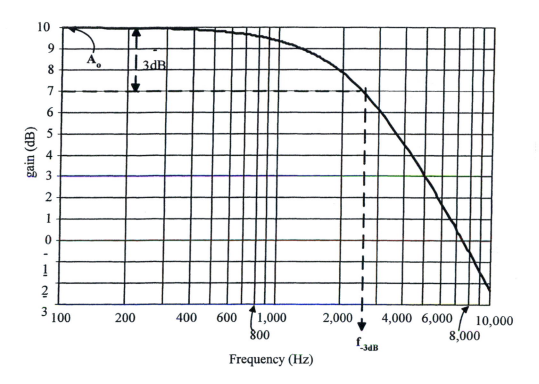

Figure 9-2 Typical low pass response curve

What does this -3dB reduction in the voltage gain mean about the power delivered to the load?

$$-3dB = 20\ \log(A_v)$$

$$A_v = 10^{-\frac{3}{20}} = 0.707$$

The voltage gain has fallen by 0.707 below its value at A_o. This means at f_{-3dB} the output voltage has fallen by 0.707.

$$v_{out\ -3dB} = 0.707 v_{out\ @Ao}$$

$$P_{out@f-3dB} = \frac{v^2}{R} = \frac{\left(0.707v_{out@Ao}\right)^2}{R}$$

$$P_{out@f-3dB} = \frac{0.5v_{out@Ao}^2}{R} = \tfrac{1}{2}P_{out@Ao}$$

Consequently, at the -3dB frequency, the power delivered to the load has been cut in half, and the voltage gain has fallen by 0.707.

As frequency changes, not only does the filter's output amplitude change, but its input-output phase relationship also shifts. At the **critical frequency**, f_o, that shift is an integer multiple of 45°. For simple filters built with a single RC pair, there is a 45° shift in phase. For each additional pair you add to the circuit, the phase shifts another 45° at f_o. Two RC pairs produce a 90° shift at the critical frequency; three pairs cause a 135° shift at f_o.

In many filters the -3dB frequency, f_{3dB}, and the critical frequency, f_o, occur at the same point. But two different effects are occurring at that frequency. The voltage gain has dropped -3dB (0.707), f_{3dB}, and the phase has shifted the correct number of 45°, f_o. However, when you use an op amp to add gain and positive feedback to a filter, these two frequencies no longer occur at the same point.

A typical phase plot combined with the gain plot is given in Figure 9-3. The combination of the gain variation and the phase shift as a function of frequency forms a complete frequency response plot. Not only is plotting both functions on the same graph convenient, but it also allows you to determine if the filter (or any system) you anticipate building will be stable (i.e., not oscillate).

Analysis of circuits containing several reactive components is most easily handled by using Laplace transforms. It is far beyond the scope of this text to present a rigorous coverage of Laplace transform circuit analysis. However, a simplified touch is useful.

When the Laplace transform of an equation (or a circuit) is used, differential and integral terms are replaced by s and 1/s. You can then manipulate and simplify the equation using the rules of algebra rather than those of differential equations.

As far as active filters are concerned, the primary term is

$$i_C = C\frac{dv}{dt}$$

To take the Laplace transform of this, just replace the derivative with s. Uppercase letters are used to indicate that they belong to the Laplace domain. Lower case letters are used for variables in the time (normal) domain.

$$I = CsV$$

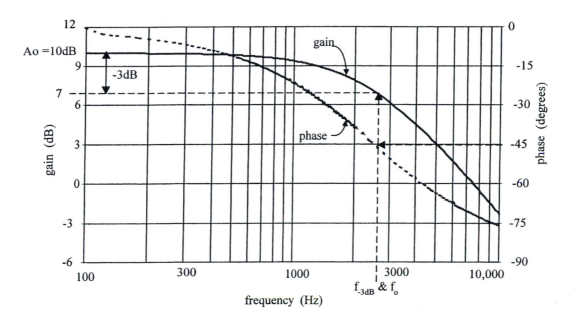

Figure 9-3 Composite gain and phase frequency response

The Laplace impedance, then, for a capacitor is

$$Z = \frac{V}{I} = \frac{1}{Cs}$$

The term s contains both the amplitude and the phase information about an equation's or a circuit's response. To obtain the frequency response from a Laplace equation, make the substitution

$$s = j\omega$$

where j is the imaginary number $\sqrt{-1}$ and $\omega = 2\pi f$.

Example 9-1
Verify that the Laplace form of capacitive impedance correctly converts back to the frequency domain.

Solution

$$Z = \frac{1}{Cs}$$

To convert from the Laplace domain (equation containing s) to the frequency domain (equation containing f or ω):

$$s = j\omega$$

$$\overline{Z} = \frac{1}{j\omega C}$$

$$\overline{Z} = \frac{1}{j2\pi fC}$$

$$\overline{Z} = 0 - j\frac{1}{2\pi fC}$$

$$\overline{Z} = 0 - jX_C$$

This is the phasor impedance of a capacitor, as you learned from AC circuits.

Using Laplace functions, you can determine the frequency response (both gain and phase) of a circuit. You must first determine the circuit's transfer function. The transfer function is just the gain (V_{out}/E_{in}) expressed in Laplace terms.

Example 9-2

    a. Determine the transfer function of the circuit in Figure 9-4.

    b. Calculate and plot the frequency response (both magnitude and phase) for $R = 637\Omega$ and $C = 0.1\mu F$.

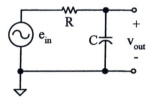

Figure 9-4 Circuit for Example 9-2

Solution

    a. To determine the transfer function, apply the voltage divider law.

$$\overline{V_{out}} = \frac{\overline{Z_C}}{R + \overline{Z_C}} \overline{e_{in}}$$

Substitution:

$$\overline{e_{in}} \rightarrow E_{in} \qquad R \rightarrow R \qquad \overline{Z_C} \rightarrow \frac{1}{Cs} \qquad \overline{V_{out}} \rightarrow V_{out}$$

$$V_{out} = \frac{\dfrac{1}{Cs}}{R + \dfrac{1}{Cs}} E_{in}$$

$$\frac{V_{out}}{E_{in}} = \frac{\dfrac{1}{Cs}}{\dfrac{RCs + 1}{Cs}}$$

$$\frac{V_{out}}{E_{in}} = \frac{1}{RCs + 1}$$

    b. To determine the frequency response, substitute

$$s = j\omega$$

$$\frac{V_{out}}{E_{in}} = \frac{1}{Rj\omega C + 1} = \frac{1}{1 + j\omega RC}$$

This is an equation with real and imaginary parts in its denominator. To separate these parts, multiply both the numerator and the denominator by the complex conjugate of the denominator.

$$\frac{V_{out}}{E_{in}} = \frac{1}{1 + j\omega RC} \times \frac{1 - j\omega RC}{1 - j\omega RC}$$

$$= \frac{1 - j\omega RC}{1 - j^2 \omega^2 R^2 C^2}$$

$$= \frac{1 - j\omega RC}{1 + \omega^2 R^2 C^2}$$

$$= \frac{1}{1+\omega^2 R^2 C^2} - j\frac{\omega RC}{1+\omega^2 R^2 C^2}$$

$$\frac{V_{out}}{E_{in}} = \text{real} + j \text{ imaginary}$$

$$\text{real} = \frac{1}{1+\omega^2 R^2 C^2} \qquad \text{imaginary} = \frac{-\omega RC}{1+\omega^2 R^2 C^2}$$

$$|\overline{G}| = \text{magnitude} = \sqrt{\text{real}^2 + \text{imaginary}^2}$$

$$= \sqrt{\left(\frac{1}{1+\omega^2 R^2 C^2}\right)^2 + \left(\frac{\omega RC}{1+\omega^2 R^2 C^2}\right)^2}$$

$$= \sqrt{\frac{1+\omega^2 R^2 C^2}{\left(1+\omega^2 R^2 C^2\right)^2}}$$

$$|\overline{G}| = \frac{1}{\sqrt{1+\omega^2 R^2 C^2}}$$

$$\phi = \text{phase shift} = \arctan\left(\frac{\text{imaginary}}{\text{real}}\right)$$

$$= \arctan\left(\frac{\dfrac{-\omega RC}{1+\omega^2 R^2 C^2}}{\dfrac{1}{1+\omega^2 R^2 C^2}}\right)$$

$$\phi = -\arctan(\omega RC)$$

Now that you have the two equations, the simplest next step is to use a spreadsheet to tabulate frequency, magnitude, and phase; then create the plot. It is similar to Figure 9-3. However, this circuit has a pass band gain, A_o, of 0dB, not the 10dB shown in Figure 9-3.

The techniques of Example 9-2 can be used to determine the frequency response of any network. However, each RC combination in the circuit increases the order of the denominator by 1. Two RC pairs cause $s^2 + bs + c$; three RC pairs cause $s^3 + bs^2 + cs + d$. Fortunately, the mathematics involved in breaking down and solving these higher-order equations has already been worked out for the popular, more useful circuits. Your job is to obtain the transfer function and recognize and extract key parameters. Actually, the major mathematical effort to convert back from the Laplace domain into the frequency domain may not have to be done at all.

9-1.2 Characteristics and Terminology

The frequency response that you have seen so far is for a low pass filter. The purpose of a low pass filter is to pass low frequency signals while stopping high frequency signals. Ideally, a low pass filter would have a frequency response as shown in Figure 9-5. All frequencies below the critical frequency would be uniformly passed. Any frequency above f_o would be completely stopped.

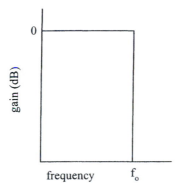

Figure 9-5 Ideal low pass filter

Of course, such a filter cannot be built. The response of a practical filter is divided into two bands as shown in Figure 9-6. The -3dB frequency, f_{-3dB}, forms the boundary between the pass band and the stop band. For some filters, gain may vary up and down (or ripple) in the pass band or in the stop band, or both. The amount of pass band ripple allowable (and, to a lesser degree, stop band ripple) is an important parameter to keep in mind when you are designing a filter. How rapidly the gain falls as the stop band is entered is called the roll-off. The first six roll-off rates are illustrated in Figure 9-7. The roll-off rate is determined by the filter's order (1, 2, 3, ...). Each increase in order increases the roll-off by 20dB/decade. In turn, as was mentioned in the Laplace transform section, the order of the filter (and its transfer function's denominator) equals the number of resistor/capacitor pairs in the circuit. So increasing the circuit's

complexity by adding an RC pair increases the circuit's order, and the difficulties of the
mathematics, but also increases the roll-off rate by 20dB/decade.

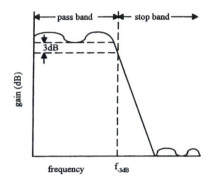

Figure 9-6 Practical low pass filter

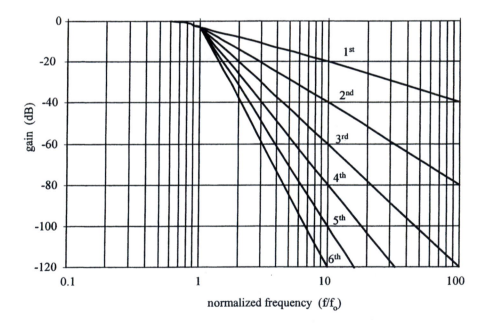

Figure 9-7 Roll-off rate comparison

You will hear roll off specified in dB/decade and dB/octave. A decade increase in frequency means that the frequency has changed by a factor of ten. An octave increase means that the frequency has doubled. Decibels per octave ratings are used primarily with audio/music applications. Table 9-2 correlates filter order (number of RC pairs and transfer function denominator order) with the roll off rate in dB/decade and dB/octave.

Table 9-2 Roll-off rate comparison

Order	dB/decade	dB/octave
1	20	6
2	40	12
3	60	18
4	80	24
5	100	30

The opposite of the low pass filter is the high pass filter. A high pass filter is illustrated in Figure 9-8. Low frequency signals and DC (which is 0Hz) are blocked, while high frequency signals are passed. Specifications, analysis, and design of high pass filters are closely analogous to those you have already seen for low pass filters.

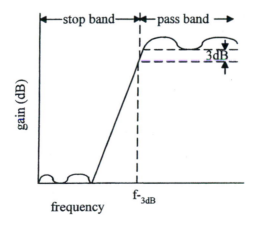

Figure 9-8 High pass filter

You have already used a simple high pass filter. The couplers used in single supply op amp amplifiers block the DC bias from previous stages or the signal generator, while passing the signal whose frequency is of interest. This RC coupler is the complement of the circuit analyzed in Example 9-2. With a single RC pair, the RC coupler has a 20dB/decade roll-off and a 45° phase shift at the critical frequency.

The band pass filter passes only those signals within a given band. Signals above and below that band are blocked. Figure 9-9(a) is the response of an ideal band pass filter, while Figure 9-9(b) is a more realistic response. Since the response rises, peaks, and then falls, there are three frequencies of interest. The **center frequency** is f_c. Depending on the component configuration and values, there may be considerable gain at the center frequency (even for filters with no built-in amplifier). The **cut-off frequencies** (f_l and f_h) occur where the gain has fallen 3dB below the center frequency gain.

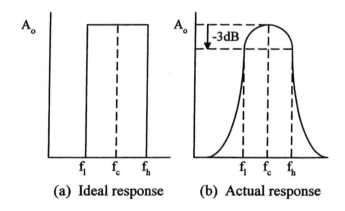

(a) Ideal response   (b) Actual response

Figure 9-9 Band pass filter response

Instead of specifying roll off-rate, **bandwidth** and **Q** are given. Bandwidth is the distance between the low frequency and high frequency cut-offs.

$$\Delta f = f_h - f_l$$

Q is the ratio of center frequency to bandwidth.

$$Q = \frac{f_c}{\Delta f}$$

This gives a measure of the sharpness or narrowness of the band pass filter. The higher the Q, the more selective the filter.

Simple band pass filters can be made with two RC pairs. This makes the transfer function second-order. However, since one pair creates the roll-off at high frequencies and the other pair handles the low frequencies, the eventual roll-off rate is half as steep as the same-order low pass filter or high pass filter.

Band pass filters are used in audio, communications, and instrumentation circuits. Equalizers and speech filters are audio band pass filters. Station tuning in radio and television uses band pass filters. Spectrum analyzers measure a circuit's frequency response with a band pass filter.

The notch filter is the complement of the band pass filter. It rejects those signals in a given band of frequencies and passes all others. The response of a notch filter is illustrated in Figure 9-10. As with the band pass filter, center frequency, low frequency cut-off, and high frequency cut-off are specified. Both bandwidth and Q are defined for the notch as they were for the band pass filter. The other specification needed is **notch depth**, which indicates how severely the signal at the center frequency is rejected.

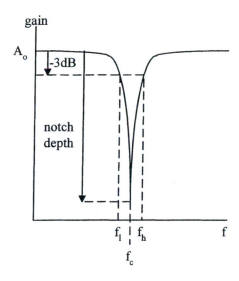

Figure 9-10 Notch filter response

9-1.3 Active versus Passive Filters

The descriptions of Sections 9-1.1 and 9-1.2 apply, more or less, to filters in general, independent of how they are built. The simplest approach to building a filter is with passive components (resistors, capacitors, and inductors). In the radio frequency range this works well. However, as the frequency comes down, inductors begin to have problems. Audio frequency inductors are physically large, heavy, and therefore expensive. To increase inductance for the lower frequency applications, more turns of wire must be added. This adds to the series resistance, degrading the inductor's performance.

Both input and output impedances of passive filters are a problem. The input impedance may be low, which loads down the source, and varies with frequency. The output impedance may be high, which limits the load impedance that the filter can drive. There is no isolation between the load impedance and the passive filter. This means that the load must be considered as a component of the filter and must be taken into consideration when you determine filter response or design. Any change in load impedance may significantly alter one or more of the filter's response characteristics.

Active filters incorporate an amplifier with resistor/capacitor networks to overcome these problems. Originally built with vacuum tubes and then with transistors, active filters now are normally centered around op amps. By enclosing a capacitor in a positive feedback loop, the inductor (with all of its low frequency problems) can be eliminated. If the op amp is properly configured, the input impedance can be increased. The load is driven from the output of the op amp, giving a very low output impedance. Not only does this improve load drive capability, but the load is now isolated from the frequency determining network. Variations in load have no effect on the active filter's characteristics.

The amplifier allows you to specify and easily adjust pass band gain, pass band ripple, cut-off frequency, and **initial** roll-off. Because of the high input impedance of the op amp, large value resistors can be used. This allows you to reduce the value (size, cost, and nonideal behavior) of the capacitors. By selecting a quad op amp IC, you can build steep roll-offs in very little space and for very little money.

Active filters also have limitations. High frequency response is limited by the gain bandwidth and slew rate of the op amp. High frequency op amps are more expensive, making passive filters a more economical choice for many rf applications. An op amp adds noise to any signal passing through it. So high quality audio applications may avoid those filter configurations that place the op amp in series with the main signal path. Active filters require a power supply. For op amps this may be two supplies. Variations in that power supply's output voltage show up, to some degree, in the signal output from the active filter. In multiple-stage applications, the common power supply provides a bus for high frequency signals. Feedback along these power supply lines can cause oscillations. Active devices, and therefore active filters, are much more susceptible to radio frequency interference and ionization than are passive RLC filters. Practical considerations limit the Q of the band pass and notch filters to less than 20. For circuits requiring very selective (narrow) filtering, a crystal filter may be a better choice.

9-2 Low Pass Filter

There are a wide variety of ways to implement a low pass filter. In this section you will first develop the response of a general Sallen-Key second-order circuit. This introduces the concepts of damping and types of response. The equal component implementation will be studied in more detail. Finally, you will see how to determine what order filter is required by a certain applications and how to cascade first- and second-order stages to build the required filter.

9-2.1 Second-Order Sallen-Key Filter

You saw the first-order filter in Example 9-2. With only rare exception, that simple, passive implementation is appropriate when you need a first-order (20dB/decade) response. The second-order filter consists of two RC pairs and has a roll-off rate of 40dB/decade. At the critical frequency the phase is shifted -90°. The denominator of the transfer function is a quadratic in s (s^2+as+b).

The schematic for the general second-order Sallen-Key filter is given in Figure 9-11. Impedances Z1, Z2, Z3, and Z4 may be either resistors or capacitors. This general form is being used so that the results are applicable to low pass or high pass filters.

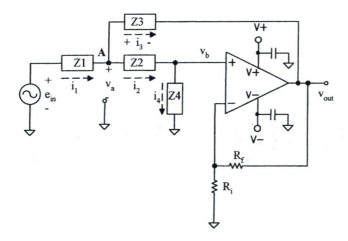

Figure 9-11 Second-order active filter model

The output, v_{out}, is determined by the signal at the op amp's noninverting input and the amplifier's gain.

$$v_{out} = A_o v_b$$

$$A_o = 1 + \frac{R_f}{R_i}$$

So
$$v_b = \frac{v_{out}}{A_o}$$

Assuming that no current flows into the op amp,

$$i_4 = \frac{v_b}{Z4}$$

$$i_2 = i_4 = \frac{v_b}{Z4}$$

At node A,
$$v_a = i_4(Z2 + Z4)$$

Combining these yields $v_a = \frac{v_b}{Z4}(Z2 + Z4)$

Current into the filter, i_1, is the difference in potential across Z1 divided by Z1.

$$i_1 = \frac{e_{in} - v_a}{Z1} = \frac{e_{in}}{Z1} - \frac{v_a}{Z1}$$

Substitute for v_a.
$$i_1 = \frac{e_{in}}{Z1} - \frac{v_b(Z2 + Z4)}{Z1Z4}$$

The current through the feedback impedance, i_3, can be calculated by summing the currents at node A.

$$i_3 = i_1 - i_2$$

Combine the equations for i_1 and i_2.

$$i_3 = \frac{e_{in}}{Z1} - \frac{v_b(Z2 + Z4)}{Z1Z4} - \frac{v_b}{Z4}$$

Summing the loop from node A, Z3, and the output yields

$$v_a - i_3 Z3 - v_{out} = 0$$

$$v_{out} = v_a - i_3 Z3$$

Substitute for i_3.

$$V_{out} = \frac{V_b}{Z4}(Z2 + Z4) - \left[\frac{e_{in}}{Z1} - \frac{V_b(Z2 + Z4)}{Z1Z4} - \frac{V_b}{Z4}\right]Z3$$

Combine this with the initial relationship for the input to output voltages of the op amp.

$$V_{out} = \frac{V_{out}}{A_o Z4}(Z2 + Z4) - \frac{e_{in}Z3}{Z1} + \frac{V_{out}(Z2 + Z4)Z3}{A_o Z1Z4} + \frac{V_{out}Z3}{A_o Z4}$$

This is an expression in v_{out}, e_{in}, and circuit components. The circuit analysis is complete. To obtain the transfer function, you have to manipulate this equation to group and separate terms, isolating v_{out}/e_{in} on the left side of the equation. When this is done, you have

$$\frac{V_{out}}{e_{in}} = \frac{A_o Z3Z4}{Z1Z2 + Z2Z3 + Z3Z4 + Z1Z3 + Z1Z4(1 - A_o)}$$

9-2.2 Second-Order Low Pass Sallen-Key Characteristics

To convert Figure 9-11 into a second-order low pass active filter, the resistors of the RC pairs must be in series with the main signal path, and the capacitors are tied to ground (or to the output which is only a few ohms from ground). This is shown in Figure 9-12.

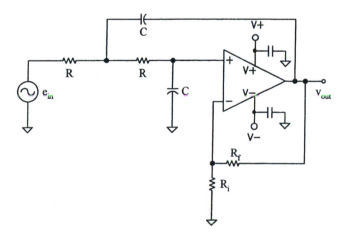

Figure 9-12 Sallen-Key, equal component, second-order, low pass, active filter

Compare Figure 9-12 to Figure 9-11. The equation developed for Figure 9-11 applies to the low pass filter of Figure 9-12 if

$$Z1 = Z2 = R \quad Z3 = Z4 = \frac{1}{Cs}$$

Making these substitutions, you have

$$\frac{V_{out}}{E_{in}} = \frac{\dfrac{A_o}{C^2 s^2}}{R^2 + \dfrac{R}{Cs} + \dfrac{1}{C^2 s^2} + \dfrac{R}{Cs} + \dfrac{R}{Cs}\left(1 - A_o\right)}$$

$$= \frac{\dfrac{A_o}{C^2 s^2}}{\dfrac{R^2 C^2 s^2 + RCs + 1 + RCs + RCs\left(1 - A_o\right)}{C^2 s^2}}$$

$$= \frac{A_o}{R^2 C^2 s^2 + 2RCs + RCs\left(1 - A_o\right) + 1}$$

$$\frac{V_{out}}{E_{in}} = \frac{A_o}{R^2 C^2 s^2 + RC\left[2 + \left(1 - A_o\right)\right]s + 1}$$

The quadratic in the denominator is more easily solved if the coefficient of s^2 is 1. Dividing numerator and denominator by $R^2 C^2$, you obtain

$$\frac{V_{out}}{E_{in}} = \frac{\dfrac{A_o}{R^2 C^2}}{s^2 + \left(\dfrac{3 - A_o}{RC}\right)s + \dfrac{1}{R^2 C^2}}$$

Second-order systems have been studied extensively. Mechanical and chemical as well as electrical second-order systems behave similarly. One transfer function is

$$\frac{A_o \omega_o^2}{s^2 + \alpha \omega_o s + \omega_o^2}$$

where

A_o = the gain
ω_o = the critical frequency in rad/s
α = the damping coefficient

This looks very similar to the second-order system response transfer function that you may have encountered in an advanced electrical networks or control systems course.

$$\frac{A_o \omega_o^2}{s^2 + 2\xi\omega_o s + \omega_o^2}$$

The damping factor (ξ) of the systems transfer function determines if the system is over-damped $(\xi > 1)$, critically damped $(\xi = 1)$, or under-damped $(\xi < 1)$. By comparing the two transfer functions, you can see that

$$\alpha = 2\xi$$

So α plays a similar role for this active filter.

Further comparisons of the Sallen-Key, equal component, low pass filter transfer function with the general second-order form reveals

$$A_o = 1 + \frac{R_f}{R_i}$$

$$\omega_o^2 = \frac{1}{R^2 C^2}$$

$$\omega_o = \frac{1}{RC}$$

$$f_o = \frac{1}{2\pi RC}$$

$$\alpha\omega_o = \frac{3 - A_o}{RC}$$

$$\alpha = 3 - A_o$$

Since it is more convenient to work with a normalized frequency, substitute

$$\omega_o = 1$$

This scales the horizontal axis to make the critical frequency occur at 1. To return to the "real" world, you just multiply the horizontal axis by ω_o.

To obtain the gain and phase relationships, substitute $s = j\omega$ into the transfer function.

$$\frac{V_{out}}{E_{in}} = \frac{A_o}{(j\omega)^2 + \alpha j\omega + 1}$$

$$= \frac{A_o}{-\omega^2 + j\alpha\omega + 1}$$

$$\frac{V_{out}}{E_{in}} = \frac{A_o}{(1-\omega^2) + j\alpha\omega}$$

Now you must separate this into a real term and an imaginary term. Start by multiplying the numerator and the denominator by the complex conjugate of the denominator.

$$\frac{V_{out}}{E_{in}} = \frac{A_o}{(1-\omega^2) + j\alpha\omega} \times \frac{(1-\omega^2) - j\alpha\omega}{(1-\omega^2) - j\alpha\omega}$$

$$= \frac{A_o\left[(1-\omega^2) - j\alpha\omega\right]}{(1-\omega^2)^2 + \alpha^2\omega^2}$$

$$\frac{V_{out}}{E_{in}} = \frac{A_o(1-\omega^2)}{(1-\omega^2)^2 + \alpha^2\omega^2} - j\frac{A_o\alpha\omega}{(1-\omega^2)^2 + \alpha^2\omega^2}$$

$$\text{Real} = \frac{A_o(1-\omega^2)}{(1-\omega^2)^2 + \alpha^2\omega^2} \qquad\qquad \text{Imaginary} = -\frac{A_o\alpha\omega}{(1-\omega^2)^2 + \alpha^2\omega^2}$$

$$|\overline{G}| = \sqrt{\text{real}^2 + \text{imaginary}^2}$$

$$= \sqrt{\frac{A_o^2(1-\omega^2)^2}{\left[(1-\omega^2)^2 + \alpha^2\omega^2\right]^2} + \frac{A_o^2\alpha^2\omega^2}{\left[(1-\omega^2)^2 + \alpha^2\omega^2\right]^2}}$$

$$= \frac{A_o \sqrt{\left[\left(1-\omega^2\right)^2 + \alpha^2\omega^2\right]}}{\left(1-\omega^2\right)^2 + \alpha^2\omega^2}$$

$$\left|\overline{G}\right| = \frac{A_o}{\sqrt{\left(1-\omega^2\right)^2 + \alpha^2\omega^2}}$$

The phase relationship is

$$\phi = \arctan\frac{\text{imaginary}}{\text{real}}$$

$$\phi = -\arctan\frac{\alpha\omega}{1-\omega^2}$$

Both the magnitude and the phase depend on the damping. And, the damping is set by the gain.

$$\alpha = 3 - A_o$$

or

$$A_o = 3 - \alpha$$

The three most commonly used damping coefficients are listed in Table 9-3. The Bessel filter has the heaviest damping and as such is both the most stable and the slowest to respond. It is often used to filter pulses because it does not overshoot or ring. It also provides the best phase (time) delay for sinusoidal signals. However, noticeable roll-off begins at $0.3\omega_o$, causing significant attenuation of the upper end of the pass band.

The Butterworth response provides the flattest frequency response. As such, it is most often the choice for audio and instrumentation circuits. It is underdamped, so there will be some overshoot when a pulse is applied. However it gives reasonable initial roll-off.

The Chebyshev filter is more lightly damped than the Butterworth. This causes the gain to increase with frequency at the upper end of the pass band, actually rising 3dB above A_o. This provides much faster initial roll-off than the Bessel or Butterworth. But this lowered damping also means that there is considerable ringing in response to a step.

Table 9-3 Second-order filter parameters

Filter Type	Damping (α)	Gain (A₀)	Correction (k$_{lp}$)
Bessel	1.732	1.268	0.785
Butterworth	1.414	1.586	1.000
3dB Chebyshev	0.766	2.234	1.390

The normalized frequency response of the magnitude is given in Figure 9-13. The damping coefficient determines the shape of the frequency response plot **near** the critical frequency. At two octaves above or below the critical frequency (1 in Figure 9-13), the gain response of each type of filter is virtually identical.

Look carefully at these three responses. The Butterworth filter has a pass band gain of 4dB. That gain falls -3dB, to 1dB, at the critical frequency. So for Butterworth damping, $f_{3dB} = f_o$. However, the Bessel filter has a pass band gain of 2dB. Its f_{-3dB} occurs early, before f_o. The Chebyshev filter ripples **up** before it falls. Its -3dB frequency occurs where the gain has dropped 3dB below the **bottom** of the pass band (at $A_o - 3dB$). So f_{-3dB} depends on the damping coefficient and is not necessarily equal to the critical frequency. That is where the k_{lp} correction factor in Table 9-3 comes in. It relates f_{-3dB} and f_o, depending on the damping coefficient.

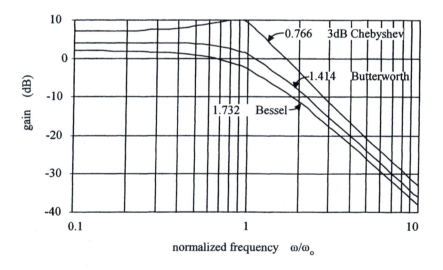

Figure 9-13 Normalized frequency response for a Sallen-Key, equal component, second-order, low pass filter

$$f_{-3dB} = k_{lp} f_o$$

Figure 9-14 shows the phase response of the second-order low pass filters for different damping coefficients. Each curve begins at 0°, passes through -90° at the critical frequency, and then asymptotes toward -180°. The heavier the damping (higher damping coefficient), the flatter the response.

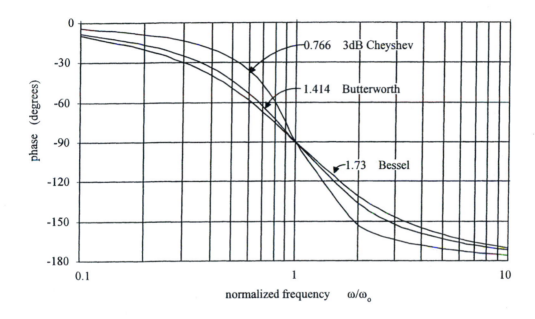

Figure 9-14 Phase shift of second-order, Sallen-Key, equal component, low-pass filter

Example 9-3

Design a Sallen-Key, equal component, second-order, low-pass filter to meet the following specifications:

$f_{-3dB} = 2kHz$ $A_o = 5$ flattest possible pass band

Solution

Since the flattest possible pass band is required, you must use a Butterworth implementation.

$$f_o = \frac{1}{2\pi RC}$$

$$f_{-3dB} = k_{lp}f_o = \frac{k_{lp}}{2\pi RC}$$

$$R = \frac{k_{lp}}{2\pi f_{-3dB}C}$$

Pick $C = 0.01\mu F$.

$$R = \frac{1}{2\pi \times 2kHz \times 0.01\mu F} = 7.96k\Omega$$

Since this is not close to a standard value resistor, pick another standard capacitor, and repeat the calculation until you have discovered a standard capacitor and a standard resistor (in the $10k\Omega$ range) which produce $f_{-3dB} = 2kHz$.
Pick $C = 0.0068\mu F$.

$$R = \frac{1}{2\pi \times 2kHz \times 0.0068\mu F} = 11.7k\Omega$$

Pick $R = 12k\Omega$.
For the Butterworth filter,

$$A_o = 3 - \alpha = 3 - 1.414 = 1.586$$

$$A_o = 1 + \frac{R_f}{R_i} = 1.586$$

$$\frac{R_f}{R_i} = 0.586 \quad \text{or} \quad R_f = 0.586R_i$$

However, to minimize the effect of offset current (which may be important in a **low** pass filter), the resistance at the inverting terminal must equal the resistance at the noninverting terminal.

$$2R = \frac{R_f R_i}{R_f + R_i}$$

Substituting for R_f, you have

$$2R = \frac{0.586R_i^2}{1.586R_i}$$

$$2 \times 12k\Omega = 0.369R_i$$

$$R_i = 65k\Omega$$

Pick $R_i = 68k\Omega$.

$$R_f = 0.586R_i = 39.8k\Omega$$

Pick $R_f = 39k\Omega$ with a series 820Ω resistor.

To obtain an overall gain of 5, you must add an amplifier after the filter. You can**not** just set the gain of the filter's amplifier to 5. The gain of the filter's amplifier must be set to 1.586 to assure that it behaves as a Butterworth filter. Any additional gain (or attenuation) must then be obtained from another amplifier.

$$A_{amp} = \frac{5}{1.586} = 3.15$$

So add a noninverting amplifier after the filter. Set its $R_{f\ amp} = 2.2k\Omega$ and $R_{i\ amp} = 1k\Omega$.

Example 9-4

Given a Sallen-Key, equal component, second-order, low pass filter, with the following values

$$R = 22k\Omega \qquad C = 0.47\mu F \qquad R_f = 12k\Omega \qquad R_i = 10k\Omega$$

determine:

$$f_{-3dB} \qquad A_o \qquad \text{the response type}$$

Also, plot the magnitude's frequency response.

Solution

The simulation of this circuit using *Electronics Workbench*^{TM} is shown in Figure 9-15. The initial run indicates a Chebyshev type of response with a pass band gain of 6.8dB. Moving the cursor up the curve reveals a 2.7dB peak, not far from the 3dB Chebyshev. The f_{-3dB} point occurs at a little over 21Hz, and there is a -40dB/decade roll-off.

Now, let's look at what theory indicates that you can expect.

$$A_o = 1 + \frac{R_f}{R_i} = 1 + \frac{12k\Omega}{10k\Omega} = 2.2$$

$$A_{o\ dB} = 20\ \log(2.2) = 6.8dB$$

$$f_o = \frac{1}{2\pi RC} = \frac{1}{2\pi \times 22k\Omega \times 0.47\mu F} = 15.4Hz$$

$$f_{-3dB} = k_{lp}f_o = 1.39 \times 15.4Hz = 21.4Hz$$

This correlates well with the simulation.

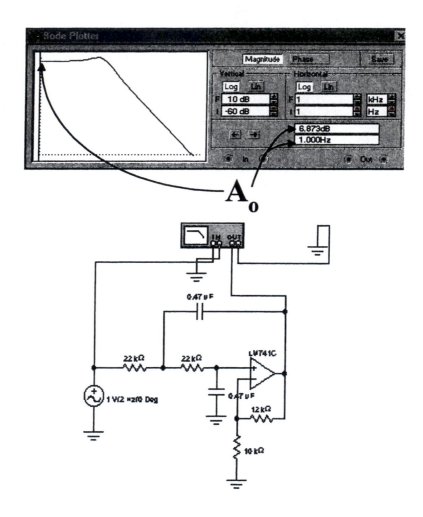

Figure 9-15 *Electronics Workbench™* simulation results for Example 9-4

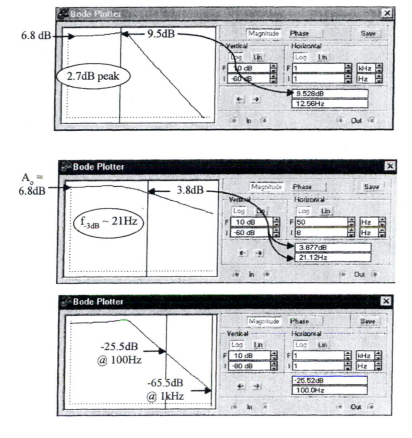

Figure 9-15 (cont.) *Electronics Workbench^{TM}* simulation results for Example 9-4

A second-order filter should have a roll-off of -40dB/decade. The simulation shows the gain dropping from -25.5dB at 100Hz to -65.5dB at 1kHz. That is -40dB/decade.

9-2.3 Higher-Order Filters

The first-order and second-order filters, though relatively easy to build, may not provide an adequate roll-off rate. The only way to improve the roll-off rate is to increase the order of the filter as illustrated in Figure 9-7. Each increase in order produces a 20dB/decade or 6dB/octave increase in the roll off-rate.

To determine what order you need, look at Figure 9-16. Figure 9-16 and the equation for n_B are for a Butterworth filter. The second equation, for n_C, is for a

Chebyshev filter. The Bessel rolls off more slowly than the Butterworth, so you may need two orders higher.

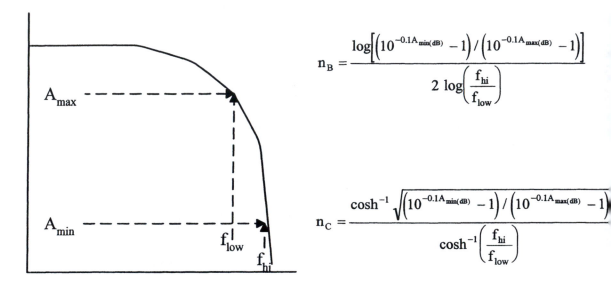

$$n_B = \frac{\log\left[\left(10^{-0.1A_{min(dB)}} - 1\right)/\left(10^{-0.1A_{max(dB)}} - 1\right)\right]}{2 \log\left(\frac{f_{hi}}{f_{low}}\right)}$$

$$n_C = \frac{\cosh^{-1}\sqrt{\left(10^{-0.1A_{min(dB)}} - 1\right)/\left(10^{-0.1A_{max(dB)}} - 1\right)}}{\cosh^{-1}\left(\frac{f_{hi}}{f_{low}}\right)}$$

Figure 9-16 Order selection

Example 9-5

It is necessary to build a filter with a -3dB frequency of 3kHz. At 10kHz the gain must be -40dB below the pass band gain. Determine the order of the filter.

Solution

You have to look carefully at the way the specifications are given. Gains are indicated with respect to the pass band gain. So we can assign an arbitrary value to it, $A_o = 0dB$. This means that

$$A_{max} = -3dB \qquad f_{low} = 3kHz \qquad\qquad A_{min} = -40dB \qquad f_{hi} = 10kHz$$

Apply the order selection equation:

$$n_B = \frac{\log\left[\left(10^{-0.1A_{min(dB)}} - 1\right) / \left(10^{-0.1A_{max(dB)}} - 1\right)\right]}{2 \log\left(\dfrac{f_{hi}}{f_{low}}\right)}$$

$$n_B = \frac{\log\left[\left(10^{-0.1 \times -40dB} - 1\right) / \left(10^{-0.1 \times -3dB} - 1\right)\right]}{2 \log\left(\dfrac{10kHz}{3kHz}\right)}$$

$$n_B = \frac{\log(9999/0.995)}{1.0458} = 3.83$$

Since you cannot make a 3.83-order filter, use a fourth-order.

Higher-order filters can be built by cascading the proper number of first- and second-order filter sections. For a fifth-order filter, this technique results in a transfer

$$\underbrace{\frac{A_o}{\left(s^2 + \alpha_1 s + \omega_1^2\right)}}_{\substack{\text{second-order} \\ \text{section}}} \underbrace{\frac{}{\left(s^2 + \alpha_2 s + \omega_2^2\right)}}_{\substack{\text{another second-} \\ \text{order section}}} \underbrace{\frac{}{\left(s + \omega_3\right)}}_{\substack{\text{first-order} \\ \text{section}}}$$

Each term in the denominator has its own damping coefficient and critical frequency. To obtain a given, well-defined response (Bessel, Butterworth, or Chebyshev), the transfer function, **as a whole,** must be solved and the appropriate coefficients determined.

$$\frac{A_o}{s^5 + as^4 + bs^3 + cs^2 + ds + e}$$

It is unreasonable to expect the α's and ω's of the cascaded filter transfer function to correlate in a simple way with the coefficients of the overall filter. You do **not** get a fifth-order, 1kHz Bessel filter by cascading two 1kHz, second-order Bessel filters and a first -order, RC passive stage.

The mathematics used to solve these higher-order polynomials is beyond the scope of this book. The results are presented in Table 9-4.

Example 9-6

Design a fourth-order Bessel filter with f_{3dB} = 3kHz. Does its roll-off meet the requirements set by Example 9-5?

Table 9-4 Higher order damping and frequency correction factors

Filter Order	Section		Bessel	Butterworth	3dB Cheby
2	2	α	1.732	1.414	0.766
		k_{lp}	0.785	1.000	1.390
3	1	α	-	-	-
		k_{lp}	0.753	1.000	3.591
	2	α	1.447	1.000	0.326
		k_{lp}	0.687	1.000	1.172
4	2	α	1.916	1.848	0.929
		k_{lp}	0.696	1.000	2.349
	2	α	1.242	0.765	0.179
		k_{lp}	0.621	1.000	1.095
5	1	α	-	-	-
		k_{lp}	0.665	1.000	5.762
	2	α	1.775	1.618	0.468
		k_{lp}	0.641	1.000	1.670
	2	α	1.091	0.618	0.113
		k_{lp}	0.569	1.000	1.061
6	2	α	1.959	1.932	0.958
		k_{lp}	0.621	1.000	3.412
	2	α	1.636	1.414	0.289
		k_{lp}	0.590	1.000	1.408
	2	α	0.977	0.518	0.078
		k_{lp}	0.523	1.000	1.042

Solution

You can build a fourth-order filter by cascading two second-order stages. From Table 9-4, the first stage must have a damping coefficient of

$$\alpha_1 = 1.916$$

$$A_{o\,1} = 3 - 1.916 = 1.084$$

For that section, the frequency is set by

$$f_{-3dB} = \frac{k_{lp}}{2\pi RC}$$

$$k_{lp} = 0.696$$

Pick $C1 = 0.01\mu F$.

$$R1 = \frac{k_{lp}}{2\pi f_{-3dB} C} = \frac{0.696}{2\pi \times 3kHz \times 0.01\mu F} = 3.69k\Omega$$

You can build this with a $3.3k\Omega$ resistor in series with a 330Ω resistor.

$$A_{o\,1} = 1 + \frac{R_{f1}}{R_{i1}} = 1.084$$

$$\frac{R_{f1}}{R_{i1}} = 0.084 \quad or \quad R_{f1} = 0.084R_{i1}$$

But

$$2R1 = \frac{R_{f1}R_{i1}}{R_{f1} + R_{i1}}$$

$$7.38k\Omega = \frac{(0.084R_{i1})R_{i1}}{1.084R_{i1}}$$

$$7.38k\Omega = 0.0775R_{i1}$$

$$R_{i1} = 95.2k\Omega$$

$$R_{f1} = 0.084 \times 95.2k\Omega = 8.0k\Omega$$

The second stage is handled in the same way, by using:

$$\alpha_2 = 1.242 \qquad\qquad k_{lp2} = 0.621$$

This gives:

$$C2 = 0.01\mu F \quad R2 = 3.3k\Omega \quad R_{i2} = 15.3k\Omega \quad R_{f2} = 11.6k\Omega$$

Neither section, **alone**, exhibits a Bessel response or has the correct cut-off. However, together the proper response is produced. The overall pass band filter gain is the product of each stage:

$$A_o = A_{o\,1} \times A_{o\,2} = 1.084 \times 1.76$$

$$A_o = 1.91 = 5.62dB$$

The simulation schematic and overall Bode plot are given in Figure 9-17. The frequency response is smooth and well behaved, suggesting a Bessel (or Butterworth) response. The pass band gain from the simulation matches the manual calculations, as does f_{-3dB}. The roll-off rate is -23.4dB/octave, a close match to the -24dB/octave predicted (6dB/octave/order; fourth order). The order selection from Example 9-5 indicated that a fourth-order (Butterworth) filter would cause the gain at 10kHz to be -40dB below A_o. But this filter falls only to -23dB at 10kHz. Sluggish early roll-off is typical of a Bessel filter. If that specification is critical, either increase the order, or use a Butterworth filter.

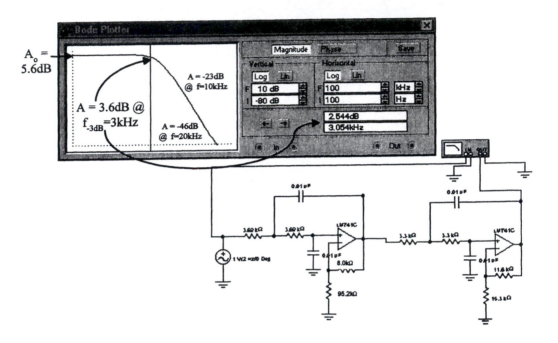

Figure 9-17 Simulation of Example 9-6

When cascading filter sections to produce higher-order filters, be sure to use the correct damping and filter correction factor from Table 9-4. Use the lowest-order filter that meets the given specifications. Damping coefficients (and therefore filter stability) become quite small as you increase the order. Use a Butterworth filter if possible. Go to Bessel if you need better transient response. But this gives poorer **initial** roll-off. Use the Chebyshev filter if initial roll-off of the Butterworth is not adequate. However, transient response and pass band flatness suffer.

9-3 High Pass Active Filter

The complement of the low pass filter is the high pass active filter. It is formed by exchanging the place of the resistors and capacitors in the frequency determining section of the filter. This is shown in Figure 9-18. Compare it carefully with Figure 9-12, the schematic of the second-order, low pass active filter. The general second order active filter is shown in Figure 9-11. Its transfer function was derived as

$$\frac{V_{out}}{E_{in}} = \frac{A_o Z3Z4}{Z1Z2 + Z2Z3 + Z3Z4 + Z1Z3 + Z1Z4(1 - A_o)}$$

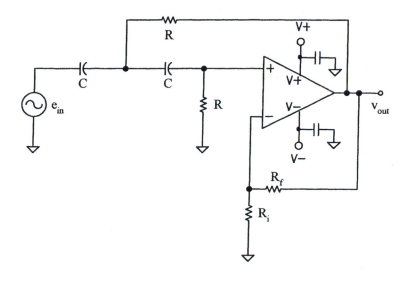

Figure 9-18 Sallen-Key, equal component, second-order,
high pass, active filter

For the Sallen-Key, equal component, second order, high pass filter of Figure 9-18,

$$Z1 = \frac{1}{Cs} \qquad Z2 = \frac{1}{Cs} \qquad Z3 = R \qquad Z4 = R$$

Substituting these into the general transfer function gives:

$$\frac{V_{out}}{E_{in}} = \frac{A_o R^2}{\dfrac{1}{C^2 s^2} + \dfrac{R}{Cs} + R^2 + \dfrac{R}{Cs} + \dfrac{R}{Cs}(1 - A_o)}$$

$$= \frac{A_o R^2 C^2 s^2}{1 + RCs + R^2 C^2 s^2 + RCs + RCs(1 - A_o)}$$

$$= \frac{A_o R^2 C^2 s^2}{R^2 C^2 s^2 + RC(3 - A_o)s + 1}$$

$$\frac{V_{out}}{E_{in}} = \frac{A_o s^2}{s^2 + \frac{(3 - A_o)}{RC}s + \frac{1}{R^2 C^2}}$$

Second-order physical systems have been studied extensively for many years. Mechanical, hydraulic, and chemical, as well as electrical, second-order systems behave similarly. The transfer function for one group of such second-order systems is

$$\frac{A_o s^2}{s^2 + \alpha \omega_o s + \omega_o^2}$$

where A_o = the gain
ω_o = the critical frequency
α = the damping coefficient

Compare the general second-order transfer function to that for the Sallen-Key, equal component, second-order, high pass filter.

$$A_o = 1 + \frac{R_f}{R_i}$$

$$\omega_o = \frac{1}{RC} \qquad f_o = \frac{1}{2\pi RC}$$

$$\alpha = 3 - A_o$$

This is **identical** to the relationships developed for the Sallen-Key, equal component, second-order, low pass filter.

Normalizing the transfer function sets $\omega_o = 1$. It then becomes

$$\frac{A_o s^2}{s^2 + \alpha s + 1}$$

To determine the gain and phase relationships in the frequency domain, substitute $s = j\omega$ into the transfer function.

$$\frac{V_{out}}{E_{in}} = \frac{-A_o\omega^2}{-\omega^2 + j\alpha\omega + 1}$$

$$= \frac{-A_o\omega^2}{\left(1-\omega^2\right) + j\alpha\omega} \times \frac{\left(1-\omega^2\right) - j\alpha\omega}{\left(1-\omega^2\right) - j\alpha\omega}$$

$$= \frac{-A_o\omega^2\left(1-\omega^2\right) + jA_o\alpha\omega^3}{\left(1-\omega^2\right)^2 + \alpha^2\omega^2}$$

$$\text{Real} = \frac{-A_o\omega^2\left(1-\omega^2\right)}{\left(1-\omega^2\right)^2 + \alpha^2\omega^2}$$

$$\text{Imaginary} = \frac{A_o\alpha\omega^3}{\left(1-\omega^2\right)^2 + \alpha^2\omega^2}$$

$$\left|\overline{G}\right| = \sqrt{\text{Real}^2 + \text{Imaginary}^2}$$

$$= \frac{\sqrt{A_o^2\omega^4\left(1-\omega^2\right)^2 + A_o^2\alpha^2\omega^6}}{\left(1-\omega^2\right)^2 + \alpha^2\omega^2}$$

$$= \frac{A_o\omega^2\sqrt{\left(1-\omega^2\right)^2 + \alpha^2\omega^2}}{\left(1-\omega^2\right)^2 + \alpha^2\omega^2}$$

$$\left|\overline{G}\right| = \frac{A_o\omega^2}{\sqrt{\left(1-\omega^2\right)^2 + \alpha^2\omega^2}}$$

Compare this to the gain magnitude of the Sallen-Key, second-order, low pass filter. The denominators are identical. However, for the high pass filter, the magnitude is **directly** proportional to the square of the frequency. An increase in frequency causes an increase in the gain. This is plotted in Figure 9-19. Compare these curves to the three for the low pass filter in Figure 9-13. The curves have just been mirrored around the $\omega = 1$ axis.

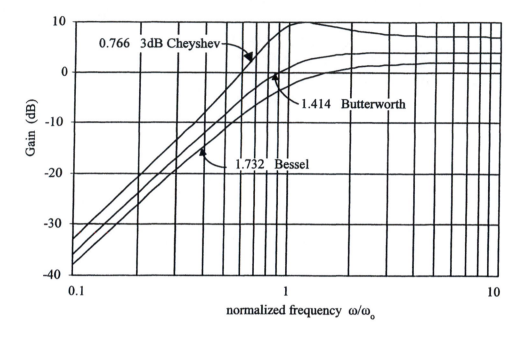

Figure 9-19 Frequency response of the Sallen-Key, equal component, second-order, high pass filter

The phase shift is

$$\phi = \arctan \frac{\text{imaginary}}{\text{real}}$$

$$\phi = -\arctan \frac{\alpha\omega}{1-\omega^2}$$

This is the same as the equation developed for the Sallen-Key, equal component, second-order, low pass filter. The plot is shown in Figure 9-14.

Since the gain magnitude plots are rotated around the normalized axis, compared to the low pass plots, the correction factors of Table 9-3 must be changed for the high pass filter.

$$k_{hp} = \frac{1}{k_{lp}}$$

Example 9-7

For a high pass filter with the components listed below, calculate f_{-3dB}, the filter type, and A_o.

$$R = 10k\Omega \qquad C = 0.1\mu F \qquad R_f = 5.8k\Omega \qquad R_i = 10k\Omega$$

Solution

$$A_o = 1 + \frac{R_f}{R_i} = 1 + \frac{5.8k\Omega}{10k\Omega}$$

$$A_o = 1.58$$

$$\alpha = 3 - A_o = 3 - 1.58$$

$$\alpha = 1.414$$

This is a Butterworth filter.

$$k_{hp} = \frac{1}{k_{lp}} = \frac{1}{1} = 1$$

$$f_{-3dB} = \frac{k_{hp}}{2\pi RC} = \frac{1}{2\pi \times 10k\Omega \times 0.1\mu F}$$

$$f_{-3dB} = 159Hz$$

The gain bandwidth for a 741C op amp is 1MHz. With a pass band gain of 1.58 (4dB), the filter's gain falls off at high frequencies.

$$f_h = \frac{GBW}{A_o} = \frac{1MHz}{1.58}$$

$$f_h = 633kHz$$

In addition, the slew rate limits the amplitude of the high frequency outputs. For full power, maximum output amplitude, the highest frequency is

$$f_{max} = \frac{slew\ rate}{2\pi V_{out\,P}}$$

The higher-order calculation that you saw for the low pass filter can be applied equally well to high pass filters. Just exchange f_{low} and f_{hi}, and A_{max} and A_{min}.

9-4 Band Pass Filter

The band pass filter was introduced in Section 9-1.2. Its frequency responses, both ideal and actual, are given in Figure 9-9. Please refer back to that section and figure.

The parameters of importance in a band pass filter are the high and low frequency cut-offs (f_h and f_l), the bandwidth (Δf), the center frequency (f_c), the center frequency gain (A_o), and the selectivity (Q). Several of these parameters were introduced in Section 9-1.2. The bandwidth is the distance between the high frequency and low frequency -3dB points.

$$\Delta f = f_h - f_l$$

The center frequency is at the **geometric** mean of these cut-off frequencies. It is also the critical frequency.

$$f_o = f_c = \sqrt{f_h f_l}$$

On a **log** plot, this puts f_c halfway between f_h and f_l. This is **not** true on a linear graph. You must be careful. With a low frequency cut-off of 100Hz, and a high frequency cut-off of 3kHz, the center frequency is at 548Hz, not 1.55kHz.

The selectivity, Q, is the ratio of the center frequency to the bandwidth.

$$Q = \frac{f_c}{\Delta f}$$

As such, it gives a measure of the relative narrowness of the filter. For a particular center frequency, the higher the Q, the narrower the bandwidth, and therefore the more selective the filter. In addition, Q gives a more usable number than bandwidth alone. A bandwidth of 1kHz is very broad for an audio filter with a center of 500Hz (Q = 0.5). But that same 1kHz bandwidth is very selective if the center frequency is 100kHz (Q = 100).

Conversely, knowing the circuit's Q and center frequency, you can calculate the upper and the lower cut-off frequencies.

$$f_l = f_c\left(\sqrt{\frac{1}{4Q^2}+1} - \frac{1}{2Q}\right)$$

$$f_h = f_c\left(\sqrt{\frac{1}{4Q^2}+1} + \frac{1}{2Q}\right)$$

9-4.1 Single Op Amp Band Pass Filter

The transfer function of a second-order band pass filter is

$$\frac{V_{out}}{E_{in}} = \frac{A_o \alpha \omega_o s}{s^2 + \alpha \omega_o s + \omega_o^2}$$

where

A_o = the gain at the center frequency
α = the damping coefficient
ω_o = the critical frequency = the center frequency

It has the same denominator as the second-order low pass filter and second-order high pass filter. The numerator of the band pass filter contains the first power of s. The numerator of the low pass filter's transfer function has no s term, while the numerator of the high pass filter has s^2.

To obtain a plot of gain versus frequency, you may first normalize the transfer function by setting $\omega_o = 1$, just as was done for the low pass and the high pass filters. Next multiply the numerator and the denominator by the complex conjugate of the denominator. Then simplify. This allows you to separate the real and the imaginary parts of the equation. The gain's magnitude, then, is

$$|\overline{G}| = \sqrt{\text{Real}^2 + \text{Imaginary}^2}$$

$$|\overline{G}| = \frac{A_o \alpha \omega}{\sqrt{\omega^4 + (\alpha^2 - 2)\omega^2 + 1}}$$

The Q, defined by center frequency and bandwidth, is set by the damping.

$$Q = \frac{1}{\alpha}$$

Figure 9-20 is a plot of the frequency response (gain magnitude) of a second-order band pass filter. The horizontal axis and the vertical axis have been normalized for convenience. To convert to a particular center frequency, f_c, multiply the horizontal axis by f_c. Add A_o (in dB) to the vertical axis to scale it for a particular A_o (dB).

There are two points you should notice. The higher the Q, the sharper (and more selective) the filter. This corresponds to lowering the damping. However, this difference appears only between approximately $0.5f_c$ and $2f_c$. Below $0.5f_c$ and above $2f_c$, the filters all roll off at 20dB/decade (6dB/octave) independent of the Q. This is because this is a second-order filter (two RC pairs). One RC pair causes 20dB/decade roll-off at low frequencies. The other governs high frequencies. A sharper roll off away from the center frequency calls for a higher order, using multiple op amps.

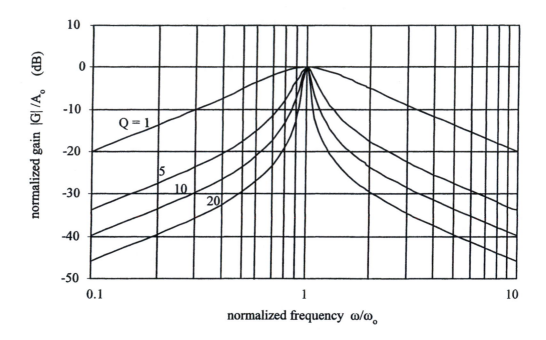

Figure 9-20 Single op amp band pass filter response
(gain and frequency both normalized)

The circuit in Figure 9-21 can be analyzed mathematically to determine how component values affect each of these parameters. The capacitors are equal to each other. As with the analysis of the other op amp circuits, because of high open loop gain and negative feedback, there is no significant difference in potential between the op amp's inverting and noninverting inputs. Also, no signal current flows into either op amp input.

Therefore, I_3 flows through both the capacitor and R3. The left end of R3 is at virtual ground. This means that all of V_{out} is dropped across R3 by I_3.

$$I_3 = -\frac{V_{out}}{R3}$$

The voltage at node A with respect to ground is the voltage dropped across the capacitor by I_3.

$$V_A = \frac{I_3}{Cs}$$

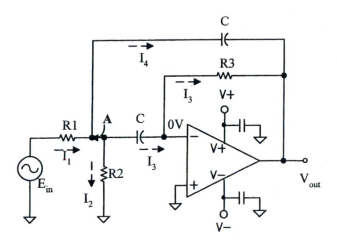

Figure 9-21 Single op amp band pass filter

$$V_A = -\frac{V_{out}}{R3\ Cs}$$

The current through R2 is produced by V_A.

$$I_2 = \frac{V_A}{R2} = -\frac{V_{out}}{R2\ R3\ Cs}$$

The current through the upper feedback capacitor, I_4, is

$$I_4 = \frac{V_A - V_{out}}{\dfrac{1}{Cs}} = (V_A - V_{out})Cs$$

After some algebraic manipulation, this becomes

$$I_4 = -\frac{V_{out}(R3\ Cs + 1)}{R3}$$

The input current, I_1, is

$$I_1 = \frac{E_{in} - V_A}{R1}$$

$$I_1 = \frac{E_{in}}{R1} + \frac{V_{out}}{R1 \; R3 \; Cs}$$

The current entering node A (I_1) must equal the current leaving the node (I_2, I_3, and I_4).

$$I_1 = I_2 + I_3 + I_4$$

Combining all of these equations yields

$$\frac{E_{in}}{R1} + \frac{V_{out}}{R1 \; R3 \; Cs} = -\frac{V_{out}}{R2 \; R3 \; Cs} - \frac{V_{out}}{R3} - \frac{V_{out}(R3 \; Cs + 1)}{R3}$$

With diligence and some luck, this can be simplified to

$$\frac{V_{out}}{E_{in}} = -\frac{R2 \; R3 \; Cs}{\left(R1 \; R2 \; R3 \; C^2\right)s^2 + \left(2 \; R1 \; R2 \; C\right)s + \left(R1 + R2\right)}$$

or

$$\frac{V_{out}}{E_{in}} = -\frac{\dfrac{1}{R1 \; C}s}{s^2 + \dfrac{2}{R3 \; C}s + \dfrac{R1 + R2}{R1 \; R2 \; R3 \; C^2}}$$

Now, compare this to the general transfer function for a band pass filter.

$$\frac{V_{out}}{E_{in}} = \frac{A_o \alpha \omega_o s}{s^2 + \alpha \omega_o s + \omega_o^2}$$

$$\omega_o^2 = \frac{R1 + R2}{R1 \; R2 \; R3 \; C^2}$$

$$f_o = f_c = \frac{1}{2\pi}\sqrt{\frac{R1 + R2}{R1 \; R2 \; R3 \; C^2}}$$

So all components combine to set the center frequency. This is a bit inconvenient.

$$\alpha \omega_o = \frac{2}{R3 \; C}$$

Equating the numerators of the two equations, you get

$$A_o \alpha \omega_o = -\frac{1}{R1 \; C}$$

Dividing these two equations gives

$$\mathbf{A_o} = -\frac{\mathbf{R3}}{\mathbf{2R1}}$$

This shows that at the center frequency, the gain is set by the ratio of the feedback resistor and the input resistor, similar to a simple inverting amplifier.

$$\alpha\omega_o = \frac{2}{R3\ C}$$

$$\alpha = \frac{2}{R3\ \omega_o C}$$

$$Q = \frac{1}{\alpha} = \frac{R3\ \omega_o C}{2}$$

$$= \frac{2\pi f_o R3\ C}{2}$$

$$Q = \frac{f_o}{\Delta f}$$

$$\frac{f_o}{\Delta f} = \frac{2\pi f_o R3\ C}{2}$$

$$\mathbf{\Delta f} = \frac{\mathbf{2}}{\mathbf{2\pi R3\ C}}$$

The bandwidth is set **solely** by R3 and C. Tuning R1 or R2 does not alter the bandwidth.

      The three highlighted equations allow you to analyze a given single op amp active filter to determine its center frequency, pass band gain, and bandwidth. Then Q, low frequency cut-off, and high frequency cut-off can also be calculated.

      For design, this same group of equations must be manipulated a bit and applied in a specific order as illustrated in Example 9-8.

Example 9-8

      Design a second-order band pass active filter to meet the following specifications:

$f_l = 3\text{kHz}$ $f_h = 3.5\text{kHz}$ $A_o = -5$

Solution

The center frequency is

$$f_c = \sqrt{f_l f_h}$$

$$f_c = \sqrt{3kHz \times 3.5kHz} = 3240Hz$$

The bandwidth is

$$\Delta f = f_h - f_l$$

$$\Delta f = 3.5kHz - 3kHz = 500Hz$$

$$Q = \frac{f_c}{\Delta f}$$

$$Q = \frac{3240Hz}{500Hz} = 6.5$$

This is within the limits set for a single op amp band pass filter.

$$1 < Q < 10$$

To calculate component values, first pick a convenient size capacitor. Let

$$C = 0.027\mu F$$

Next, R3, which alone sets the bandwidth, must be determined.

$$\mathbf{R3} = \frac{\mathbf{2}}{\mathbf{2\pi\Delta fC}}$$

$$R3 = \frac{2}{2\pi \times 500Hz \times 0.027\mu F} = 23.6k\Omega$$

Pick   R3 = 24kΩ.

The gain is set by both R1 and R3. Since you now have R3, R1 can be calculated.

$$A_o = -\frac{R3}{2R1}$$

$$\mathbf{R1} = -\frac{\mathbf{R3}}{\mathbf{2A_o}}$$

$$R1 = -\frac{24k\Omega}{2(-5)} = 2.4k\Omega$$

Finally, the value of R2 must be calculated to set the center frequency.

$$f_c = \frac{1}{2\pi}\sqrt{\frac{R1 + R2}{R1\ R2\ R3\ C^2}}$$

But, you need an equation for R2. After some more algebra, the result is

$$R2 = \frac{R1}{4\pi^2 R1\ R3\ f_c^2 C^2 - 1}$$

$$R2 = \frac{2.4k\Omega}{4\pi^2 (2.4k\Omega)(24k\Omega)(3240Hz)^2 (0.027\mu F)^2 - 1} = 146\Omega$$

The **order** in which you adjust these values is critical. Since bandwidth is affected only by R3, alter its value first to obtain the desired bandwidth. Gain is affected by R3 and R1, so, without changing R3 (bandwidth adjust), next tweak R1 to set the center frequency gain. Finally, the center frequency is affected by all three resistors. Last, without changing R1 or R3, adjust R2 to obtain the desired center frequency.

Only center frequency is altered by R2. So you can move the center frequency about by changing R2 without affecting either the bandwidth or the center frequency gain. If you need to adjust the gain (with R1), you must also retune the center frequency. Changing the bandwidth with R3 requires that you readjust both the gain and the center frequency.

There are two component limitations to this single op amp band pass filter. Look at the equation for R2. To get a real value for R2, the denominator may not be zero or go negative. That is,

$$4\pi^2 R1\ R3\ f_c^2 C^2 - 1 > 0$$

or

$$f_c > \frac{1}{2\pi\sqrt{R1\ R3}\ C}$$

With a little more algebra, this same condition can be expressed in terms of the filter's initial specifications. This allows you to decide, at the beginning, if this configuration can be used to implement the band pass filter.

$$2Q^2 > |A_o|$$

If you need less gain to assure that the inequality is met, set the filter's gain to a lower level, and add an amplifier after the filter to set the overall circuit's gain correctly.

The second component limitation is the op amp's gain bandwidth. At the center frequency, the op amp's open loop gain must be at least $20Q^2$ to insure less than 10% gain error. In terms of the op amp's gain bandwidth,

$$\mathbf{GBW} \geq \mathbf{20Q^2} \times \mathbf{f_c}$$

One final word of warning: remember that $Q = 1/\alpha$. As you increase the filter's Q to make it more and more selective, you are lowering the circuit's damping. For any $\alpha < 2$ ($\xi < 1$) the circuit is underdamped. This means that when the circuit is hit with a pulse, the output rings. It is possible that the circuit may oscillate at f_c if you set the damping too low. So choose the highest damping, the lowest Q, that works for your application. If you need more rejection than this single op amp band pass filter can provide without excessive ringing, then you may need to move to a multistage band pass filter.

9-4.2 Multistage Bandpass Filters

The single op amp band pass filter works reasonably well for moderate values of Q if 20dB/decade (6dB/octave) roll-off is adequate once you move away from the center frequency. However, if you need wide band operation (Q < 1), very narrow band operation (Q > 20), or steeper roll-off, you must cascade several filters.

Cascading two single op amp band pass filters together increases the roll-off rate from 20dB/decade to 40dB/decade (12dB/octave). A third stage sets the roll-off rate to 60dB/decade. However, as with cascading low pass or high pass filters, the overall resultant damping coefficient, and therefore Q, change.

For a multistage band pass filter to produce maximum sharpness, set the center frequency of all stages equal to the desired overall center frequency.

$$f_c = f_{c1} = f_{c2} = f_{c3}$$

Set the Q of each stage equal to the desired overall Q corrected by a correction factor.

$$Q_{each\ stage:2\ stages} = 0.644 Q_{cascaded}$$

$$Q_{each\ stage:3\ stages} = 0.510 Q_{cascaded}$$

$$Q_{each\ stage:n\ stages} = \sqrt{2^{\frac{1}{n}} - 1}\ Q_{cascaded}$$

The overall cascaded filter gain is the product of the gain of each stage, or the sum of the dB gains of each stage.

$$A_{o\ overall} = A_{o1} \times A_{o2} \times A_{o3} \times ... \times A_{on}$$

Example 9-9

a. Design a circuit to meet the following specifications:

$$A_o = -9 \qquad Q = 20 \qquad f_c = 60Hz \qquad 60dB/decade\ roll\text{-}off$$

b. Confirm the design's performance with a simulation.

Solution

To obtain a 60dB/decade roll off, you must cascade three single op amp band pass filters. $\qquad f_c = 60Hz = f_{c1} = f_{c2} = f_{c3}$

$$Q1 = Q2 = Q3 = 0.51Q_{cascaded} \qquad\qquad Q_{each\ stage} = 0.51 \times 20 = 10.2$$

$$A_{o\ overall} = A_{o1} \times A_{o2} \times A_{o3} = -9 \qquad A_{o\ each\ stage} = \sqrt[3]{-9} = -2.08$$

Now, you simply design a single op amp band pass filter with

$$f_c = 60Hz \qquad A_o = -2.08 \qquad Q = 10.2$$

The result uses R1 = 27.6kΩ, R2 = 280Ω, R3 = 115kΩ, and C = 0.47μF.

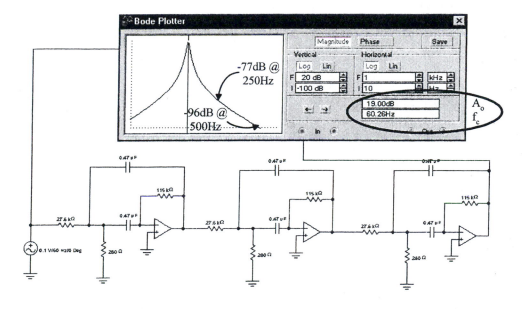

(a) Schematic setup and overall frequency response simulation

Figure 9-22 *Electronics Workbench^{TM} simulation of Example 9-9*

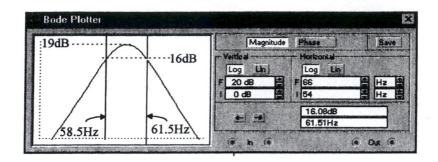

(b) Expanded view of the frequency response around f_c to verify Q

Figure 9-22 (cont.) *Electronics Workbench*™ simulation of Example 9-9

From the Bode plot in Figure 9-22(a), you can see that the center frequency is 60Hz and the pass band gain is 19dB (ratio gain of 8.9). Notice that because of this high gain, it is necessary to reduce the input amplitude to 0.1V. The roll-off is so steep that it is necessary to look at an octave instead of a decade. From 250Hz to 500Hz, the gain drops from -77dB to -96dB. This is a roll-off of 19dB/octave. Theory predicts 18dB/octave for a third-order band pass filter.

In Figure 9-22(b) the response has been expanded around the peak. The bandwidth is

$$\Delta f = 61.5Hz - 58.5Hz = 3Hz$$

This gives an overall Q of

$$Q = \frac{f_c}{\Delta f} = \frac{60Hz}{3Hz} = 20$$

You can alter the shape of two cascaded filter stages by keeping the Q of each stage equal and slightly staggering the center frequencies. To produce a symmetrically shaped response, one filter's frequency should be shifted up by the same proportion that the other filter is shifted down. This is easily done by

$$f_{c\ low} = mf_c$$

$$f_{c\ high} = \frac{f_c}{m}$$

m = stagger factor

With m = 1, a **maximum peak** frequency response is produced, as you just saw in Example 9-9. As **m** falls, the top of the pass band spreads out. However, the initial and final roll-off rates are not affected.

Maximum flatness results with

$$m = 0.946 \quad (2 \text{ stages})$$

$$m = 0.877 \quad (3 \text{ stages})$$

To stagger tune three stages,

$$f_{c \text{ low}} = mf_c$$

$$f_{c \text{ mid}} = f_c$$

$$f_{c \text{ high}} = \frac{f_c}{m}$$

Also, when you stagger tune a three-stage filter, best shape is obtained if

$$Q_{low} = Q_{high}$$

$$Q_{mid} = \tfrac{1}{2} Q_{low}$$

Decreasing the stagger factor more causes the pass band to ripple across the top.

A wide band pass band filter requires a low Q. If you build this with a single op amp band pass filter, the roll-off rate is only 20dB/decade, right up to the center frequency. This does not produce a selective filter.

However, in many audio and speech processing applications, low Q filters (wide band) are needed with flat tops and sharp edges. You can build such a filter by cascading a Butterworth low pass filter with a Butterworth high pass filter. The relative effects are shown in Figure 9-23.

The high frequency cut-off of the band pass filter is the -3dB frequency of the low pass Butterworth filter. Conversely, the low frequency cut-off is the -3dB frequency of the high pass Butterworth filter. Fourth-order low and high pass filters have been shown. You should use the lowest-order filters capable of providing adequate performance. The gain of all three filter has been normalized to make comparisons easier. The actual center frequency gain of the cascaded low pass - high pass filter is the product of the ratio gains of the two cascaded filters.

By keeping the critical frequencies of the two cascaded filters well separated, there is very little interaction between the damping coefficients. This means that you need not apply additional frequency correction factors (beyond what is needed to form each separate filter). However, this restricts this technique of band pass filter construction to wide band (low Q) applications.

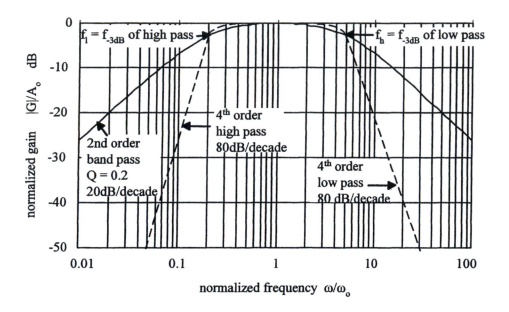

Figure 9-23 Wide band pass filter response of fourth-order low pass filter cascaded with a fourth-order high pass filter

9-5 State Variable Filter

The filters you have seen so far have been relatively simple single op amp circuits or several single op amp circuits cascaded. The state variable, however, uses three or four op amps and two feedback paths. Though a bit more complicated, the state variable configuration offers several features not available with the simpler Sallen-Key filters. First, all three filter types are available **simultaneously**. By properly summing these outputs, some very interesting responses can be made. Band pass filters with Q's up to 50 can be built. The pass band gain for low pass and high pass can be set to 1 and independent of damping. The damping and critical frequencies can be **electronically** tuned.

Figure 9-24 is the schematic of a three op amp, unity gain, state variable filter. Op amps U2 and U3 are integrators. Op amp U1 sums the input with the low pass output and a portion of the band pass output. The circuit is actually a small analog computer, designed to solve the differential equation (transfer function) for each filter type.

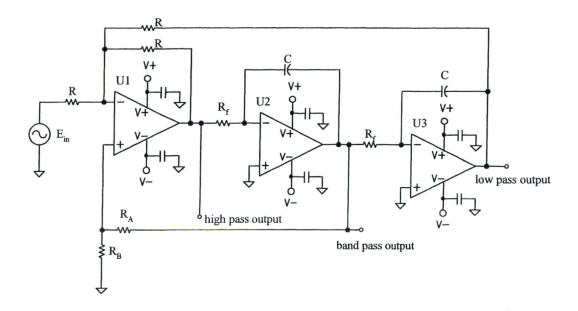

Figure 9-24 Three op amp, state variable filter

You will see from the following derivations that the critical frequencies of each of the three filters are equal.

$$f_o = \frac{1}{2\pi R_f C}$$

The pass band gain for the low pass and high pass filter is 1.

$$A_{o\,low\,pass} = A_{o\,high\,pass} = 1$$

The damping is set by the positive feedback through R_A and R_B. This determines the type of low pass and high pass response (Bessel, Butterworth, or Chebyshev).

$$\alpha = 3\frac{R_B}{R_A + R_B}$$

This also sets the Q and the gain of the pass filter.

$$Q = \frac{1}{\alpha}$$

$$A_{o \text{ band pass}} = Q$$

The detailed analysis of the state variable filter is done in several steps. First, to determine the effect that U2 has on the output of U1, notice that the inverting input pin of U2 is at virtual ground. So the current flowing through U2's input resistor, R_f, is

$$I_{R_f U2} = \frac{V_{HP}}{R_f}$$

Since none of this current flows into U2, it must all go through U2's feedback capacitor.

$$V_C = \frac{I}{Cs}$$

$$V_{BP} = -\frac{I_{R_f U2}}{Cs}$$

$$V_{BP} = -\frac{\dfrac{V_{HP}}{R_f}}{Cs} = -\frac{V_{HP}}{R_f Cs}$$

Applying the same technique to U3 yields

$$V_{LP} = -\frac{V_{BP}}{R_f Cs}$$

Op amp U1 is a summer with three inputs. To determine its output, V_{HP}, you can use superposition. When considering E_{in}, or V_{LP}, the other inputs are all grounded, leaving a simple inverter with unity gain.

$$V_{HP \text{ from } E \text{ in}} = -\frac{R}{R} E_{in} = -E_{in}$$

$$V_{HP \text{ from } V_{LP}} = -\frac{R}{R} V_{LP} = -V_{LP}$$

The band pass output, V_{BP}, is voltage divided by a ratio of

$$\beta = \frac{R_B}{R_A + R_B}$$

and then applied to U1 as a noninverting amplifier. Since the input resistors from E_{in} and V_{LP} are both grounded when you do superposition, the gain (as a noninverting amp for the signal βV_{BP}) is

$$V_{HP \text{ from BP}} = \left(1 + \frac{R}{\frac{R}{2}}\right)\beta V_{BP} = 3\beta V_{BP}$$

By superposition, the output is the sum of the effects of each input.

$$V_{HP} = -E_{in} - V_{LP} + 3\beta V_{BP}$$

You now have three simultaneous equations that describe the behavior of the state variable filter.

$$V_{BP} = -\frac{V_{HP}}{R_f Cs}$$

$$V_{LP} = -\frac{V_{BP}}{R_f Cs}$$

$$V_{HP} = -E_{in} - V_{LP} + 3\beta V_{BP}$$

For the high pass output, substitute the equation for V_{LP},

$$V_{HP} = -E_{in} + \frac{V_{BP}}{R_f Cs} + 3\beta V_{BP}$$

Now substitute the equation for V_{BP},

$$V_{HP} = -E_{in} + \frac{-\dfrac{V_{HP}}{R_f Cs}}{R_f Cs} + 3\beta\left(-\frac{V_{HP}}{R_f Cs}\right)$$

After a bit algebraic manipulation,

$$V_{HP}\left(\frac{R_f^2 C^2 s^2 + 3\beta R_f Cs + 1}{R_f^2 C^2 s^2}\right) = -E_{in}$$

$$V_{HP} = -\frac{s^2 E_{in}}{s^2 + \dfrac{3\beta}{R_f C}s + \dfrac{1}{R_f^2 C^2}}$$

$$\frac{V_{HP}}{E_{in}} = -\frac{s^2}{s^2 + \dfrac{3\beta}{R_f C}s + \dfrac{1}{R_f^2 C^2}}$$

Compare this with the standard high pass transfer function.

$$\frac{A_o s^2}{s^2 + \alpha \omega_o s + \omega_o^2}$$

Equating coefficients, for the high pass output,

$$A_o = -1 \qquad\qquad \omega_o = \frac{1}{R_f C} \qquad\qquad \alpha = 3\beta = \frac{3R_B}{R_A + R_B}$$

For the band pass output,

$$V_{BP} = -\frac{V_{HP}}{R_f C s}$$

Substitute the equation for V_{HP},

$$V_{BP} = -\frac{\dfrac{-s^2 E_{in}}{s^2 + \dfrac{3\beta}{R_f C}s + \dfrac{1}{R_f^2 C^2}}}{R_f C s}$$

$$V_{BP} = \frac{s^2 E_{in}}{s^2 + \dfrac{3\beta}{R_f C}s + \dfrac{1}{R_f^2 C^2}} \times \frac{1}{R_f C s}$$

$$\frac{V_{BP}}{E_{in}} = \frac{\dfrac{1}{R_f C}s}{s^2 + \dfrac{3\beta}{R_f C}s + \dfrac{1}{R_f^2 C^2}}$$

Compare this to the standard band pass transfer function.

$$\frac{A_o \alpha \omega_o s}{s^2 + \alpha \omega_o s + \omega_o^2}$$

Equating coefficients, for the band pass output,

$$\omega_o = \frac{1}{R_f C} \qquad \alpha = 3\beta = \frac{3R_B}{R_A + R_B}$$

These are the same as for the high pass output. Equate the numerators.

$$A_o \alpha \omega_o = \frac{1}{R_f C}$$

$$A_o = \frac{1}{R_f C} \times \frac{1}{\omega_o} \times \frac{1}{\alpha}$$

$$A_o = \frac{1}{R_f C} \times R_f C \times Q$$

$$A_o = Q$$

Now for the low pass section,

$$V_{LP} = -\frac{V_{BP}}{R_f C s}$$

Substitute the relationship for V_{BP},

$$V_{LP} = -\frac{\dfrac{E_{in}}{R_f C} s}{s^2 + \dfrac{3\beta}{R_f C} s + \dfrac{1}{R_f^2 C^2}} \times \frac{1}{R_f C s}$$

$$\frac{V_{LP}}{E_{in}} = \frac{-\dfrac{1}{R_f^2 C^2}}{s^2 + \dfrac{3\beta}{R_f C} s + \dfrac{1}{R_f^2 C^2}}$$

Compare this with the standard low pass transfer function.

$$\frac{A_o \omega_o^2}{s^2 + \alpha \omega_o s + \omega_o^2}$$

Equating coefficients for the low pass output (just as for the high pass output) yields

$$A_o = -1 \qquad\qquad \omega_o = \frac{1}{R_f C} \qquad\qquad \alpha = 3\beta = \frac{3R_B}{R_A + R_B}$$

The state variable filter produces the standard second-order, low pass, band pass, and high pass responses. The critical frequencies of each are equal, and the damping is set by the feedback from the band pass output. For all three outputs this damping has precisely the same effect (at the same numerical values) as it did for the single op amp filters. For low pass and high pass, the damping coefficient of 1.414 produces a Butterworth response. Damping of 1.73 gives a Bessel response, and $\alpha = 0.766$ causes 3dB peaks (Chebyshev). Also, the -3dB frequencies of the low pass sections are shifted from the critical frequency according to the correction factor, k_{lp}, just as you saw for the Sallen-Key filters. The high pass -3dB frequency is similarly shifted by the high pass correction factor.

$$k_{hp} = \frac{1}{k_{lp}}$$

For the band pass section, changing the damping coefficient inversely alters Q and gain.

$$Q = \frac{1}{\alpha} \qquad\qquad A_{o\,BP} = Q$$

But the critical frequency is set by R_f and C. It is not altered by changes in the damping coefficient. This means that changes in damping only (and directly) affect the bandwidth. So tuning the band pass filter is very convenient. The R_f resistors adjust the center frequency (only). Resistors R_A and R_B adjust the bandwidth (only).

At this point, it is important that you realize that **optimum** performance from the three outputs cannot be obtained simultaneously. If you want maximum flatness in the pass bands of the low pass and high pass outputs, you must select a Butterworth response with $\alpha = 1.414$. But a damping coefficient of 1.414 gives a Q of

$$Q = \frac{1}{\alpha} = \frac{1}{1.414}$$

$$Q = 0.707$$

The band pass filter is not very selective and actually attenuates the center frequency by 30% ($A_{o\,BP} = 0.707$).

On the other hand, if you select Q = 20, to obtain reasonable selectivity with good center frequency gain, the low and high pass outputs have a damping coefficient of 0.05. This causes a pass band peak of over 25 dB (20 log Q). You can optimize **either** the band pass output or the low and high pass outputs, but never all three.

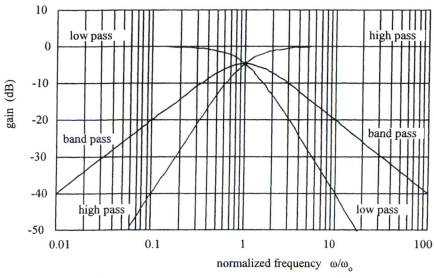

(a) State variable Bessel response

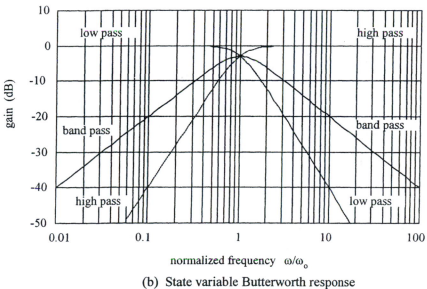

(b) State variable Butterworth response

Figure 9-25 State variable filter responses to changes in damping

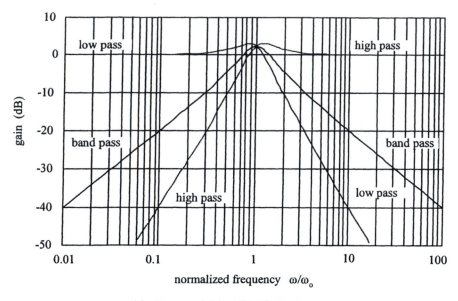

(c) State variable 3dB Chebyshev response

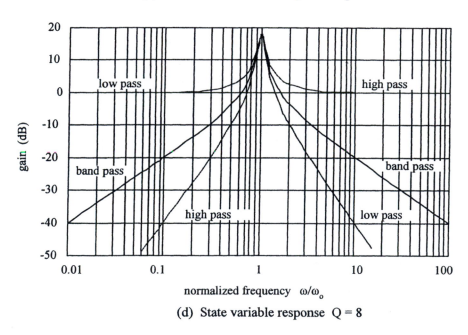

(d) State variable response Q = 8

Figure 9-25 (cont.) State variable filter responses to changes in damping

The frequency response for four different values of damping are shown in Figure 9-25. Bessel response is shown in (a). Gain and Q are low. Notice that all three responses for the Bessel filter coincide at f_o. This is also true for the Butterworth response in (b), though Q and band pass gain have increased. Decreasing the damping even further caused the curves to peak, as in the 3dB Chebyshev of (c). There is indeed a 3dB peak in the low pass and the high pass responses. Higher Q can be obtained by lowering the damping coefficient further. A filter with Q = 8 is shown in (d). Notice the large peaks in the low pass and high pass responses. In fact, the peaks of all three responses coincide at f_o and are all 20 log Q tall.

Example 9-10

Given a state variable filter with the following component values (see Figure 9-24):

$R = 10k\Omega$ $R_f = 4.7k\Omega$ $R_A = 22k\Omega$ $R_B = 7.5k\Omega$ $C = 0.01\mu F$

determine the filter response by calculating the parameters below.
a. damping coefficient and type of filter (Bessel, Butterworth, Chebyshev).
b. low pass and high pass f_{-3dB}.
c. band pass center frequency.
d. band pass bandwidth.
e. band pass gain at the center frequency.

Solution

a. The damping coefficient is

$$\alpha = \frac{3R_B}{R_A + R_B}$$

$$\alpha = \frac{3 \times 7.5k\Omega}{22k\Omega + 7.5k\Omega} = 0.763$$

The 3dB Chebyshev filter has a damping coefficient of 0.766.

b. The critical frequency is

$$f_o = \frac{1}{2\pi RC}$$

$$f_o = \frac{1}{2\pi \times 4.7k\Omega \times 0.01\mu F} = 3.39kHz$$

The low pass and high pass f_{-3dB} must be corrected by the factor from Table 9-3.

$$f_{-3dB \, low \, pass} = k_{lp} f_o$$

$$f_{-3dB \, low \, pass} = 1.39 \times 3.39kHz = 4.7kHz$$

For the high pass output,

$$k_{hp} = \frac{1}{k_{lp}}$$

$$k_{hp} = \frac{1}{1.39} = 0.719$$

$$f_{-3dB \, high \, pass} = k_{hp} f_o$$

$$f_{-3dB \, high \, pass} = 0.719 \times 3.39kHz = 2.44kHz$$

c.   The band pass center frequency is f_o.

$$f_{BP \, center \, freq} = f_o = 3.39kHz$$

d.   The pass band bandwidth comes from Q.

$$Q = \frac{1}{\alpha}$$

$$Q = \frac{1}{0.766} = 1.305$$

$$Q = \frac{f_o}{\Delta f}$$

$$\Delta f = \frac{f_o}{Q}$$

$$\Delta f = \frac{3.39kHz}{1.305} = 2.60kHz$$

e.   The band pass gain at center frequency is

$$A_o = Q$$

$$A_o = 1.305$$

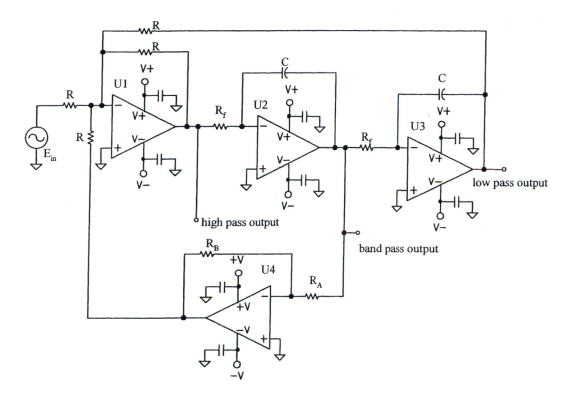

Figure 9-26 Four op amp, state variable active filter

An alternate approach to building a state variable filter is shown in Figure 9-26. A fourth op amp is needed, but since op amps come in quad packs, this does not necessarily increase your parts count. The band pass feedback, which sets the damping, is fed through an inverter and then summed at the inverting terminal of U1. The damping coefficient is

$$\alpha = \frac{R_B}{R_A}$$

This configuration now gives simple, independent control of critical frequency and damping with resistors connected to virtual ground: R_f controls f_o, and R_A controls α. This allows you to control this filter's performance electronically. Preceding each of these resistors with an analog multiplier or a voltage out multiplying DAC effectively raises their resistance. You could replace each resistor with a digital potentiometer, or with a current out multiplying DAC. In Chapter 10, you will read about a component whose resistance can be altered with a voltage (the operational transconductance amplifier). It works well when its output current is driven into virtual ground. Three of these OTAs give you analog voltage control over the performance of this filter.

Cascading state variable filters is just like cascading Sallen-Key or single stage band pass filters. The roll-off rate increases, but you have to account for changes in the damping coefficients, correction factors, and Q.

9-6 Notch Filter

The band pass filter enhances one set of frequencies while rejecting all others. The notch filter does just the opposite. It rejects a band of frequencies while passing all others. This becomes quite handy for recovering signals that are buried in line (60Hz) noise. Also, harmonic distortion is measured by notching out the fundamental. The remaining signal is the sub- and super harmonics. These are a direct indication of harmonic distortion.

9-6.1 Band Pass and Summer

Several different circuits produce notch filters. However, perhaps the most straightforward technique is to subtract the band pass filter output from its input, as illustrated in Figure 9-27. The center frequency, f_c, is the largest out of the band pass filter, subtracting the most from the input signal. This yields a very small output at f_c. Away from the center frequency, the band pass filter has only a small output. This subtracts only a small amount from the input signal, leaving a large output away from the center (or notch) frequency.

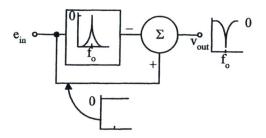

Figure 9-27 Notch filter block diagram

The single op amp band pass filter has an inverted output. This means that its gain or transfer function is already negative. Consequently, when building a circuit to implement Figure 9-27, use a summer rather than a difference amplifier. Also, the band pass filter has a gain of A_o, so the center frequency is $-A_o\, e_{in}$. To completely subtract out, the input of the summer must be precisely $A_o\, e_{in}$. A gain of A_o must be inserted between the input signal and the summer. Look at Figure 9-28.

The output of the circuit is the sum of the outputs from the two blocks.

$$V_{out} = A_o E_{in} + \frac{-A_o \alpha \omega_o s}{s^2 + \alpha \omega_o s + \omega_o^2} E_{in}$$

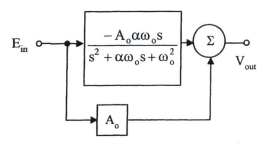

Figure 9-28 Practical notch filter block diagram

$$\frac{V_{out}}{E_{in}} = A_o - \frac{A_o \alpha \omega_o s}{s^2 + \alpha \omega_o s + \omega_o^2}$$

$$= A_o \left(1 - \frac{\alpha \omega_o s}{s^2 + \alpha \omega_o s + \omega_o^2} \right)$$

$$= A_o \left(\frac{s^2 + \alpha \omega_o s + \omega_o^2 - \alpha \omega_o s}{s^2 + \alpha \omega_o s + \omega_o^2} \right)$$

$$\frac{V_{out}}{E_{in}} = \frac{A_o \left(s^2 + \omega_o^2 \right)}{s^2 + \alpha \omega_o s + \omega_o^2}$$

This is the transfer function for a second-order notch filter. The terms are defined precisely as they are for the band pass filter.

As with the other filters, to obtain the gain equation, substitute $s = j\omega$ into the transfer function. Manipulate the result to separate the real and the imaginary parts. Then apply

$$|\overline{G}| = \sqrt{\text{real}^2 + \text{imaginary}^2}$$

The result is

$$|\overline{G}| = \frac{A_o|\omega_o^2 - \omega^2|}{\sqrt{\omega^4 + (\alpha^2 - 2)\omega_o^2\omega^2 + \omega_o^4}}$$

Remember, $\qquad Q = \dfrac{1}{\alpha}$

This is plotted in Figure 9-29 for different values of Q. The gain and the frequency have both been normalized.

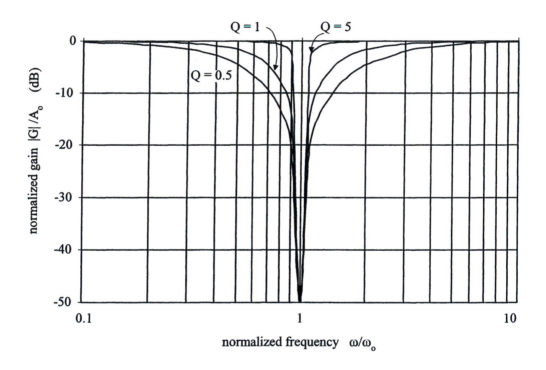

Figure 9-29 Notch filter response

To build a notch filter, you build a band pass filter with the desired Q and bandwidth. Then connect its output into a summer. The other input of the summer is connected back to the input of the band pass filter. Finally, the gain must be adjusted. The schematic of a second-order notch filter is shown in Figure 9-30. The band pass filter consists of U1, R1, R2, R3, and C. The summer consists of U2, R4, R5, and R6. Build and tune the band pass filter as you did in the previous section. To select values for the summer, you must make the gain at the center frequency for the band pass filter, through the summer, the same as the input's gain through the summer. That is,

$$\frac{R3}{2R1} \times \frac{R6}{R5} = \frac{R6}{R4}$$

Or if $R5 = R6$

then $\frac{R3}{2R1} = \frac{R6}{R4}$

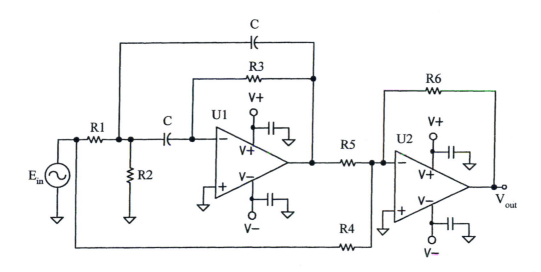

Figure 9-30 Band pass plus summer notch filter schematic

Example 9-11

Design a notch filter to meet the following specifications:

$f_l = 3kHz$

$f_h = 3.5kHz$

A_o (outside the notch) = 5

Confirm your design with a simulation.

Solution

This is the same problem as Example 9-8, so use that band pass filter.

$$C = 0.027\mu F \quad R1 = 2.4k\Omega \quad R2 = 150\Omega \quad R3 = 24k\Omega$$

Pick R5 = R6 = 12kΩ.

Rearrange the equation for the resistors around the summer.

$$R4 = \frac{2R1 \times R6}{R3}$$

$$R4 = \frac{2 \times 2.4k\Omega \times 12k\Omega}{24k\Omega} = 2.4k\Omega$$

Remember, the more closely the gains match, the deeper the notch. It's tempting to make R4 a 2.2kΩ resistor with a series 500Ω rheostat. This would allow you to tweak the gain to maximize notch depth. However, rheostats cause many problems in production. So test selecting fixed resistors, laser trimming, or just using precision resistors is usually a better solution.

The simulation schematic and results are shown in Figure 9-31. The pass band gain is 13.9dB. Specifications indicate that it should be 20 log(5) = 13.98dB. In order to obtain better resolution, the AC Analysis was modified from the default to 5000points/decade. Even with this, the closest point available to f_l is at 10.3dB, 2.97kHz. It should be at 10.9dB, 3kHz. The high frequency cut-off, f_h, occurs at 10.9dB, 3.45kHz. The notch should be at

$$f_c = \sqrt{f_l \times f_h}$$

$$= \sqrt{3kHz \times 35kHz}$$

$$f_c = 3.24kHz$$

The simulation shows it at 3.21kHz. To get a better look at the notch depth, you have to spread the Bode plot display out more than is indicated by Figure 9-31. When you do, it shows that the gain falls from 13.9dB to -41.6dB, a difference of -56dB.

9-6.2 State Variable Implementation

By adding an inverting summer to a single op amp band pass filter, you are able to build a notch filter. You can do the same with the band pass section of the state variable filter. However, Q equals the pass band gain of the state variable band pass section. So any change to Q causes a change to $A_{o\,BP}$, which then requires that you adjust the gain of the summer.

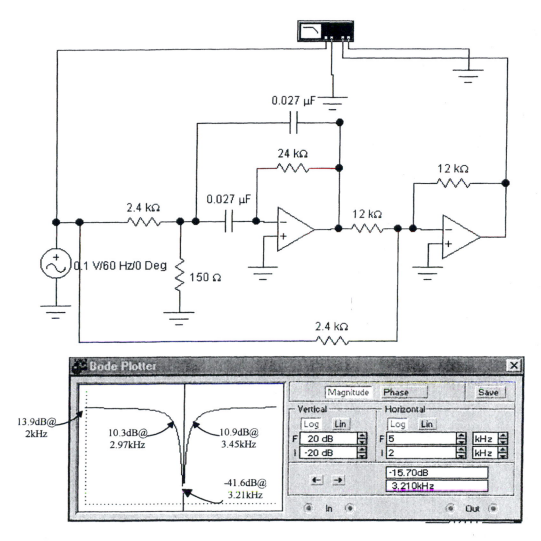

Figure 9-31 *Electronics Workbench*^{TM} simulation results for Example 9-11

A better approach is to sum the state variable's low pass and high pass outputs. With this configuration you have the advantages of the state variable filter: high, stable Q; unity gain; and independent (even electronic) adjustment of the critical frequency and Q.

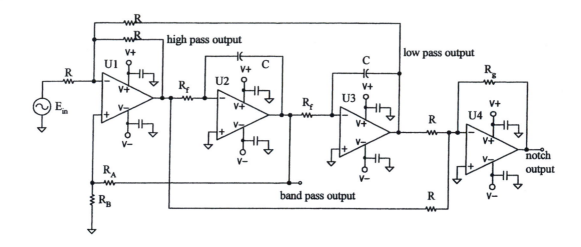

Figure 9-32 State variable implemented notch filter

The schematic is shown in Figure 9-32. Op amps U1-U3 and their associated resistors and capacitors form the state variable filter. So the critical frequency is set by C and R_f. The damping is set by R_A and R_B. The **high** pass and the **low** pass outputs are summed (and given any additional gain if desired) by the inverting summer U4.

The output from the low pass section (derived in the state variable section) is

$$V_{LP} = \frac{-\dfrac{E_{in}}{R_f^2 C^2}}{s^2 + \dfrac{3\beta}{R_f C} s + \dfrac{1}{R_f^2 C^2}}$$

The output for the high pass section is

$$V_{HP} = \frac{-s^2 E_{in}}{s^2 + \dfrac{3\beta}{R_f C} s + \dfrac{1}{R_f^2 C^2}}$$

The notch summer, U4, adds these two signals, producing

$$V_N = -\frac{R_g}{R}\left(V_{LP} + V_{HP}\right)$$

Combine these relationships.

$$V_N = -\frac{R_g}{R} E_{in} \left(\frac{-\dfrac{1}{R_f^2 C^2} - s^2}{s^2 + \dfrac{3\beta}{R_f C}s + \dfrac{1}{R_f^2 C^2}} \right)$$

Combine and rearrange terms, then divide through by E_{in}.

$$\frac{V_N}{E_{in}} = \frac{\dfrac{R_g}{R}\left(s^2 + \dfrac{1}{R_f^2 C^2}\right)}{s^2 + \dfrac{3\beta}{R_f C}s + \dfrac{1}{R_f^2 C^2}}$$

The transfer function for the notch filter is

$$\frac{V_N}{E_{in}} = \frac{A_o\left(s^2 + \omega_o^2\right)}{s^2 + \alpha\omega s + \omega_o^2}$$

Comparing terms, this means

$$A_o = \frac{R_g}{R}$$

$$\omega_o = \frac{1}{R_f C}$$

$$\alpha = 3\beta = \frac{3R_B}{R_A + R_B}$$

Example 9-12

For the state variable implementation of the notch filter shown in Figure 9-32:

$$R = R_g = 1k\Omega \quad R_f = 4.9k\Omega$$

$$C = 10nF$$

$$R_A = 18.5k\Omega \quad R_B = 1k\Omega$$

Predict the performance manually, and verify those predictions with a simulation.

Solution

$$f_o = \frac{1}{2\pi R_f C}$$

$$f_o = \frac{1}{2\pi \times 4.9k\Omega \times 10nF} = 3.25kHz$$

$$\alpha = \frac{3R_B}{R_A + R_B}$$

$$\alpha = \frac{3 \times 1k\Omega}{18.5k\Omega + 1k\Omega} = 0.1538$$

$$Q = \frac{1}{\alpha} = \frac{1}{0.1538} = 6.5$$

These are precisely the same parameters as for the single op amp, band pass filter, implemented notch of Example 9-11. The simulations provide a comparison of two implementations of the same filter. See Figure 9-33.

Initially, the Bode plot in Figure 9-33 was run from 1kHz to 5kHz. At both extremes, the gain was 0dB. This matches theory. As with Example 9-11, to get good resolution, the AC Frequency analysis was altered to provide 5000points/decade; then the range was zoomed in. The notch is at 3.25kHz. This, too, is what theory predicts. The -3dB frequencies occur at 3.0kHz and 3.5kHz, giving

$$\Delta f = 3.5kHz - 3.0kHz = 500Hz$$

$$Q = \frac{f_o}{\Delta f}$$

$$Q = \frac{3.25kHz}{500Hz} = 6.5$$

Again, theory and simulation are identical. The biggest difference between the band pass implemented notch and the state variable implemented notch is the notch depth. For the band pass version, the simulation in Example 9-11 shows a -56dB depth. The same simulation program with the same op amps at the same frequency and Q shows that the state variable implementation of the notch gives a -70dB depth. This added depth means that the signal from the state variable circuit is $\frac{1}{5}$ that of the band pass filter. In practical terms, at -70dB below 1V, the signal is just over 300μV. This is well into the inherent noise generated by many op amp circuits. You just cannot make the output much smaller.

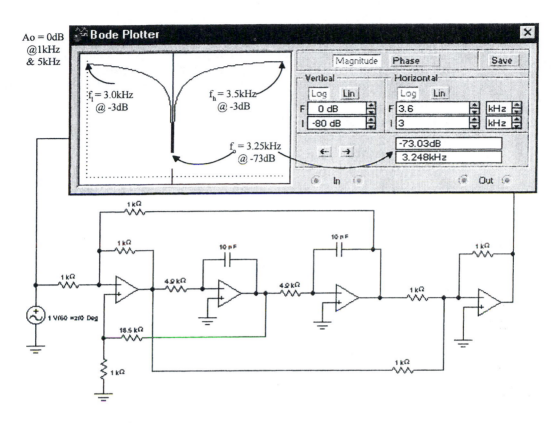

Figure 9-33 *Electronics Workbench*[TM] simulation for Example 9-12

9-7 Voltage Controlled Filter

The ability to control the critical frequency of a filter with an analog voltage or a digital command allows you to automatically and electronically alter how a signal is processed. A common example is dynamic noise reduction. Noise contaminating an audio circuit usually occurs at the higher frequencies as pop, hiss, and static. When amplifying low amplitude signals to a usable volume, this high frequency background noise is also amplified, interfering with the music. So the amplitude of the signal going into the circuit is detected and used to set the cut-off of a voltage controlled low pass filter. At low amplitudes, the cut-off frequency is low, removing the higher frequency noise. For higher amplitude inputs, the noise is too small to affect the output, so the filter's cut-off is raised to pass the high frequency notes from the cymbals and piccolos.

Voltage controlled filters may be the simplest way for a microprocessor to control an analog filter. Variation of the data into a voltage out, multiplying digital to analog converter, like the MAX512 from the third chapter, directly alters the signal within the filter, setting its critical frequency. This may be preferable to using digitally controlled potentiometers. In fact, if you use a log encoded DAC, variation in data sweeps the critical frequency exponentially across a wide frequency range.

There are two key elements within a voltage controlled filter: the integrator and the multiplier. In Chapter 5 you saw the integrator used to control the offset of an amplifier. Look at Figure 9-34.

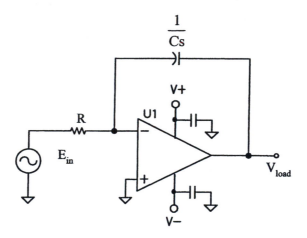

Figure 9-34 Integrator (also Figure 5-30)

In Chapter 5 you saw that

$$V_{load} = -\frac{1}{R_i C_f} \int e_{in} \, dt$$

In the Laplace domain the input current is

$$I = \frac{E_{in}}{R}$$

This current flows through the feedback capacitor, producing the output voltage.

$$V_{load} = -\frac{I}{Cs}$$

$$V_{load} = -\frac{\dfrac{E_{in}}{R}}{Cs} = -\frac{E_{in}}{RCs}$$

This gives the transfer function

$$\frac{V_{load}}{E_{in}} = -\frac{1}{RCs}$$

$$\tau = RC$$

So, this gives the transfer function of the integrator as

$$\frac{V_{out}}{E_{in}} = -\frac{1}{\tau s}$$

Three multipliers are shown in Figure 9-35.

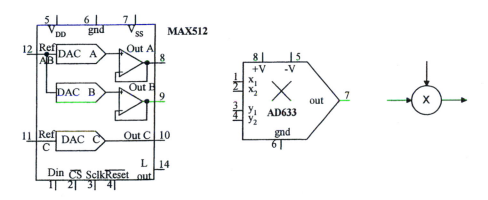

(a) Multiplying digital to   (b) Analog   (c) Multiplier simulation
    analog converter   multiplier   diagram symbol

Figure 9-35 Multiplier schematic symbols

You saw the MAX512 multiplying digital to analog converter, shown in Figure 9-35-(a), in Chapter 3. Its output is set by the product of the input to Ref AB and the digital code serially loaded.

$$v_{load} = v_{Ref}\frac{code}{256}$$

Defining the multiplier's gain as G, then its transfer function is

$$\frac{V_{load}}{E_{in}} = \frac{code}{256} = G$$

So the gain is a fraction and is set by the digital code you send to the DAC.

The AD633 is an analog multiplier that is covered more fully in Chapter 10. It has differential inputs X_1, X_2, Y_1, Y_2. For use in a simple voltage controlled filter, ground the negative inputs (X_2, Y_2), apply the signal to one input (Y_1) and a DC control voltage to the other input (X_1). With this connection, the output is

$$V_{load} = \frac{e_{X1} \times e_{Y1}}{10V}$$

The 10V is internally set by the AD633. Other analog multipliers may have a different reference, though 10V is common. Its Laplace transfer function is

$$V_{load} = \frac{E_{control}}{10V} e_{in}$$

$$\frac{V_{load}}{E_{in}} = \frac{E_{control}}{10V} = G$$

So the gain is set by the DC voltage that you apply ($E_{control}$) and is a fraction if you keep $E_{control} \le 10V$.

The schematic of a low pass, first-order, voltage controlled filter using the analog multiplier is shown in Figure 9-36.

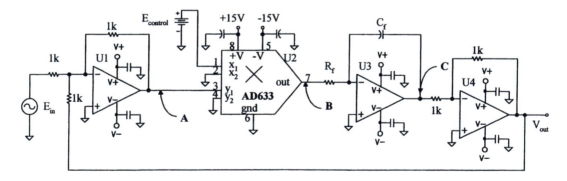

Figure 9-36 Low pass, first-order, voltage controlled filter

$$V_{out} = -V_C$$

$$V_C = -\frac{1}{\tau s} V_B \qquad\qquad \text{where } \tau = R_f C_f$$

$$V_{out} = \frac{1}{\tau s} V_B$$

$$V_B = G V_A \qquad\qquad \text{where } G = \frac{E_{control}}{10V}$$

$$V_{out} = \frac{G}{\tau s} V_A$$

$$V_A = -E_{in} - V_{out}$$

$$V_{out} = \frac{G}{\tau s}\left(-E_{in} - V_{out}\right)$$

$$V_{out} = -\frac{G}{\tau s} E_{in} - \frac{G}{\tau s} V_{out}$$

$$V_{out} + \frac{G}{\tau s} V_{out} = -\frac{G}{\tau s} E_{in}$$

$$V_{out}\left(1 + \frac{G}{\tau s}\right) = -\frac{G}{\tau s} E_{in}$$

$$V_{out}\left(\frac{\tau s + G}{\tau s}\right) = -\frac{G}{\tau s} E_{in}$$

$$\frac{V_{out}}{E_{in}} = -\frac{G}{\tau s} \times \frac{\tau s}{\tau s + G}$$

$$\frac{V_{out}}{E_{in}} = -\frac{G}{\tau s + G}$$

$$\frac{V_{out}}{E_{in}} = -\frac{\dfrac{G}{\tau}}{s + \dfrac{G}{\tau}}$$

Compare this to the classical first-order, low pass transfer function.

$$\frac{V_{load}}{E_{in}} = \frac{\omega_o}{s + \omega_o}$$

Equating coefficients

$$\omega_o = \frac{G}{\tau}$$

$$\omega_o = \frac{\dfrac{E_{control}}{10V}}{R_f C_f}$$

$$f_o = \frac{E_{control}}{2\pi RC \times 10V}$$

So the circuit is, indeed, a first-order low pass filter, and the critical frequency is directly controlled by the analog input voltage.

You could simplify the circuit (though the derivation would not be quite as direct) by grounding X_1 and driving X_2 of the AD633. This adds an inversion to the signal at point B. With that, you then do not need the inverter U4 and its two resistors.

If you replace the AD633 analog multiplier with a multiplying DAC, such as the MAX512, the G is set digitally, allowing the critical frequency to be controlled by the data from a microprocessor.

$$f_o = \frac{code}{2\pi RC \times 256}$$

Taking the output from the integrator produces a low pass filter. When you move the output to the summer, without changing any components, the same circuit is a high pass filter. The low pass filter of Figure 9-36 has been redrawn in Figure 9-37, with a few nodes relabeled.

$$V_{out} = -E_{in} - V_D$$

$$V_D = -V_C$$

$$V_{out} = -E_{in} + V_C$$

$$V_C = -\frac{V_B}{\tau s}$$

$$V_{out} = -E_{in} - \frac{V_B}{\tau s}$$

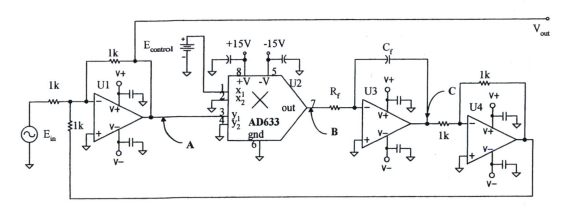

Figure 9-37 Voltage controlled first order high pass filter

$$V_B = GV_A = GV_{out} \quad \text{where} \quad G = \frac{E_{control}}{10V}$$

$$V_{out} = -E_{in} - \frac{G}{\tau s} V_{out}$$

$$V_{out} + \frac{G}{\tau s} V_{out} = -E_{in}$$

$$V_{out}\left(1 + \frac{G}{\tau s}\right) = -E_{in}$$

$$V_{out}\left(\frac{\tau s + G}{\tau s}\right) = -E_{in}$$

$$\frac{V_{out}}{E_{in}} = -\frac{\tau s}{\tau s + G}$$

$$\frac{V_{out}}{E_{in}} = -\frac{s}{s + \dfrac{G}{\tau}}$$

Compare this to the standard transfer function of a first-order high pass filter.

$$\frac{V_{out}}{E_{in}} = \frac{s}{s + \omega_o}$$

So
$$\omega_o = \frac{G}{\tau}$$

For an analog multiplier,

$$\omega_o = \frac{\dfrac{E_{control}}{10V}}{R_f C_f} \qquad f_o = \frac{E_{control}}{2\pi RC \times 10V}$$

Or, for a multiplying DAC,

$$f_o = \frac{code}{2\pi RC \times 256}$$

This is the same critical frequency as for the low pass filter.

Example 9-13

Design a first-order voltage controlled filter whose critical frequency is 1kHz, and whose input control voltage is 5V. Verify performance at control voltages of 0.1V, 5V, and 10V.

Solution

The critical frequency is set by

$$f_o = \frac{E_{control}}{2\pi RC \times 10V}$$

$$R = \frac{E_{control}}{2\pi f_o C \times 10V}$$

Pick C = 22nF.

$$R = \frac{5V}{2\pi \times 1kHz \times 22nF \times 10V} = 3.62k\Omega$$

This can be made with a 3.3kΩ resistor in series with a 330Ω resistor. The multiplier in MicroSim's *PSpice* outputs

$$v_{out} = e_x \times e_y$$

There is no 1/10V. To provide this, add a 900Ω and a 100Ω voltage divider after the multiplier symbol. The schematic of the simulation is shown in Figure 9-38, and the probe frequency responses are shown in Figure 9-39.

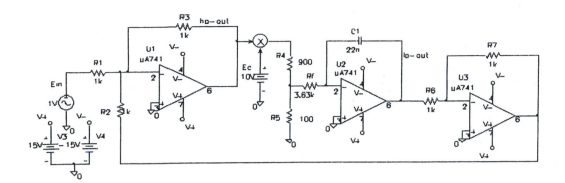

Figure 9-38 Simulation for Example 9-13

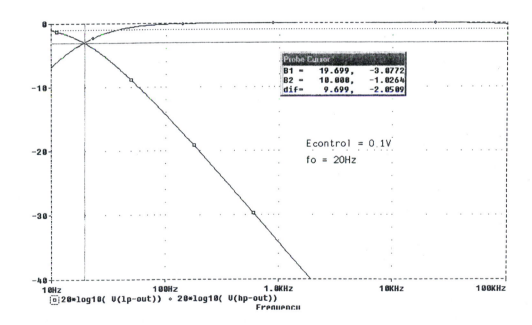

Figure 9-39 Frequency responses for Example 9-13

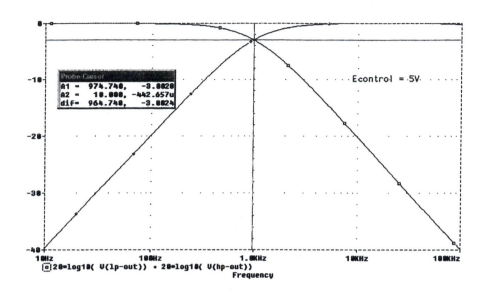

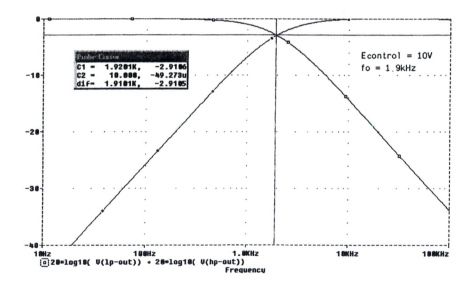

Figure 9-39 (cont.) Frequency responses for Example 9-13

9-8 Switched Capacitor Filters

The switched capacitor filter is one of the most significant advances in analog integrated circuits since the monolithic integrated circuit op amp. It combines the techniques of analog active filter design with the concepts of digital signal processing to produce filters that are compact, stable, accurate, inexpensive, easy to use, and digitally tunable.

9-8.1 Switched Capacitor Filter Fundamentals

The critical frequency of a switched capacitor filter is usually set by an external digital clock signal. This allows a digital circuit, such as a microprocessor, to control the filter's critical frequency. For simpler applications, many switched capacitor filters can generate their own clock signal with an external resistor and capacitor. The filter's critical frequency range exceeds the audio band, occasionally extending from 0.1Hz to 100kHz.

    The filter's damping and order may vary widely among ICs. The dedicated low pass filter ICs are usually internally configured to a particular type (Bessel, Butterworth, elliptical) and order. Fourth through eighth-order low pass filters are available. Most are complete within the IC. You need to provide only the input and clock (or RC pair for self-clocking).

    For greater flexibility, there is a group of state variable switched capacitor filters. Their internal configuration implements the integrators and op amps. The critical frequency is controlled by the clock. The type (low pass, band pass, or high pass), pass band gain, and damping are all determined by where you take the output and which resistors you connect around the IC. The state variable filter is inherently second-order. However, several state variable filters are available within a single IC.

    The term **switched capacitor** describes how these filters allow digital control of their critical frequency. A cell, consisting of two CMOS transistor switches and a capacitor, emulates a resistor. The switches transfer charge from the cell's input to the capacitor, and then to the cell's output. The more rapidly the switches are switched, the more charge is transferred from the input of the cell to its output. That is, changing the switching frequency alters the current through the cell, and therefore its effective resistance. This technique allows a frequency to control the resistance within the IC. Changing a resistance can be used to change the critical frequency of a filter.

    An additional benefit of this technology is a notable reduction in size. The space to integrate a switched capacitor cell (including the two CMOS switches, the control circuitry and the capacitor) which emulates a 10MΩ resistor is 1/100 the space needed to produce the same resistor using traditional analog integration techniques. This then allows the manufacture of complex analog functions under digital (clock) control in a small, easy-to-use, inexpensive package.

    Of course, the advantages of switched capacitor filters are not realized without problems. One of the problems is typical of any system in which the signal is processed in pieces (sampled), not passed through continuously. This problem is defined by the Nyquist criterion, and is called **aliasing**. The Nyquist criterion states that for a signal to be faithfully processed, that signal's highest harmonic must be sampled at least twice. Most switched capacitor filters have a fixed relationship between the clock and the critical frequency. Usually

$$f_{clock} = 50f_o$$

or

$$f_{clock} = 100f_o$$

This may depend on the IC itself or on a logic level sent to the IC. Since the sampling frequency is normally half of the clock frequency, the maximum allowable frequency into the filter is

$$f_{in\ max} = \frac{f_{clock}}{4}$$

Any signal above this frequency violates the Nyquist criterion. Aliasing occurs.

Figure 9-40 illustrates an extreme case of aliasing. The actual signal presented to the input is shown as a solid line. However, the signal is sampled at a lower frequency, only at each marked spot. So the filter perceives the signal shown by the dotted line. This phantom signal is processed as if it were an actual input, producing a significant, erroneous output. The solution to aliasing is the addition of a simple RC low pass filter connected **directly** to the input of the switched capacitor filter. The critical frequency of this first-order passive filter is

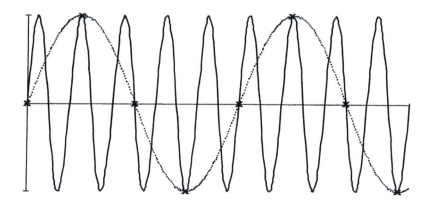

Figure 9-40 Aliasing in a sampled data system

$$f_{o\ antialiasing} = \frac{f_{sampling}}{10} = \frac{f_{clock}}{20}$$

This antialiasing filter improves the high frequency performance of low pass and band pass filters. However, for the high pass and notch configurations, the need for an antialiasing filter sets an upper limit to the frequencies that can be processed by the switched capacitor filter. If you really must pass filters above $f_{o\ antialiasing}$, then you must use an all-analog filter. The switched capacitor filter will not work.

The output signal is reconstructed from the input at the clock's frequency. So a smooth sine wave in the switched capacitor filter's pass band does not produce a smooth output. Instead, the output has stair steps, one step occurring on each cycle of the clock. These steps may be so small, or the frequency so high, that you do not even notice them. However, if they present a problem, a second, passive, low pass filter connected directly to the **output** of the switched capacitor filter removes these steps. Set the critical frequency of this output filter to be the same as the input, antialiasing filter. Of course, the series resistor of the output filter raises the output impedance of the circuit. Add a voltage follower after the output filter to lower the circuit's output impedance.

As with any circuit that uses both analog and digital signals, power supply decoupling is **critical**. A 0.1µF capacitor, connected directly from each power supply pin to circuit common, is a minimum. Separate analog and digital grounds, joined at the power supply, and a 10Ω resistor between the power bus and the decoupling capacitor(s) may also be needed. All of this added support circuitry is shown in Figure 9-41.

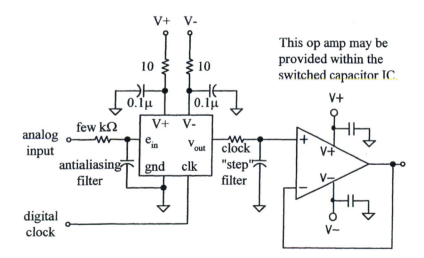

Figure 9-41 Switched capacitor support circuitry

The input impedance of many switched capacitor filters varies inversely with the clock frequency. Typically, this impedance falls no lower than 100kΩ. So, the resistor in the antialiasing filter should be on the order of a few kΩ.

The performance of a switched capacitor cell is illustrated in Figure 9-42. When the switch is connected to the input, on the left, the capacitor instantaneously charges to V_1, acquiring a charge Q_1.

$$Q_1 = CV_1$$

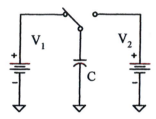

Figure 9-42 Fundamentals of a switched capacitor cell

When the clock swings the switch to the right, the capacitor is connected to the output voltage, V_2. Assuming that $V_1 > V_2$, the capacitor discharges to the new charge, Q_2.

$$Q_2 = CV_2$$

When the capacitor is connected to the input again, it again charges to Q_1. It discharges to Q_2 each time it is connected to the output. So, from cycle to cycle, charge is transferred from the input to the output of the cell. The amount of charge that is transferred is

$$\Delta Q = Q_1 - Q_2 = C(V_1 - V_2)$$

Current from the input to the output of the cell is defined as

$$I = \frac{\Delta Q}{\Delta t}$$

This transfer of charge occurs in one period of the clock, $\Delta t = T_{clock}$. Substituting for this and for ΔQ yields

$$I = \frac{C(V_1 - V_2)}{T_{clock}} = C(V_1 - V_2)f_{clock}$$

This current flows through the switched capacitor cell, from V_1 to V_2. How much current depends on not only the voltage difference, but also the size of the capacitor and the frequency of the clock. The switched capacitor cells present a resistance to the flow of this current.

$$R = \frac{V_1 - V_2}{I} = \frac{V_1 - V_2}{C\,(V_1 - V_2)f_{clock}}$$

$$R = \frac{1}{Cf_{clock}}$$

In effect, when placed between two defined voltages, the switched capacitor cell presents a resistance whose value can be changed by altering the frequency of the clock.

Figure 9-43 shows an integrator such as that used in the voltage controlled filter and the state variable filter. The input resistor has been replaced by a switched capacitor cell. The MOS switches are driven by a two-phase, non-overlapping clock.

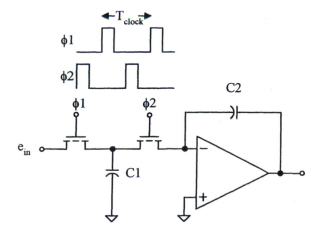

Figure 9-43 Switched capacitor integrator

When a resistor is used as the input for an integrator, its critical frequency is

$$f_o = \frac{1}{2\pi RC_2}$$

Substituting the resistive value of the switched capacitor cell for R gives

$$f_o = \frac{1}{2\pi \dfrac{1}{C_1 f_{clock}} C_2}$$

$$f_o = \frac{1}{2\pi} \frac{C_1}{C_2} f_{clock}$$

The critical frequency depends on the clock frequency and on the ratio of two capacitors. The clock frequency can be set very accurately with a crystal oscillator. The two capacitors are integrated close to each other and at the same time. Imperfections that alter the value of one capacitor also alter the value of the other proportionally. Since it is the ratio of these two values that is important, manufacturing variations cancel. Similarly, drift in one capacitor's value also appears in the other. This too cancels. Accurate, stable control of the critical frequency can be easily achieved.

9-8.2 A Low Pass, Switched Capacitor, Filter IC

The MF4 is a unity gain, fourth-order, Butterworth, low pass, switched capacitor, active filter produced by National Semiconductor. The 3dB down frequency is set by the clock. There are two models, the MF4-50 and the MF4-100. The suffix indicates the relationship between the clock frequency and the 3dB down frequency.

$$f_{-3dB} = \frac{f_{clock}}{50} \text{ for the MF4-50}$$

$$f_{-3dB} = \frac{f_{clock}}{100} \text{ for the MF4-100}$$

For a given 3dB down frequency, the MF4-100's clock is at twice the frequency as the clock for a MF4-50. There are twice as many steps in the output from the MF4-100 as in the output from the MF4-50. However, the steps from the MF4-100 are half as tall. So, the output from the MF4-100 is smoother.

The upper frequency limit for the output from the MF4 is set by the maximum clock frequency allowable for a given power supply and accuracy. These limitations are the same for both the MF4-50 and the MF4-100. For this maximum clock frequency, then, the MF4-50 has twice the 3dB down frequency that the MF4-100 has. Operating from ±5V supplies, at rated accuracy, the maximum clock frequency is 250kHz. So the

MF4-50 can be used as a low pass filter with its 3dB down frequency adjustable from 0.1Hz to 5kHz. If you choose the MF4-100, the output is smoother, but you can adjust f_{-3dB} only from 0.1Hz to 2.5kHz. For either model, the maximum clock frequency may be extended to 1MHz. However, the frequency response may peak 0.5dB as the corner frequency is reached. Lowering the supply voltage reduces the maximum allowable clock frequency.

The clock can be generated by the MF4 itself, or it can be an externally produced CMOS or TTL signal. Figure 9-44 shows how to enable self-clocking. The level shift pin (L.Sh pin 3) must be tied to V⁻. This enables the tristate buffer, turning the oscillator on. To prevent loading this tristate buffer, set

$$R_{clock} \geq 4.7k\Omega$$

This resistor and the external capacitor set the clock frequency.

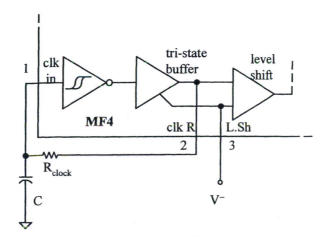

Figure 9-44 MF4 self-clocking *(Courtesy of National Semiconductor)*

$$f_{clock} = \frac{1}{1.69R_{clock}C}$$

When powering the MF4 from ±5V, either a TTL (0V to 5V) or a CMOS (-5V to +5V) external clock may be applied. To use a TTL level clock, connect the L.Sh pin to ground. This turns the tristate buffer off. Apply the TTL level clock to the clk R pin.

A CMOS level (−5V to +5V) can be used by connecting the L.Sh pin to V⁻. This enables the tristate buffer. The CMOS clock should be applied to the input of the Schmitt trigger (clk in). These two clocking schemes are shown in Figure 9-45.

When powering the MF4 from a single supply, the supply must be between 5V and 14V.

$$5V \leq V^+ \leq 14V$$

Apply the clock signal to the clk in pin. Enable the tristate buffer by connecting the L.Sh pin to the most negative voltage in the circuit (ground). The clock's high logic level must exceed 70% of the supply, and its low level must fall below 30% of the supply. If $V^+ = 5V$, a TTL clock should work as long as $V_{hi} > 3.5V$.

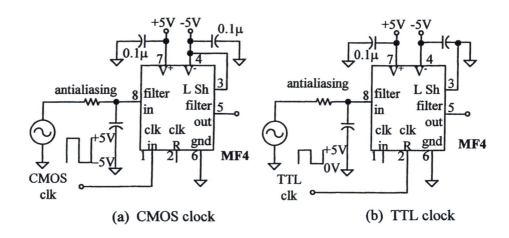

(a) CMOS clock　　　　　　　　　　　　　　(b) TTL clock

Figure 9-45 Operating the MF4 from ±5V *(Courtesy of National Semiconductor)*

If you are working from a single supply, the analog ground must be biased up to half of the supply voltage. A single supply schematic is shown in Figure 9-46. Be sure to set the input filter's cut-off well below the lowest frequency that you plan to pass into the filter. There is also a ±1V saturation voltage. So the peak signal out can be no larger than 1V below the supply and can drop no lower than 1V above ground.

The circuit in Figure 9-47 uses an MF4-50 to synthesize a sine wave. The digital input is a TTL signal whose frequency is 50 times greater than the desired sine wave frequency. This square wave is derived from a crystal oscillator, dividers, and perhaps a phase locked loop. The resulting frequency is digitally controlled, accurate, and stable.

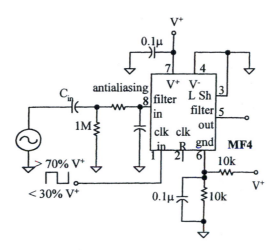

Figure 9-46 MF4 single supply operation *(Courtesy of National Semiconductor)*

That frequency sets the -3dB down frequency of the MF4-50. Any signal above this clock is severely attenuated.

This digital signal also goes to two cascaded 7490s. The first is configured to divide its input by 5. The second divides by 10. However, it is connected as a divide-by-5 followed by a divide-by-2. The result is a 50% duty cycle output at 1/50 of the clock frequency. That input to the filter is at the 3dB down frequency set by the clock.

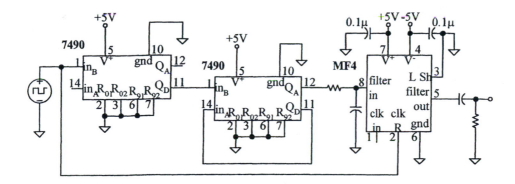

Figure 9-47 Digital sine wave generator

The MF4 converts the square wave at its input into a sine wave by rejecting the square wave's upper frequency harmonics. With a steep enough roll-off, only the fundamental passes through the filter, becoming the sine wave output.

When the input digital signal changes frequency, the filter's frequency is altered. The two dividers provide an input square wave to the filter that is always at f_{3dB}. So the output frequency always tracks the clock frequency, but it is a sine wave at 1/50 the frequency of the signal from the digital input. At a maximum clock frequency of 1MHz, a 20kHz output sine wave is produced. If the output is at all distorted, a sixth-order or an eighth-order filter may replace the MF4; or you could just add a second MF4 after the first. Give both filters the same clock signal.

The TTL signal into the MF4 goes from 0V to 5V (more or less), so its average value is 2.5V. This is a DC level that passes through the low pass filter. The output sine wave rides on this DC level. The output RC coupler blocks this DC, putting the sine wave evenly above and below ground. Set the coupler's cut-off a decade below the lowest output frequency you expect. Also, that resistor contributes to the circuit's output impedance. You may want to follow the coupler with a voltage follower.

9-8.3 A State Variable, Switched Capacitor, Filter IC

There are integrated circuits that have one, two, three, or four switched capacitor, controlled state variable filters within a single package. Some allow f_oQ to be as high as 1.8MHz. These ICs offer you the improved performance of the state variable filter configuration; digital (clock) control of the critical frequency; and low pass, band pass, high pass, all pass, and notch outputs. Q and each of the gains are adjustable with external resistors. Finally, configurations other than the classical state variable topology are possible.

Although not as densely packed as some, nor as fast as others, the MF10 from National Semiconductor is a widely used example of the switched capacitor, state variable, filter IC. The critical frequency is adjustable from 0.1Hz to over 20kHz with an externally generated clock. There is no facility to internally generate a clock. The clock-to-critical-frequency ratio is pin selectable at 50 or 100. This ratio is also dependent on the ratio of two external resistors. So, you can compensate for an inconvenient base clock frequency. The maximum clock frequency is

$$f_{clock\ max} = 1MHz$$

So, with a clock divider ratio of 50, the maximum critical frequency is

$$f_{o\ max} = 20kHz$$

However, there is also a limit on the product of the critical frequency and Q.

$$f_o \times Q < 200kHz$$

The MF10 can be powered either from a dual supply ($\pm$V) or from a single supply. In either case, the total difference in potential between the IC's power pins must remain within

$$8V < V^+ - V^- < 14V$$

This allows the popular $\pm$5V supplies. The maximum supply current is 12mA. So the $-$5V supply could easily be produced from a 7660 inverter from a single $+$5V supply.

Each output of the MF10 requires 1V headroom. Therefore, when powered from $\pm$5V, the maximum output is $\pm$4V. Each output can source 3mA but can sink only 1.5mA. So be sure to keep load resistance large enough to prevent loading.

$$R_{load} > \frac{V_{supply} - 1V}{1.5mA}$$

A considerably simplified block diagram of the MF10 is shown in Figure 9-48. It contains two second-order state variable filters. The control circuitry is shared, and connections are shown at the right end of the diagram. Notice that the MF10's integrators are **non**inverting. This change requires that the feedback from the band pass output of the MF10 be connected to the inverting input of the summer.

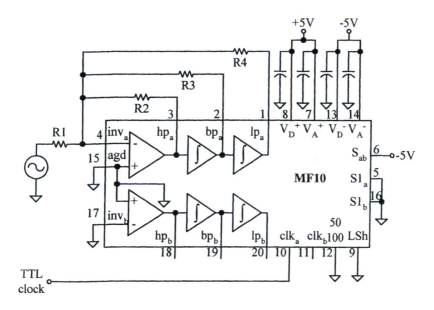

Figure 9-48 Simplified block diagram of the MF10 as a state variable filter

This alters the transfer function, setting

$$Q = \frac{R3}{R4} \sqrt{\frac{R2}{R4}}$$

$$\alpha = \frac{1}{Q}$$

The critical frequency is set by both the clock and the resistors.

$$f_o = \frac{f_{clock}}{100} \sqrt{\frac{R2}{R4}} \quad \text{or} \quad \frac{f_{clock}}{50} \sqrt{\frac{R2}{R4}}$$

Careful use of these external resistors provides quite a bit of flexibility. They can be used to correct for a clock frequency that is available, but not precisely the value you need. Or, when cascading the two stages of the MF10 to make a fourth-order filter, you can shift each filter's critical frequency to set the overall f_{-3dB} correctly. These shifts can be accomplished with the resistors, allowing you to use the same clock for each stage and still get an overall fourth-order Bessel or Butterworth or Chebyshev response at the desired f_{-3dB}.

The pass band gains are also affected by these resistors.

$$A_{o\ LP} = -\frac{R4}{R1}$$

$$A_{o\ BP} = -\frac{R3}{R1}$$

$$A_{o\ HP} = -\frac{R2}{R1} \quad @ \ f = \tfrac{1}{2} f_{clock}$$

To produce a unity gain, state variable response, set $R = R1 = R2 = R4$. This simplifies all of the relationships (although it removes some of the flexibility). The unity gain, state variable equations become

$$Q = \frac{R3}{R}$$

$$f_o = \frac{f_{clock}}{100} \quad \text{or} \quad \frac{f_{clock}}{50}$$

$$A_{o\ LP} = -1 \quad A_{o\ HP} = -1 \quad A_{o\ BP} = -Q$$

The MF10 has separate $V_A{}^+$ and $V_D{}^+$ positive power input pins. These are connected together internally, so, you must connect each to the same voltage level.

However, separating the functions allows (encourages) you to use a separate bypass capacitor on each pin. The same is true for the V_A^- and V_D^- pins. When you are using split supplies, you should connect the analog ground pin (pin 15) to a clean (analog) ground. When possible, keep your analog and digital power supply lines separate, connecting the two busses only at the IC power pins.

    The MF10 can be run from a single supply between 8V and 14V. As with the MF4, tie V^- to analog ground, divide the V^+ supply in half with a voltage divider, and tie the agnd pin to that point. The resistors in the voltage divider should be between 2kΩ and 100kΩ. Also, bypass this point to ground with a capacitor between 4.7μF and 470μF. Finally, add a 0.1μF capacitor directly from the agnd pin to analog ground.

    There are four control pins that you must properly connect. The 50/100 pin (pin 12) sets the ratio between the clock frequency and the critical frequency. When this pin is connected to V^+, the ratio is 50; when it is connected to agnd, the ratio is 100. Connecting this pin to V^- turns the filter off and limits the supply current to 2.5mA.

    The level shift pin, L.Sh (pin 9) determines the logic levels for the clock input. Normally, connect this pin to analog ground. This allows CMOS clock operation with either dual or single supplies, and TTL operation with dual supplies. When the MF10 is powered from a single supply, the agnd pin is biased to half of the supply. To accept a TTL clock when powered from a single supply, connect the L.Sh pin to the V^- pin (which is tied to the system ground).

    To configure the MF10 for state variable operation, connect S1 (pins 5 and 16) to agnd. Also, set S_A and S_B to V^-.

    As with the other switched capacitor filters, there are several other practical considerations. Power supply decoupling must be rigorously followed. An antialiasing filter is usually in order, since one of the filters is high pass. There are steps in each of the output signals. Using a clock divider ratio of 100 makes the steps smaller but also limits the maximum critical frequency.

Summary

Filter behavior is usually described in a frequency response plot. Gain (and often phase) is plotted on the vertical axis, while frequency is plotted logarithmically on the horizontal axis. Gain is expressed in decibels, which is a log ratio of powers. Assuming that the circuit's input and load impedances are equal allows you to calculate the dB gain in terms of input and load voltage. The f_{-3dB} frequency occurs where the gain has dropped by -3dB below the pass band gain. At this point there has been a 30% reduction in voltage and a 50% reduction in power delivered to the load.

    The -3dB frequency serves as the dividing line between the stop band and the pass band. Low pass filters pass low frequency and stop high frequency signals. High pass filters do just the opposite. The band pass filter passes only a certain range of frequencies. All lower and higher frequency signals are stopped. The notch filter stops the band of frequencies while passing higher and lower frequencies.

Laplace transforms allow you to easily analyze the response of circuits containing differential and integral elements. Capacitors' impedances are replaced by 1/Cs, and inductors' with Ls. You then analyze the circuit algebraically, obtaining a "gain" in terms of s. This "gain" is the transfer function and contains the full magnitude and phase response information. To decode the transfer function, substitute s = jω, separate real and imaginary parts, and then solve.

$$gain = \sqrt{real^2 + imaginary^2}$$

$$\phi = \arctan\left(\frac{imaginary}{real}\right)$$

The order of the denominator of the transfer function determines the roll-off rate in the stop band. For low pass and high pass filters, the roll-off rate several octaves from the critical frequency is 20dB/decade/order. This is the same as 6dB/octave/order.

For band pass and notch filters, the critical frequency lies at the geometric mean of the low frequency and high frequency cut-offs. This is called the center frequency. Bandwidth is the distance between the low and the high frequency cut-offs. Q is the ratio of the center frequency to bandwidth. It is a measure of the filter's selectivity.

A passive filter usually has low input impedance and high output impedance and is easily loaded and frequency shifted by any resistance connected to the output. There is poor isolation between the load and the source. To obtain second-order or higher response, with control over the response shape, you must use inductors as well as capacitors. At audio frequencies and below, these inductors are expensive and bulky and have rather poor electrical characteristics.

Active filters, especially those built with op amps, have none of these problems. However, they require power supplies and may be more expensive than simple passive filters. They are limited by the op amp's amplitude and high frequency responses. Parasitic feedback can cause oscillations in active filters. They are also subject to the same environmental hazards and limitations (heat, radiation, etc.) as any other semiconductor.

Several transfer functions were derived in the chapter. The most important ones are listed below.

First-order low pass
$$\frac{A_o \omega_o}{s + \omega_o}$$

First-order high pass
$$\frac{A_o s}{s + \omega_o}$$

Second-order low pass $$\frac{A_o \omega_o^2}{s^2 + \alpha \omega_o s + \omega_o^2}$$

Second-order high pass $$\frac{A_o s^2}{s^2 + \alpha \omega_o s + \omega_o^2}$$

Second-order band pass $$\frac{A_o \alpha \omega_o s}{s^2 + \alpha \omega_o s + \omega_o^2}$$

Second-order notch $$\frac{A_o \left(s^2 + \omega_o^2\right)}{s^2 + \alpha \omega_o s + \omega_o^2}$$

where

A_o = the pass band gain

ω_o = the critical frequency

α = the damping coefficient

The critical frequency is determined by input and feedback resistors and capacitors. Highly damped filters (Bessel) give good transient response but have slow initial roll-off. The Butterworth response gives optimum flatness in the pass band. Less damping allows pass band ripple and poor transient response, but it gives fast initial roll-off (Chebyshev). For Sallen-Key, equal component filters, the damping is set by adjusting the pass band gain. This allows damping and critical frequency to be set independently. However, the -3dB frequency is affected by damping, so you must use appropriate correction factors.

To obtain higher-order response, several single op amp filters can be cascaded. The orders add. However, each stage's damping and frequency correction factors are often considerably different from those of a single op amp filter and from those of the other stages of the filter.

Wide band, band pass filters are best obtained by cascading a low pass with a high pass filter. For $1 < Q < 10$, you can build a single op amp filter with $Q = 1/\alpha$. For this filter, several parameters are interdependent. Be sure to follow the recommended design and tuning sequence. A one op amp band pass filter gives an ultimate roll-off rate of 20dB/decade. Higher roll-off rates can be obtained by cascading. As with other cascaded filters, adjustments in α (or Q) and overall gain are required.

Combining an integrator (or two) with a summer allows you to build two very stable and flexible filters. The state variable filter is inherently second-order, providing low pass, high pass, and band pass responses simultaneously. Critical frequency and damping are independently set. Pass band gain for low and high pass are often set to unity. The band pass gain is Q. By adding a multiplier (either analog or digital) in front of the resistor in an integrator, the critical frequency can be controlled by an external analog voltage or digital data.

Two notch filters were shown. You may use a single op amp, band pass filter, combined with a summer. The output from the band pass filter is subtracted from the input, leaving a notch response at the circuit's output. The high pass and low pass outputs from a state variable can be summed to give a notch. Though requiring more parts than the first technique, this approach allows independent (and electronic) adjustment of the filter's parameters.

The switched capacitor technique provides a wide variety of sophisticated filters in a single IC package. Fourth- through eighth-order low pass filters and several state variable filters are available. The switched capacitor filter uses two CMOS switches and an on-board capacitor to emulate a resistor. Varying the frequency that you throw the switch varies the effective resistance. Placed in an integrator configuration, this resistance then allows an external frequency to control the filter's critical frequency. Because the signal is sampled, the Nyquist criterion, antialiasing, and steps in the output all must be handled. However, the ability to alter the filter's critical frequency electronically allows some unique applications. Most switched capacitor filters are limited to the audio frequency range (or just above) and work best from ±5V supplies. Two examples were investigated.

Problems

9-1   a. Convert the following ratio gains to dB: 0.03, 0.125, 8, and 50.
       b. Convert the following dB voltage gains to ratio: -22dB, 18dB, -65dB, and 45dB.

9-2   For the circuit in Figure 9-49:
       a. Draw the Laplace transformed circuit.
       b. Analyze the circuit to obtain its transfer function.
       c. Substitute $s = j\omega$ into the transfer function, and derive the gain magnitude equation.
       d. Derive the gain phase equation from the transfer function.
       e. For $R = 2.2k\Omega$ and $C = 0.1\mu F$, plot the magnitude and the phase frequency response.

9-3   Sketch the frequency response of each of the following filters:
       a. First-order low pass with $A_o = 10$ and $f_{3dB} = 300Hz$.
       b. Third-order high pass with $A_o = 12dB$ and $f_{3dB} = 1kHz$.
       c. Fourth-order band pass (second-order on each edge) with $A_o = 20dB$, $f_o = 800$, and $Q = 12$.
       d. Notch with $A_o = 6dB$, $f_h = 30kHz$, $f_l = 25kHz$, and notch depth $= -30dB$.

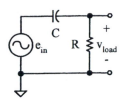

Figure 9-49 Schematic for Problem 9-2

9-4   Express the roll-off rate for Problem 9-3 (a-c) in dB/decade and dB/octave.

9-5   Describe two **specific** uses for a:
    a.   low pass active filter.
    b.   high pass passive filter.
    c.   band pass active filter.

9-6   List four disadvantages of passive filters and explain how active filters overcome each.

9-7   List four disadvantages of active filters.

9-8   For the circuit in Figure 9-50:
    a.   Analyze the circuit to obtain its transfer function in the Laplace domain.
    b.   Substitute $s = j\omega$ into the transfer function, and derive the gain magnitude and phase equations.

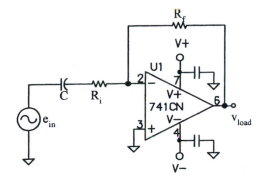

Figure 9-50 Schematic for Problem 9-8

9-9   Repeat Problem 9-8 for the circuit in Figure 9-51.

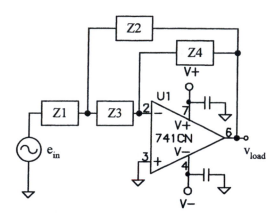

Figure 9-51 Schematic for Problem 9-9

9-10   Complete the derivation for the circuit in Figure 9-51, determining A_o, α, and ω_o in terms of R1, R2, C1, and C2, if:

$$Z1 = R1 \quad Z2 = \frac{1}{sC1} \quad Z3 = R2 \quad Z4 = \frac{1}{sC2}$$

9-11   Given a second-order, low pass, Sallen-Key, equal component, active filter with:

$$R_i = 10k\Omega, \quad R_f = 2.7k\Omega, \quad R1 = R2 = 1.1k\Omega, \quad C = 0.022\mu F$$

calculate the critical frequency, f_{-3dB}; A_o; damping coefficient; and response type.

9-12   Design a second-order, low pass, Sallen-Key, equal component, 3dB Chebyshev, active filter. Set $f_{-3dB} = 5kHz$.

9-13   For the circuit in Figure 9-52, calculate the following:
a. critical frequency for each stage.
b. damping for each stage.
c. response type.
d. overall filter f_{-3dB}.
e. overall filter roll-off rate.
f. overall filter pass band gain.

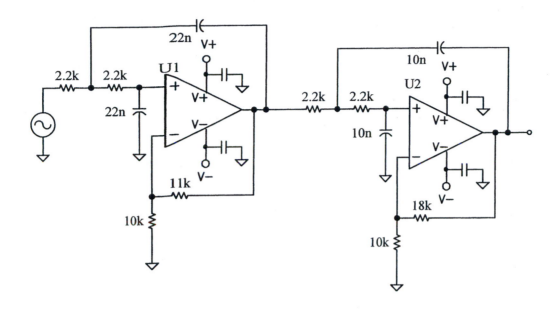

Figure 9-52 Schematic for Problem 9-13

9-14 Design a low pass Butterworth filter with a -3dB frequency of 100Hz. At 200Hz, the gain must be at least −18dB below the pass band gain.

9-15 Given a second-order, high pass, Sallen-Key, equal component, active filter with:

$$R_i = 10k\Omega \quad R_f = 2.7k\Omega \quad R1 = R2 = 1.1k\Omega \quad C = 0.022\mu F$$

   a. Calculate the critical frequency, f_{-3dB}, A_o, damping coefficient, and type.
   b. Compare the results with those obtained in Problem 9-11.

9-16 **a.** Design a second-order, high pass, Sallen-Key, equal component, 3dB Chebyshev, active filter. Set $f_{-3dB} = 5kHz$.
   b. Compare the results to those obtained in Problem 9-12.

9-17 **a.** Design a high pass Butterworth filter with a -3dB frequency of 100Hz. At 50Hz, the gain must be at least −18dB below the pass band gain.
   b. Compare the results to those obtained in Problem 9-14.
   c. Calculate the gain bandwidth and slew rate requirements for the op amps.

9-18   Derive the transfer function for the band pass filter in Figure 9-53. Compare the results of your derivation to the transfer function of Figure 9-21.

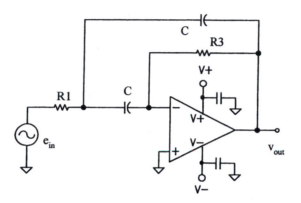

Figure 9-53 Two-resistor, single op amp, band pass filter for Problem 9-18

9-19   Derive the transfer function for the circuit in Figure 9-54.

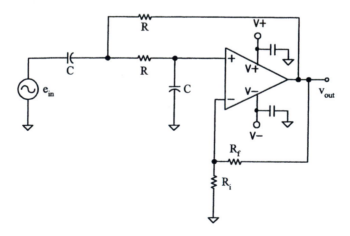

Figure 9-54 Schematic for Problem 9-19

9-20   For the circuit in Figure 9-55, calculate f_o, Δf, Q, f_l, f_h, and A_o.

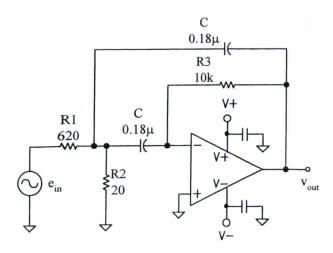

Figure 9-55 Schematic for Problem 9-20

9-21 Design a band pass filter with a center frequency of 400Hz, Q = 20, A_o = +10, and a 40dB/decade roll-off rate. What are the gain bandwidth and slew rate requirements of the op amps?

9-22 Design a band pass filter with f_l = 500Hz, f_h = 2000Hz, and a roll-off rate of 24dB/octave.

9-23 Design a 3dB Chebyshev active filter using a three op amp, state variable configuration. Set f_o = 5kHz. Calculate $f_{-3dB\ low\ pass}$ and $f_{-3dB\ high\ pass}$.

9-24 **a.** Design a band pass filter with a center frequency of 400Hz and Q = 20, using a three op amp, state variable filter.
  b. Sketch the low pass and high pass outputs.

9-25 Analyze the circuit in Figure 9-56.
  a. Calculate A_o for each output, α, Q, and f_o.
  b. Plot the frequency response for each output.

9-26 Derive the three transfer functions for the state variable filter in Figure 9-26. Define all of the appropriate parameters.

9-27 Design a 60Hz notch filter with a pass band gain of 0dB and Q = 10 using the band pass and summer topology.

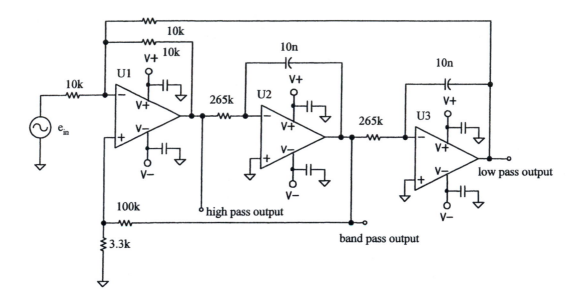

Figure 9-56 Schematic for Problem 9-25

9-28   Analyze the circuit in Figure 9-57. Calculate f_o, A_o, and Q.

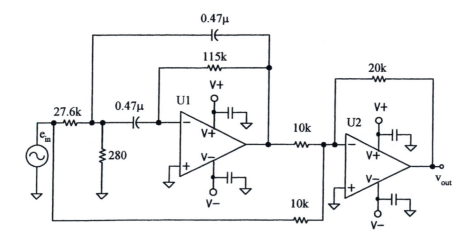

Figure 9-57 Schematic for Problem 9-28

9-29　Design a 60Hz notch filter with a pass band gain of 0dB and Q = 10 using the state variable topology.

9-30　Analyze the circuit in Figure 9-58.
　　a.　Calculate f_o, A_o, and Q.
　　b.　Plot the frequency response.

9-31　Design a voltage controlled band pass filter. Control voltage E1 sets f_l, and control voltage E2 sets f_h. When each control voltage is set to 5V (midrange), $f_l = 500Hz$ and $f_h = 2000Hz$.

9-32　For the circuit that you designed in Problem 9-31, calculate f_o, Δf, and Q:
　　a.　E1 = 5V, and E2 = 5V.
　　b.　E1 = 7.5V, and E2 = 3.3V.
　　c.　E1 = 7.5V, and E2 = 7.5V.

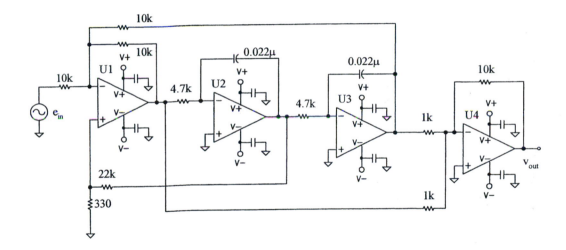

Figure 9-58 Schematic for Problem 9-32

9-33　Design a low pass filter using the MF4 with a −3dB frequency that is sweepable from 0.1Hz to 1kHz by a TTL clock. Include an antialiasing filter and rigorous power supply decoupling, but omit the output step filter.
　　a.　Implement the design using ±5V power supplies.
　　b.　Implement the design using a single +12V power supply with a 3V offset.

9-34   Alter the designs of Problem 9-33 to allow the MF4 to self-clock with −3dB
frequencies of 0.1Hz, 10Hz, and 1kHz.   Calculate appropriate timing
components for each −3dB frequency.   Indicate any other required changes to
the schematic.

9-35   Analyze the circuit in Figure 9-59.
   a.   Calculate α, Q, and A_o for each output; f_o; and the peak gain for the low and
high pass outputs.
   b.   Sketch the frequency response of each output.

9-36   Design a 60Hz notch filter with a pass band gain of 0dB and a Q of 10 using the
MF10 and the state variable topology.   Be sure to provide the clock circuit
design.   Use ±5V power supplies.   Be sure to select a clock design and an
external op amp that operate well with these supplies.   Don't forget decoupling
and antialiasing.   What is the circuit's upper frequency limit?   Explain.

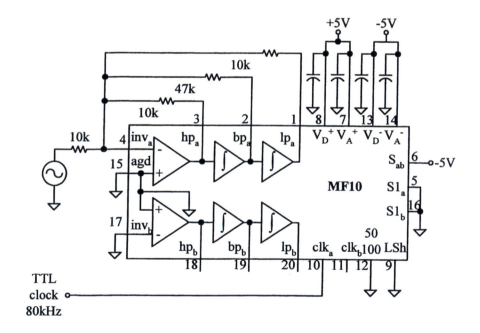

Figure 9-59 Schematic for Problem 9-35

Sallen-Key, Low Pass, Active Filter Lab Exercise

A. Setup

1. Design a second-order, low pass, Sallen-Key, equal component, Butterworth filter. Set $f_{-3dB} = 1kHz$.

2. Build the circuit that you have designed.

3. Measure the capacitors. Their values must match within 5%.

4. Adjust the frequency determining resistances (potentiometers) for the values of the capacitors that you have. These resistors must also match within 5%.

5. Apply power, and then a $1V_{RMS}$, 100Hz sine wave to the filter.

6. Verify that the pass band gain is correct. If it is not, alter the gain setting resistors to produce the correct gain.

B. Butterworth Evaluation

1. Monitor the input and the output on the oscilloscope. Also measure the output amplitude with the digital voltmeter.

2. Complete the following "quick look" table:

   $A_o = $ _____ at _____ Hz.

   $f_{-3dB} = $ _____ Hz at _____ gain. (Find this point accurately.)

   $f_o = $ _____ Hz at _____ ° phase shift.

   A at 4kHz = _____ dB

   A at 8kHz = _____ dB

   Roll-off rate = _____ dB/octave

3. Compare each of these values to its theoretical value.

4. When the filter is working correctly, adjust the frequency as indicated below. Measure and record the filter's gain at each frequency: 100, 200, 300, 400, 500, 600, 700, 800, 900, 1000, 2000, 3000, 4000, 5000, 6000, 7000, 8000, 9000, 10,000, and 20,000.

5. Plot the dB gain versus frequency on a semilog graph.

6. Apply a 100Hz, $1V_P$, $0V_{DC}$ square wave. Record the output wave shape. Discuss ringing and circuit damping.

C. Chebyshev Filter Evaluation

Repeat **all** steps of Sections A and B for a 3dB Chebyshev filter with $f_{-3dB} = 600Hz$.

State Variable, Active Filter Lab Exercise

A. Design

Design a unity gain, state variable active filter with $f_o = 800Hz$ and $Q = 5$.

B. Evaluation

1. Build the circuit that you have designed. Apply power.

2. Set the input signal to $1V_{RMS}$, 100Hz. Monitor the outputs with a multimeter and one channel of the oscilloscope. Monitor the input with the other channel.

3. Verify and record that the signals at the low pass, band pass, and high pass outputs are correct.

4. Set the input signal's frequency to 10kHz.

5. Verify and record that the signals at the low pass, band pass, and high pass outputs are correct.

6. Compare the $A_{low\ pass\ @\ 100\ Hz}$ and the $A_{high\ pass\ @\ 10kHz}$ with the theoretical values.

7. Monitoring the band pass output, alter the input signal frequency to determine its center frequency, gain at the center frequency, low frequency cut-off, high frequency cut-off, bandwidth, and Q. Record each of these parameters.

8. Compare the $A_{o\ band\ pass}$, $f_{o\ band\ pass}$, and Q that you just measured with the theoretical values of these parameters.

9. Monitor the low pass output. Alter the input signal's frequency to determine and record the following: peak gain, frequency at the peak gain, and f_o (frequency where there is 90° phase shift).

10. Compare the $f_{o\ low\ pass}$ with the theoretical value for this parameter.

11. Monitor the high pass output. Alter the input signal's frequency to determine and record the following: peak gain, frequency at the peak gain, and f_o (frequency where there is 90° phase shift).

12. Compare the $f_{o\ high\ pass}$ with the theoretical value for this parameter.

13. When you are satisfied that your filter is functioning properly, complete a frequency response test. Measure the gain at the low pass, band pass, and high pass outputs for each of the following frequencies: 10, 40, 80, 100, 200, 300, 400, 500, 600, 700, 800, 850, 900, 1000, 2000, 3000, 4000, 6000, 8000, and 10,000.

14. Plot all three dB gains on the same semilog graph and determine each roll-off rate of each output.

10

Nonlinear Circuits

Most of the circuits you have seen so far have had a linear relationship between their input and their output. In fact, considerable effort was often expended to overcome nonideal characteristics and to linearize the circuits' responses.

However, there is a class of circuits that intentionally has a highly nonlinear response. Ideal rectifiers, peak detectors, transfer curve synthesizers, log and antilog amps, multipliers, and variable gain amplifiers are all nonlinear circuits.

These nonlinear circuits find diverse applications. Industrial instrumentation, general signal processing, communications, and audio electronics all use nonlinear circuits to enhance their overall performance.

Objectives

After studying this chapter, you should be able to do the following:
1. Explain how enclosing a diode in an op amp's negative feedback loop eliminates the diode's offset.
2. Analyze a given half wave and full wave ideal rectifier circuit, calculating Z_{in} and V_{load}.
3. Design a half wave and a full wave ideal rectifier to meet given specifications.
4. Explain why two diodes are necessary in the ideal rectifier circuit.
5. Qualitatively explain the operation of a peak detector.
6. Design a peak detector.
7. List three sources of peak detector error in the tracking mode and three sources of error in the hold mode.
8. Analyze a given transfer curve synthesizer, calculating each slope and each breakpoint voltage and drawing the transfer curve.
9. Given a transfer function, design a synthesizer that produces that function.
10. Explain the significance of the root mean squared (RMS) value of a voltage.
11. Calculate the true RMS value of a variety of wave shapes.
12. Illustrate the proper use of a true RMS converter IC.
13. Draw the schematic of a log amp and an antilog amp.
14. For each, describe the operation, derive the output voltage equation, and discuss the advantages and limitations.
15. Describe two different techniques of IC multiplication, including the advantages and disadvantages of each.

16. Draw the schematic and explain the operation of each of the following applications of the multiplier:
    a. frequency doubler
    b. true power converter
    c. voltage controlled amplifier
    d. balanced and conventional amp
    e. divider
    f. square root converter
17. Explain the difference between a transconductance amplifier and an operational amplifier.
18. Discuss the effects of programming on an operational transconductance amp (OTA).
19. Analyze and design the following OTA circuits:
    a. electronic parameter control
    b. voltage amplifier
    c. programmable resistor
    d. variable gain amp
    e. tunable active filter
    f. multiplier

10-1 Diode Feedback Circuits

One of the simplest ways of producing a nonlinear response in an amplifier is to include a diode in its negative feedback loop. Properly positioned, diodes, resistors, and voltage references can produce an ideal rectifier (without the traditional 600mV loss), a circuit that finds and holds the peak value of its input, and a circuit whose input-to-output (transfer) relationship you can customize.

10-1.1 Ideal Rectifier

In electrical measurements and control, it is often necessary to convert an AC signal to an equivalent DC. The simplest solution is to use a diode rectifier. However, to turn on the rectifier, 200mV to 600mV of the signal must be dropped across the diode. This portion of the input never appears at the output. Even for a $10V_p$ signal, that loss across the diode causes a 2% error. Converting AC signals in the millivolt range cannot be done with a simple diode.

      The circuit in Figure 10-1 allows AC to DC conversion of small signals. It is a conventional amp with a diode added within the negative feedback loop. Notice, however, that the output is taken from the diode (in the middle of the feedback loop), **not** at the op amp's output.

      When the input voltage is zero, the op amp's output is zero (ignoring offsets), and the diode is off. This means that the negative feedback loop has been broken, and the op amp is operating open loop. In this open loop condition the input signal is inverted and amplified by the op amp's open loop gain (typically 2×10^5 for a 741). A drop of $-3\mu V$ at the inverting input of the op amp sends its output to 0.6V. This turns the diode on. Consequently, the offset of the diode is provided by the op amp, requiring only $3\mu V$ of the input signal.

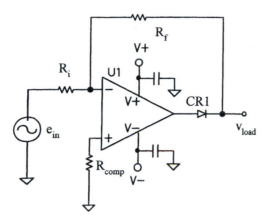

Figure 10-1 Simple ideal rectifier

Once the diode is on, the circuit functions as a regular inverting amplifier, with an output of

$$V_{load} = -\frac{R_f}{R_i} e_{in}$$

for the rest of the negative input cycle. The op amp's output follows v_{load}, but is approximately 600mV higher.

When the input voltage goes positive, the op amp is rapidly driven into negative saturation. The diode is off. At first you may be tempted to conclude that the circuit's output must go to zero. However, since the op amp has no negative feedback, its inverting terminal is no longer held at virtual ground. Resistors R_f and R_i are the only things between the input and the output. They form a voltage divider with the input's source impedance and the load impedance. During the negative input cycle,

$$V_{load} = \frac{R_{load}}{R_{source} + R_i + R_f + R_{load}} e_{in}$$

This is quite different from the zero output desired.

This simple ideal rectifier has one other problem. During the positive input cycle, with CR1 off, the op amp is saturated. When the cycles switch, it may take a significant time to pull the op amp out of negative saturation and to slew to 0.6V. This delay significantly limits the high frequency response.

Both of these problems are solved by adding a second diode and feedback resistor as shown in Figure 10-2. During the negative input cycle, the op amp drives CR1 on and then functions as an inverting amplifier, as you have already seen. This produces the positive output. When the input goes positive, the op amp's output swings negative to -0.6V. This turns on CR2 and turns off CR1. The negative feedback loop is now provided through R_{f1} and CR2. The negative cycle is produced at CR2's anode, just as the positive half cycles are produced through CR1.

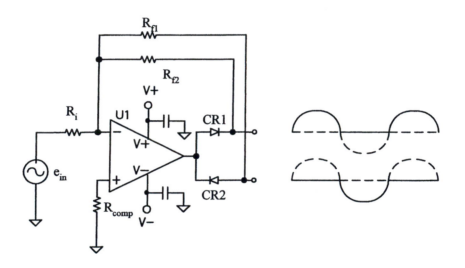

Figure 10-2 Improved ideal rectifier

Since the negative feedback loop is now complete for both half cycles, the inverting input remains at virtual ground. When CR1 is off, there is no current flow through R_{f2}, so there is no voltage across it. With the left end of R_{f2} at virtual ground, and no voltage dropped across R_{f2}, the output at CR1 must be at 0V during the positive half cycle of the input. Also, with both cycles providing negative feedback, the op amp never saturates. This improves speed.

You do not have to use both outputs. Use only the one that provides the polarity voltage needed. However, be sure to include both diodes and feedback resistors. Otherwise, during the off output cycle, the op amp saturates, and the output voltage does not go to zero.

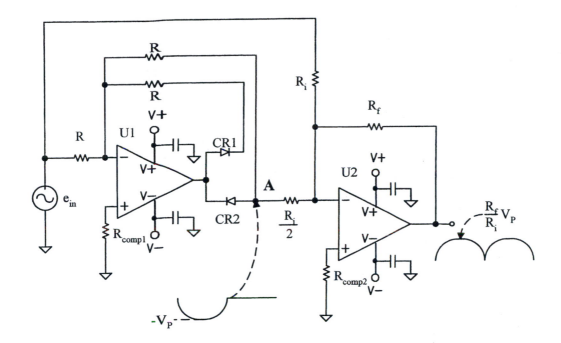

Figure 10-3 Full wave ideal rectifier

The circuit in Figure 10-2 provides both **half** wave outputs. You can obtain a **full** wave rectified output by adding an inverting summer, as shown in Figure 10-3. The half wave rectifier built around U1 works as you have already seen, producing a negative half wave at point A.

Op amp U2 is configured as an inverting summer. The input from the ideal rectifier is given a gain of

$$\text{gain from point A} = -\frac{R_f}{\dfrac{R_i}{2}} = -2\frac{R_f}{R_i}$$

The signal from the input around the top of U1 receives a gain of

$$\text{gain of input} = -\frac{R_f}{R_i}$$

So the output from the summer is

$$V_{load} = -\frac{R_f}{R_i}\left(v_{input} + 2v_{half\,wave}\right)$$

During the positive input half cycle, the rectifier produces a negative output half wave cycle. The output from the summer is

$$V_{load} = -\frac{R_f}{R_i}\left[+e_{in} + 2(-e_{in})\right] = \frac{R_f}{R_i}e_{in}$$

A positive half cycle is produced at the output.

During the negative input half cycle, the ideal rectifier holds point A at 0V. The output from the summer is

$$V_{load} = -\frac{R_f}{R_i}\left[-e_{in} + 0\right] = \frac{R_f}{R_i}e_{in}$$

Again, a positive half cycle has been produced at the output. The full wave rectifier of Figure 10-3 actually takes the absolute value of the input.

$$V_{load} = \frac{R_f}{R_i}|e_{in}|$$

When you are building analog computational circuits, this could be quite handy.

10-1.2 Peak Detector

The ideal rectifiers of the previous section yield an output proportional to the average value of a repetitive waveform. There are several applications where the **peak** value of a signal must be detected and then held for later analysis. The peak of a pulse wave train is often of more interest than its average value. When you are performing destructive testing, it is critical to track and hold the maximum signal. Spectral and mass spectrometer analysis also requires the use of peak detectors.

The principle of including a diode in the feedback loop to overcome its offset is used in building peak detectors. Look at Figure 10-4. The op amp is configured as a voltage follower. Diode CR1 is used to keep the op amp from going into negative saturation for negative inputs. This improves the circuit's speed of response and lowers the reverse voltage rating needed for CR2. Diode CR2 is the ideal rectifier, enclosed in the op amp's negative feedback loop.

With no charge on the capacitor, $+3\mu V$ at the input causes $+0.6V$ at the op amp's output. This turns the diode on. When the diode goes on, the op amp functions as a voltage follower. The output tries to track the input. Capacitor C charges through Z_{out} of the op amp and $R_{diode\,on}$ to the input, e_{in}.

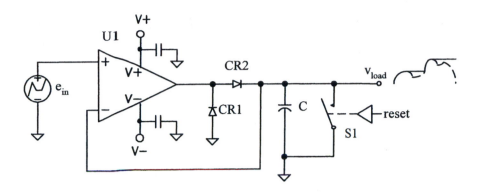

Figure 10-4 Peak detector

The charge rate is limited by C, $R_{\text{diode on}}$, $Z_{\text{op amp out}}$, and $I_{\text{max op amp}}$. If the input voltage drops below the output voltage, diode CR2 goes off. Then the op amp heads to negative saturation but is clamped at about ground when diode CR1 goes on. The output voltage is held at its highest value by the capacitor since it has no place to discharge.

There are several sources of error in the tracking mode. The op amp must be able to operate stably when driving a capacitive load. Also, the circuit must not be underdamped. This means that when you apply a square wave, the circuit must not overshoot. Such overshoots would be erroneously captured as peaks. To test for proper damping, short out diode CR2 and apply a square wave input. Observe the output for overshoots. The op amp must have a slew rate quick enough to track the fastest rise that you expect. But be careful. Faster op amps tend to be slightly underdamped, so breadboard testing may be necessary to find the right op amp.

Leakage is the largest source of error in the hold mode. Any current drawn by the load discharges the capacitor, causing an error called **droop**. You may need to add a high impedance buffer (op amp voltage follower) after the capacitor and its discharge switch to isolate them from the load. The bias current into that buffer op amp discharges the capacitor, just as does the bias current into the inverting input of U1. For precision peak detectors, carefully select the op amps to have low bias currents, low drift, and high slew rate.

The reverse bias leakage current through the diode and the capacitor's leakage current may cause noticeable droop over long hold intervals. Choose a high-quality **film** capacitor. Finally, to discharge the capacitor and reset the circuit so that it can detect a new peak, a switch is normally used. A transistor or an analog switch is convenient, but causes more leakage and droop than a relay.

One tempting solution to these hold mode droop problems is simply increasing the value of the capacitor. Increasing C provides more charge and therefore lowers the effect of leakage currents. However, the larger C is, the more slowly it accepts charge during the tracking mode. The capacitor may not charge to a peak before the voltage has fallen.

Example 10-1

Explain the operation of the circuit in Figure 10-5.

Solution

Op amp U1, CR1, CR2, and C form a positive peak detector. The LM311, U2, is a noninverting comparator. The output from the peak detector is divided by R2 and R3, placing $0.9V_P$ as the reference voltage to the comparator.

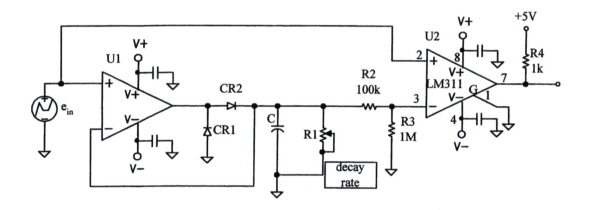

Figure 10-5 Schematic for Example 10-1

This circuit is an adaptive comparator. It provides a logic level transition for every cycle at the input. Switching occurs near the peak of the input, rather than at the zero crossing. This minimizes false switching caused by noise on the ground line or by other harmonics on the input Both of these are problems in sensitive systems. It also allows reliable switching to occur as the input varies in amplitude, providing one TTL compatible output cycle for each input wave, even as the input gradually fades. This is particularly a problem in audio applications, since the volume of the music or the speaker varies considerably. You may want to include a little hysteresis by adding both a large resistor from the output to the comparator's noninverting input and a small resistor between that input and the source voltage.

10-1.3 Transfer Curve Synthesizer

The transfer curve of a circuit is a plot of the circuit's input voltage or current on the horizontal axis versus its output voltage or current on the vertical axis. For the bulk of circuits it is desirable for this transfer curve to be a straight line. This means that changes on the output are directly proportional to changes on the input.

There are times, however, when a linear transfer curve is **not** desirable. If a transducer's response is itself nonlinear, running its output through a circuit with a complementary, nonlinear transfer curve results in an overall linear response. Squaring and taking the square root of analog signals both can be done with nonlinear transfer curves. The same is true for taking the log or antilog of analog signals. Wave shaping, from a triangle wave input to a sine wave output, is easily done with a circuit whose transfer curve is nonlinear.

Figure 10-6(a) is the transfer curve of a squarer. It is obviously nonlinear. However, it can be reasonably well approximated by a series of straight lines, each with a progressively steeper slope. This piece-wise linear approximation of the squarer is shown in Figure 10-6(b). It has only five segments and forms a good fit. The more line segments (pieces) that you use, the closer the linear approximation is to the actual nonlinear transfer curve.

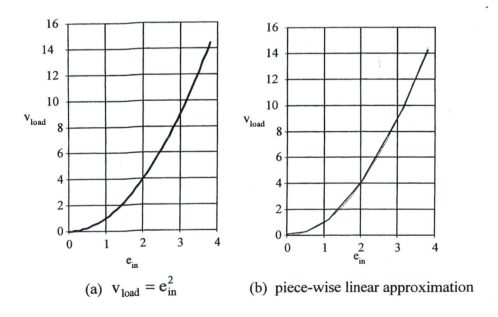

(a) $V_{load} = e_{in}^2$ (b) piece-wise linear approximation

Figure 10-6 $V_{load} = e_{in}{}^2$ transfer curve

The circuit in Figure 10-7 has a two-segment, piece-wise linear transfer curve. For small values of negative input voltage, V_{ref} holds the cathode of the diode positive. This keeps it off. Resistor R2 sees an open and does not affect the input circuit. The result is a simple inverting amplifier with R_f as the feedback resistor and R1 as the input resistor. The gain is

$$\frac{V_{load}}{e_{in}} = -\frac{R_f}{R1}$$

This is also the slope for segment 1 of the transfer curve.

$$S1 = -\frac{R_f}{R1}$$

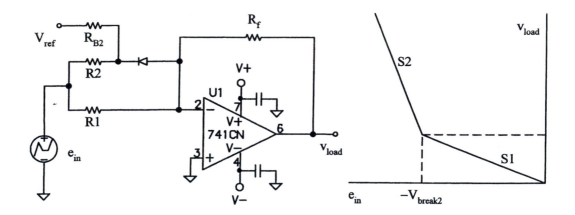

(a) Two segment piece-wise linear circuit   (b) Transfer curve

Figure 10-7 Two-segment, piece-wise linear transfer curve synthesizer

As the input voltage becomes more and more negative, the voltage at the cathode of the diode falls. At -$V_{break\ 2}$, the diode goes on. Ignoring the 0.6V offset of the diode and its on resistance, you can see that this places R2 in parallel with R1. The amplifier's gain changes to

$$\frac{V_{load}}{e_{in}} = -\frac{R_f}{\dfrac{R1 \times R2}{R1 + R2}}$$

which is also the slope of the second line segment.

$$S2 = -\frac{R_f}{\dfrac{R1 \times R2}{R1 + R2}}$$

$$= -R_f \times \frac{R1 + R2}{R1 \times R2}$$

$$= -\left(\frac{R_f R1}{R1 + R2} + \frac{R_f R2}{R1 + R2}\right)$$

$$S2 = -\left(\frac{R_f}{R1} + \frac{R_f}{R2}\right)$$

The slope has automatically increased. Consequently, you can increase the slope of the circuit's transfer curve by causing diodes to switch in resistance to parallel the input resistor, lowering R_i, and increasing gain.

For a three-segment curve, another input resistor, R3, diode, and bias resistor, R_{B3}, network must be added in parallel with R1. When that network switches in, the slope rises to

$$S3 = -\left(\frac{R_f}{R1} + \frac{R_f}{R2} + \frac{R_f}{R3}\right)$$

This can be generalized for as many segments as you need. Each additional segment requires an input resistor, a bias resistor, and a diode network added in parallel with R1.

$$S_n = -\left(\frac{R_f}{R1} + \frac{R_f}{R2} + \frac{R_f}{R3} + ... + \frac{R_f}{R_n}\right)$$

The input resistors set the slopes. The break points are determined by V_{ref}, R_B, and the input resistors. Look again at Figure 10-7(a). The inverting input, and therefore the diode's anode, are held at virtual ground by the op amp's negative feedback. Resistor R2, R_{B2}, the diode, and V_{ref} can be redrawn as shown in Figure 10-8.

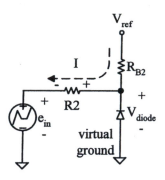

Figure 10-8 Break point analysis schematic

As long as the voltage across the diode is positive, the diode is off. When V_{diode} is driven to ground (actually -0.6V) by e_{in}, the diode goes on and the next line segment begins.

$$I = \frac{V_{ref} - e_{in}}{R2 + R_{B2}}$$

Summing the bottom loop gives

$$V_{diode} = I\,R2 + e_{in}$$

$$= \frac{V_{ref} - e_{in}}{R2 + R_{B2}} R2 + e_{in}$$

For simplicity, assume that $V_{diode} = 0V$ when it goes on.

$$0 = \frac{V_{ref} - e_{in}}{R2 + R_{B2}} R2 + e_{in}$$

After a bit of algebraic manipulation, this equation can be solved for the value of e_{in} that drives V_{diode} to zero, switching it on.

$$V_{break2} = e_{in} = -V_{ref}\frac{R2}{R_{B2}}$$

$$R_{B2} = -R2\frac{V_{ref}}{V_{break2}}$$

You should be aware of several observations and precautions. Placing the diode network in the input circuit allows the slope of the transfer curve to **increase** at each break point. If the slope must decrease with increasing input voltage, you must use a different circuit. Second, the circuit is inherently an inverter. This puts the transfer curve in the second quadrant for negative inputs (positive outputs) and in the fourth quadrant for positive inputs (negative outputs). Adding an inverting amplifier either before or after this circuit gives you an overall noninverting circuit (first and third quadrants).

Each time a diode switches on (at each break point), the input impedance is lowered. Be careful that the circuit does not load the source at high input voltage levels.

The derivation above assumed that

$$V_{diode\,on} = 0V$$

$$R_{diode\,on} = 0\Omega$$

These approximations should cause no problems as long as you insure that

$$V_{break} \gg 0.6V$$

$$R1, R2, R3, \ldots \gg R_{diode\,on}$$

Finally, the circuit in Figure 10-7(a) accommodates only negative inputs. To produce a similar transfer curve for positive inputs, you must add input resistors, diodes, and bias networks in parallel with R1. For these positive networks, reverse the diodes and use a negative reference to hold the diodes off. A seven-segment transfer curve synthesizer for both positive and negative inputs is shown in Figure 10-9. The resulting transfer curve with appropriate parameters is given in Figure 10-10.

Example 10-2

Analyze the circuit in Figure 10-11.

a. Calculate each slope and each break voltage.

b. Verify your calculations with a computer simulation.

Solution

a.   The slope around the origin, 0V input and output is

$$S1 = -\frac{R_f}{R_i} = -\frac{10k\Omega}{20k\Omega} = -0.5$$

When the input swings negative enough, D1 is turned on. The slope then becomes

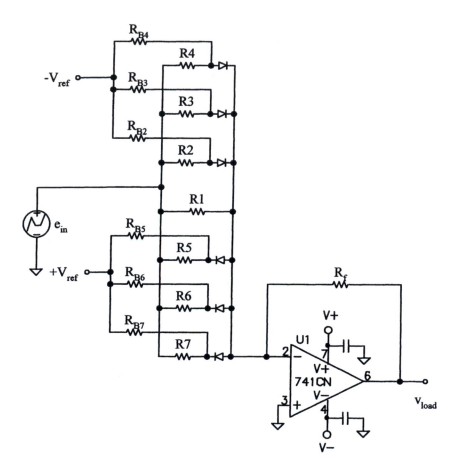

Figure 10-9 Bipolar transfer curve synthesizer

$$S2 = -\left(\frac{R_f}{R1} + \frac{R_f}{R2}\right) = -\left(\frac{10k\Omega}{20k\Omega} + \frac{10k\Omega}{20k\Omega}\right) = -1$$

When the input swings positive enough, D2 is turned on. The slope then becomes

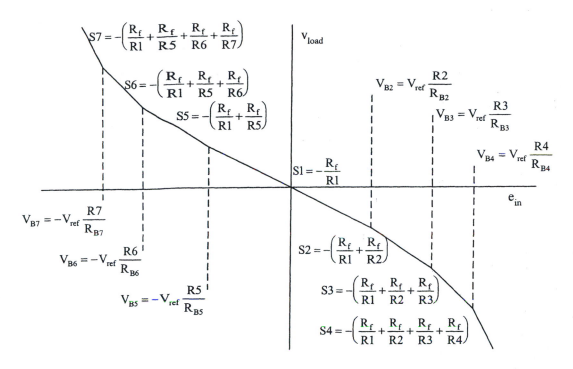

Figure 10-10 Bipolar piece-wise linear transfer curve

$$S3 = -\left(\frac{R_f}{R1} + \frac{R_f}{R3}\right) = -\left(\frac{10k\Omega}{20k\Omega} + \frac{10k\Omega}{6.67k\Omega}\right) = -2$$

The positive break voltage is

$$V_{break} = -V_{ref}\frac{R3}{R_{B3}} = -(-15V)\frac{6.67k\Omega}{20k\Omega} = 5V$$

The negative break voltage is

$$V_{break} = -V_{ref}\frac{R2}{R_{B2}} = -(15V)\frac{20k\Omega}{60k\Omega} = -5V$$

b. The schematic and the analysis setup in PSpice are shown in Figure 10-12. Figure 10-13 gives the results of the DC Sweep, with the break points and slopes.

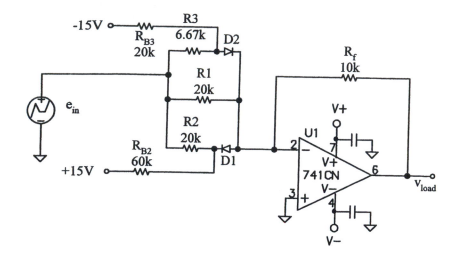

Figure 10-11 Schematic for Example 10-2

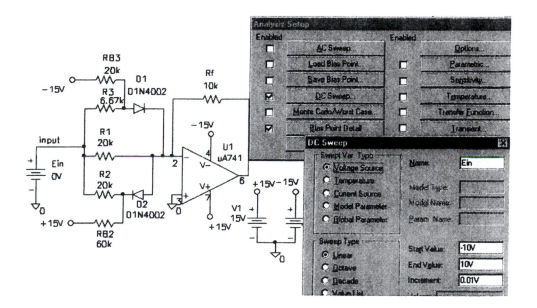

Figure 10-12 Simulation setup for Example 10-2

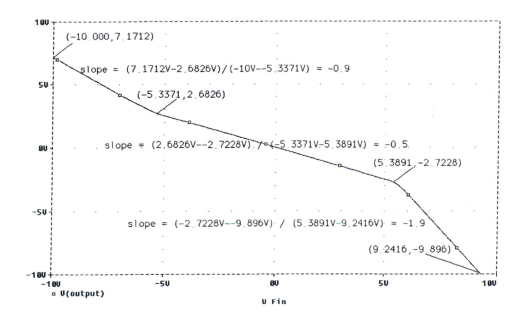

Figure 10-13 DC sweep of simulation for Example 10-2

Placing the resistor diode network in the **feedback** network produces a transfer curve whose slope **decreases** with an increase in input voltage. The schematic is given in Figure 10-14(a), and the two-segment transfer curve in Figure 10-14(b). For proper operation, the diode must be connected to the inverting input, which provides a virtual ground.

With a small positive output (negative input), the negative reference voltage holds the diode off. The gain, and the slope S1, are determined by

$$S1 = -\frac{R1}{R_i}$$

When the input becomes more negative, the output is driven more positive. This makes the anode of the diode more positive (or less negative). At some point, the diode goes on. Resistor R2 now parallels R1, decreasing the feedback resistance, the gain, and the slope.

$$S2 = -\frac{R1 // R2}{R_i}$$

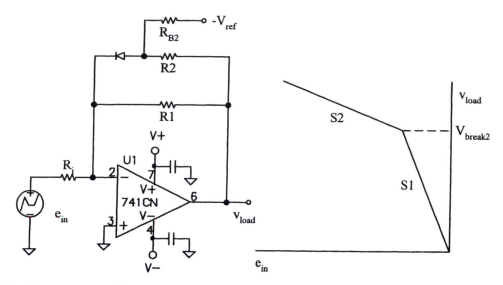

(a) Two-segment piece-wise linear circuit   (b) Transfer curve

Figure 10-14 Two-segment piece-wise linear circuit with decreasing slope

If a third and a fourth resistor diode network are added,

$$S3 = -\frac{R1 \, // \, R2 \, // \, R3}{R_i}$$

and

$$S4 = -\frac{R1 \, // \, R2 \, // \, R3 \, // \, R4}{R_i}$$

Although straightforward, these equations are a bit difficult to manipulate for analysis or design. However, if you reorganize them, you have

$$S1 = -\frac{1}{\dfrac{R_i}{R1}}$$

$$S2 = -\frac{1}{\left(\dfrac{R_i}{R1} + \dfrac{R_i}{R2}\right)}$$

$$S3 = -\cfrac{1}{\left(\cfrac{R_i}{R1} + \cfrac{R_i}{R2} + \cfrac{R_i}{R3}\right)}$$

$$S_n = -\cfrac{1}{\left(\cfrac{R_i}{R1} + \cfrac{R_i}{R2} + \cfrac{R_i}{R3} + ... + \cfrac{R_i}{R_n}\right)}$$

These equations are directly analogous to the slope equations for the circuit with the network in the input.

The voltage at the anode of the diode in Figure 10-14(a) is

$$v_{anode} = v_{load} - \left(V_{ref} - v_{load}\right)\frac{R2}{R2 + R_{B2}}$$

Consequently, for this configuration, the diode voltage is controlled by the output voltage, not the input. If you assume that the diode turns on at

$$V_{anode} \approx 0V$$

then a little algebra leads to

$$v_{load} = V_{break2} = -V_{ref}\frac{R2}{R_{B2}}$$

This is the output voltage at the break point between S1 and S2. It is the same as the equations for the break point voltage in the increasing slope circuit. However, for the circuit with the diode network in the **input**, the break voltage refers to **input** voltage. For the circuit with the diode network in the **feedback**, the break voltage refers to **output** voltage.

Example 10-3

Design a circuit that shapes a $6V_P$ input triangle wave to a $6V_P$ sine wave. Use $\pm 12V$ supplies.

Solution

You must first determine the transfer curve required. This is shown in Figure 10-15(a).

$$v_{load} = 6V_P \sin\theta$$

Expressing this in terms of the input voltage yields

$$v_{load} = 6V_P \sin\left(e_{in} \times \frac{90°}{6V}\right)$$

As the input voltage increases linearly, the output voltage changes sinusoidally.

Next, this must be converted to a piece-wise linear approximation that can be implemented by the transfer curve synthesizer. After careful examination of Figure 10-15(a), four segments were selected. These are plotted in Figure 10-15(b).

The slope decreases with increasing input, so the resistor diode network must be placed in the feedback loop. The overall circuit diagram is shown in Figure 10-16. The slopes are as follows:

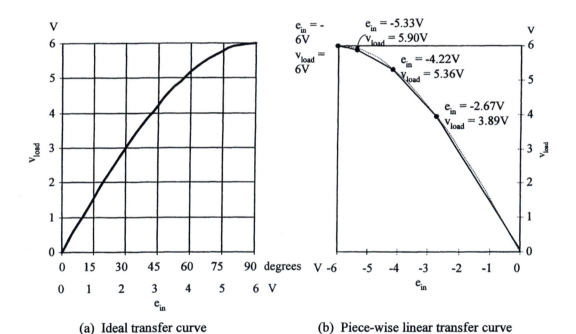

(a) Ideal transfer curve   (b) Piece-wise linear transfer curve

Figure 10-15 Sine shaper

$$S1 = \frac{3.89V - 0V}{-2.67V - 0V} = -1.46 \qquad\qquad \frac{1}{S1} = -0.686$$

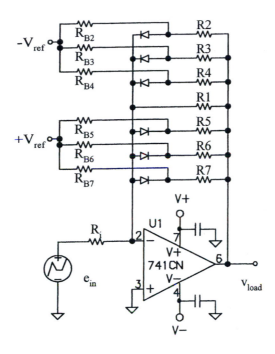

Figure 10-16 Sine shaper schematic for Example 10-3

$$S2 = \frac{5.36V - 3.89V}{-4.22V - (-2.67V)} = -0.948 \qquad \frac{1}{S2} = -1.05$$

$$S3 = \frac{5.90V - 5.36V}{-5.33V - (-4.22V)} = -0.486 \qquad \frac{1}{S3} = -2.06$$

$$S4 = \frac{6V - 5.90V}{-6V - (-5.33V)} = -0.149 \qquad \frac{1}{S4} = -6.70$$

$$V_{break2} = 3.89V$$

$$V_{break3} = 5.36V$$

$$V_{break4} = 5.90V$$

Pick $R_i = 10k\Omega$.

$$\frac{1}{S1} = -\frac{R_i}{R1} = -0.685$$

$$R1 = -\frac{10k\Omega}{-0.685} = 14.6k\Omega$$

$$\frac{1}{S2} = -\frac{R_i}{R1} - \frac{R_i}{R2}$$

$$-1.05 = -0.685 - \frac{R_i}{R2}$$

$$\frac{R_i}{R2} = 0.365 \qquad\qquad R2 = \frac{10k\Omega}{0.365} = 27.4k\Omega$$

$$\frac{1}{S3} = -\frac{R_i}{R1} - \frac{R_i}{R2} - \frac{R_i}{R3}$$

$$-2.06 = -1.05 - \frac{R_i}{R3}$$

$$\frac{R_i}{R3} = 1.01 \qquad\qquad R3 = 9.9k\Omega$$

$$\frac{1}{S4} = -\frac{R_i}{R1} - \frac{R_i}{R2} - \frac{R_i}{R3} - \frac{R_i}{R4}$$

$$-6.71 = -2.06 - \frac{R_i}{R4}$$

$$\frac{R_i}{R4} = 4.65 \qquad\qquad R4 = 2.2k\Omega$$

$$V_{break2} = 3.89V = -(V_{ref})\frac{R2}{R_{B2}}$$

$$R_{B2} = -\frac{-V_{ref}}{V_{break2}}R2 = -\frac{-12V}{3.89V}27k\Omega = 83.3k\Omega$$

$$R_{B3} = -\frac{-V_{ref}}{V_{break3}}R3 = -\frac{-12V}{5.36V}10k\Omega = 22.4k\Omega$$

$$R_{B4} = -\frac{-V_{ref}}{V_{break4}}R4 = -\frac{-12V}{5.90V}2.2k\Omega = 4.47k\Omega$$

Resistors R2-R4 and R_{B2}-R_{B4} alter the slope of the positive output cycle. To cause the negative cycle to be similarly shaped,

R5=R2,   R6=R3,   R7=R4, R_{B5}=R_{B2}, R_{B6}=R_{B3}, R_{B7}=R_{B4} +V_{ref}=12V

10-2 True RMS Converters

The root mean squared (RMS) value of a signal is a measure of the amount of power it can deliver to a load. As such, it is the most valid way to compare signals of widely different shapes. For example, a $1V_P$ steady DC signal dissipates the same power as a $1.414V_P$ sine wave, a $1.73V_P$ triangle wave, a $2V_P$ 50% rectangular pulse, or a $4.7V_P$ 25% SCR waveform. The root mean squared values of all of these signals is $1V_{RMS}$.

Mathematically, the root mean squared value is defined as

$$v_{RMS} = \sqrt{\frac{1}{T}\int_0^T v(t)^2\,dt} = \sqrt{ave\left[v(t)^2\right]}$$

That is, the square **root** is taken of the **mean** (i.e., average) of the **square** of the voltage. This definition assures that numerically equal RMS signals produce the same power.

When rectifying a signal, even with the ideal rectifiers of the previous section, you produce the absolute value of the signal. Following that rectifier with a filter provides the average of that absolute value. The result is called the **mean absolute deviation**, the v_{MAD}.

$$v_{MAD} = ave\left[|v(t)|\right]$$

Certainly these two equations are not the same. In simple applications, gain may be added to the mean absolute deviation to force that signal to equal the RMS value. But each different wave shape requires its own, different gain constant. Traditionally, the gain is set to make the output of the MAD circuit indicate the RMS value of a pure sine wave. However, the output from such a circuit does not properly indicate either v_{RMS} or v_{MAD} for any other shape, as illustrated in Table 10-1.

Pure sine waves are present in only the simplest of laboratory experiments. In practice, virtually every circuit produces some degree of distortion; and many applications use signals radically different in shape from sinusoids. So, some other circuitry must be used to produce a measure of the true RMS of a signal, regardless of its

Table 10-1 Error introduced by mean absolute (average) responding circuits when measuring common waveforms *(Courtesy of Analog Devices)*

Waveform Type $1V_P$ Amplitude	Crest Factor $\dfrac{V_P}{V_{RMS}}$	True RMS Value V_{RMS}	Average Responding Circuit Calibrated to Read RMS Value of a Sine Wave Will Read:	% of Reading Error of Ave Responding Circuit (%)
Sine wave	1.41	0.707	0.707	0
Symmetrical Square wave	1.00	1.00	1.111	+11
Triangle wave	1.73	0.590	0.555	−2
Gaussian noised ($98\% < 1V_P$)	3.00	0.333	0.266	−20
Rectangular pulses	2.00	0.500	0.250	−50

SCR
   50% duty cycle
   25% duty cycle
	2.02
	4.71
		0.495
		0.212
			0.354
			0.150
				−28
				−30

wave shape.

Analog Devices manufactures a family of true RMS converter ICs. Each member outputs a DC voltage that is numerically equal to the root mean squared value of the IC's input signal, regardless of the signal's wave shape. Table 10-2 lists these ICs. The AD536A is one of the first converters built. It is a general purpose converter that accepts a wide range of input amplitudes. It also provides an internal log amp. This allows the output to be encoded in dB. The next IC is the AD636. It is similar to the AD536A. However, the AD636 has been optimized for operation at low power supply voltages and quiescent current requirements. The maximum input is restricted to $200mV_{RMS}$, providing a $200mV_{DC}$ output. This is ideal for battery operation and matches the input requirements of a variety of digital multimeter/LCD analog to digital converters. For precise instrumentation, the AD637 provides the best accuracy across a wide range of amplitudes and frequencies. Of course, this increased performance is more expensive. The AD736 and AD737 are second-generation parts. Implementing the RMS conversion differently than the other ICs, these two converters are less expensive and use notably less power than the other ICs.

Table 10-2 True RMS converter ICs from Analog Devices

Model	Full Scale V_{RMS}	Accuracy %	Bandwidth kHz (100mV)	Log	Comments
AD536A	7.0	0.2	450	yes	general purpose
AD636	0.2	0.5	1000	yes	low power, DMM
AD637	7.0	0.02	600	yes	most accurate, high freq
AD736	0.2	0.3	170	no	general purpose, low power
AD737	0.2	0.3	170	no	AD736 without buffer

The block diagram of the AD736 is given in Figure 10-17. The AD736 has an output buffer, while the AD737 does not. The buffer allows the AD736 to drive impedances as low as 2kΩ, while the AD737 is specified only when driving no load. However, the buffer does increase the DC error, and increases the power supply current by 40μA. So the AD737 is the choice when driving high input impedance loads (such as most A-D converters), while the AD736 provides rated performance into 2kΩ.

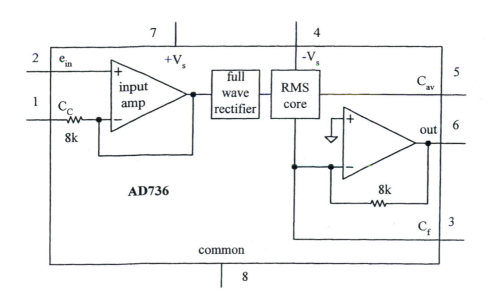

Figure 10-17 Block diagram of the AD736 *(Courtesy of Analog Devices)*

A dual supply, $\pm V$, is required. Typically this is $\pm 5V$. However, it may be as large as $\pm 16V$ but must be at least $+2.8V$, $-3.2V$. With no input signal, the AD736 draws about 200μA from the supplies. While processing an input, this quiescent current rises to 270μA.

To avoid damage to the IC, you must keep the input amplitude within the supply voltages, $\pm V$. To achieve the rated accuracy of 0.3%, the input must be limited to 200mV$_{RMS}$. For this same accuracy, the maximum transient (i.e., peak) voltage is 1V$_P$. That means that the crest factor (V_P/V_{RMS}) must be 5 or less. However, SCR phase angle fired signals and music waveforms often have crest factors greater than 5. For these signals you should expect errors to exceed the rated 0.3%.

The minimum resolution is 100μV. This implies that the AD736 can be used with a $3\frac{1}{2}$-digit A-D converter (1999), but not a $4\frac{1}{2}$-digit (19999) DVM.

There are two inputs. Pin 2 provides $10^{12}\Omega$ of input resistance, more than enough to assure that even a high impedance input voltage divider is not loaded down. An inverting input is also provided at pin 1. The 8kΩ resistor between the input pin and the connection to its amplifier sets this input impedance to 8kΩ. Be careful. Pin 1 is an **inverting** input.

The frequency response is directly proportional to the signal amplitude. Smaller signals have a significantly reduced high frequency cut-off. This is different from any other circuit you have seen in this text. So beware. The -3dB down frequency for a full scale input with an input of 200mV$_{RMS}$ is 190kHz. However, for a 1mV$_{RMS}$ input, this drops to 5kHz.

If you are aiming for a highly accurate conversion (0.3%), then the -3dB frequency is not a valid parameter. At the -3dB frequency the output of the converter has fallen to 0.707 of its low frequency value. That is a 30% error! Analog Devices also specifies the 1% cut-off. For a 200mV$_{RMS}$ signal into the high impedance input, $f_{1\%} = 33$kHz. If you can tolerate the 8kΩ input impedance and inversion of pin 1, then $f_{1\%}$ rises to 90kHz. For both inputs, $f_{1\%}$ falls to 1kHz when the input amplitude drops to 1mV$_{RMS}$.

To use the AD736, you only need to add one external capacitor. The schematic is shown in Figure 10-18. The averaging capacitor, C_{ave}, is needed to allow the IC's RMS core to produce the **mean** portion of the calculation. The larger this capacitor, the smaller the error in the output DC. However, increasing C_{ave} increases the time you must wait after the amplitude changes until the output has settled to a steady value. So, you must deal with a trade-off between steady state error and settling time. Also, as with frequency response, input amplitude affects speed. However, the smaller the input amplitude, the more quickly the output settles. This is the opposite of the relationship between amplitude and frequency.

There will be some ripple in the DC output at the frequency of the input signal. The capacitor, C_f, in Figure 10-18 is used to remove AC variations from the output signal. The proper values of C_{ave} and C_f depend on application, signal amplitude, and frequency. The manufacturer's recommended values and the resulting settling times are listed in Table 10-3.

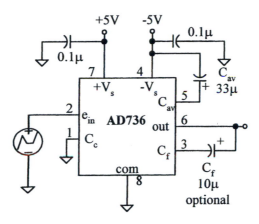

Figure 10-18 AD736 with external components *(Courtesy of Analog Devices)*

Table 10-3 Capacitor selection chart *(Courtesy of Analog Devices)*

Application	Input Level V_{RMS}	Low Frequency Cutoff (-3dB) Hz	Max Crest Factor	C_{av} µF	C_f µF	Settling* to 1% s
General purpose	0-1	20 200	5 5	150 15	10 1	0.36 0.04
RMS	0-0.2	20 200	5 5	33 3.3	10 1	0.36 0.04
SCR waveform	0-0.2	50 60	5 5	100 82	33 27	1.200 1.000
	0-0.1	50 60	5 5	50 47	33 27	1.200 1.000

Audio
  speech
  music
	0-0.2
	0-0.1
		300
		20
			3
			10
				1.5
				100
					0.5
					68
						0.018
						2.400

* Settling time is specified over the stated RMS input level with the input signal increasing from zero. Settling times are greater for decreasing amplitude input signals.

A DC component added to a signal changes the RMS value of that signal. To measure the true RMS value of the AC portion only of such a composite signal, you must remove the DC. This requires the addition of C_{in} and C_C. Look at Figure 10-19. Each capacitor, when combined with resistance within the IC, forms a high pass filter, passing the AC signal but blocking the DC. Since the input capacitor is connected to $10^{12}\Omega$, any practical value produces an adequate low frequency cut-off. The second coupling capacitor, C_C, combines with the $8k\Omega$ from the input amplifier to set the low frequency cut-off. Remember, however, that at this frequency, the AC signal is attenuated by 30% (i.e., 0.707). So a good rule is to set f_l at no more than 0.1 of the lowest frequency signal that you want to measure. This should reduce the filter's error to 0.5%.

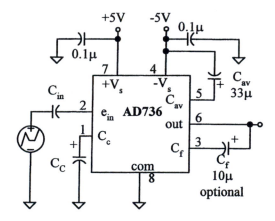

Figure 10-19 AC coupling the AD736

$$f_L = \frac{1}{2\pi \times 8k\Omega \times C_C} \le 0.1f_{lowest\,signal}$$

The AD736 and AD737 were designed with battery power in mind. But you must use a supply splitter, as shown in Figure 10-21, to provide ±V. Notice that circuit common is set in the middle of the 9V battery. This must be the common reference for previous and subsequent circuits as well. So evaluate your entire circuit and its power requirements carefully.

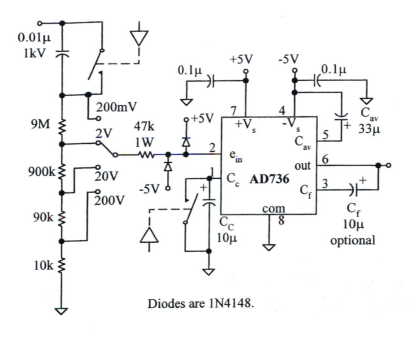

Diodes are 1N4148.

Figure 10-20 AD736 with high impedance input attenuator
(Courtesy of Analog Devices)

The full scale input is 200mV$_{RMS}$ at rated accuracy. To provide a variety of ranges, you can add an input voltage divider and a selector switch as shown in Figure 10-20. The selector switch may be manual, or it may be digitally controlled. The diodes provide overvoltage protection. Place a short around the two AC coupling capacitors if you want to pass the DC part of the input.

Many transducers produce a differential output signal. A bridge circuit is a classical example. Information is also often transmitted over wires as the difference in potential between two signals. This way, any noise picked up along the way appears common to both lines and can be subtracted out. Finally, most analog control techniques require that the input (measurement of what the process is actually doing) be subtracted from the set point (what you want the process to do). Further electronics then attempts to reduce this error. In all of these applications, two voltages must be subtracted. Certainly, you could use the differential amplifier you saw in Chapter 2. However, if you also need an RMS converter, you can use the much simpler circuit shown in Figure 10-22. The voltage follower must be added to the inverting input to raise its input impedance to $10^{12}\Omega$ to match the noninverting amp.

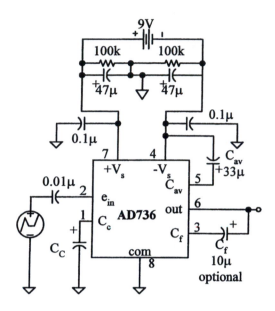

Figure 10-21 Battery powered operation *(Courtesy of Analog Devices)*

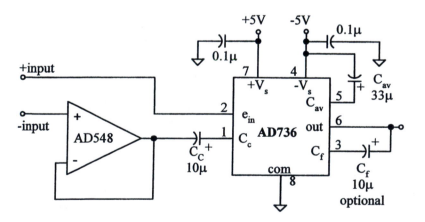

Figure 10-22 AD736 differential input connection *(Courtesy of Analog Devices)*

10-3 Logarithmic and Antilogarithmic Amplifiers

Logarithmic amplifiers have several areas of applications. Analog computation may require ln x, log x, or sinh x. These can all be performed continuously with analog log amps. It is very convenient to have direct dB displays on digital voltmeters and spectrum analyzers. Log amps are used in these instruments. Signal dynamic range compression enhances computer signal processing as well as audio recording. Log amps can be used to compress a signal's dynamic range.

Two major areas of applications for log amps are in complex analog calculations and signal processing. To raise a signal to a power, or multiply two signals, or divide two signals, you could first take the log of the signal. Then exponentiation to any power comes from amplification (multiplication by a constant). Multiplication of two log "encoded" signals is performed by a summer, while division of the two log "encoded" signals is performed by a difference amp.

To process a signal whose range varies from tens of microvolts to tens of volts requires very high quality, expensive equipment, to have both the resolution (μV) and range (tens of volts). If you first take the log of the signal, each decade of input signal change requires only 1V of output signal change. This log compression of the signal significantly lowers the demands on the processing circuits for both tight resolution and wide range.

The fundamental log amp is formed by placing a transistor (or two) in the negative feedback path of an op amp, as shown in Figure 10-23. For a positive input, current flows from the source, left to right, through R_i, and through the npn transistor, into the output of the op amp. A negative input causes current to flow from the output of the op amp, through the pnp transistor, and into the input source. If your signal is unipolar, then you need only one of the transistors. For signals that may go either positive or negative, both the npn and the pnp transistors are needed.

The classical relationship of current through a bipolar junction transistor is

$$I_C = \alpha I_s \left(e^{-qV_{be}/kT} - 1 \right)$$

where

$\alpha \sim 1$ and constant with I and T

I_S = the emitter saturation current $\sim 10^{-12}$A

q = one electron's charge = 1.60219×10^{-19} C

k = Boltzmann's constant = 1.38062×10^{-23} J/°K

T = temperature = 296°K (room temperature)

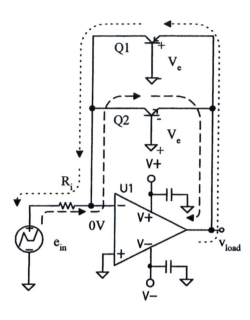

Figure 10-23 Basic log amp

Since the base of each transistor is grounded and the emitter is being driven,

$$V_{be} = -V_e$$

Substituting for the constant in the exponential gives

$$\frac{kT}{q} = 26mV$$

Assuming that $\alpha = 1$, then the transistor collector current becomes

$$I_C = I_s\left(e^{\frac{V_e}{26mV}} - 1\right)$$

$$e^{\frac{V_e}{26mV}} = \frac{I_C}{I_s} + 1$$

Since I_s is very small, the right side of this equation is $>>1$, so the 1 can be ignored.

$$e^{\frac{V_e}{26mV}} = \frac{I_C}{I_s}$$

Taking the natural logarithm of both sides

$$\frac{V_e}{26mV} = \ln\frac{I_C}{I_s}$$

$$V_e = 26mV \ln\frac{I_C}{I_s}$$

Now, look at Figure 10-23 again.

$$v_{load} = -V_e$$

$$I_C = \frac{e_{in}}{R_i}$$

Combining these gives the output of the circuit.

$$v_{load} = -26mV \ln\frac{e_{in}}{R_i I_s}$$

Although this equation was developed with the natural log, it holds equally well for $\log_{10}$, by proper scaling.

$$\log_{10}(x) = 0.4343\ln(x)$$

The emitter saturation current, I_s, though on the order of 10^{-12}A varies significantly from one transistor to another, and with temperature. So, the output becomes random. The circuit in Figure 10-24 eliminates this problem.

The input is applied to one log amp, while a reference voltage is applied to another. The three transistors are integrated closely together in the silicon wafer. This provides a close match of saturation currents and ensures good thermal tracking. For this reason,

$$I_{sQ1} = I_{sQ2} = I_{sQ3} = I_s$$

The output from U1 is

$$e_1 = -26mV \ln\frac{e_{in}}{R_i I_s}$$

The output from U2 is

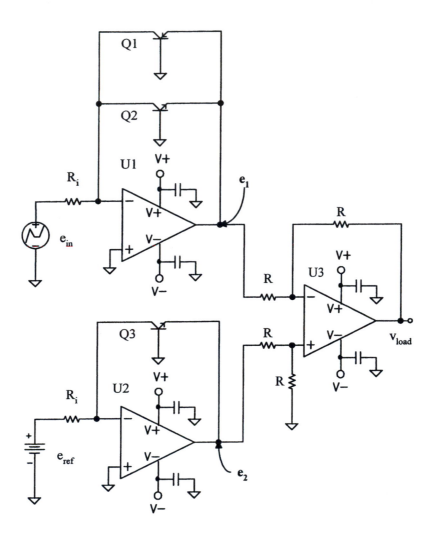

Figure 10-24 Log amp with I_s compensation *(Courtesy of Analog Devices)*

$$e_2 = -26\text{mV} \ln \frac{e_{\text{ref}}}{R_i I_s}$$

Op amp U3 subtracts these two signals

$$v_{\text{load}} = -26\text{mV} \ln \frac{e_{\text{ref}}}{R_i I_s} - \left(-26\text{mV} \ln \frac{e_{\text{in}}}{R_i I_s} \right)$$

$$V_{load} = 26mV \left(\ln \frac{e_{in}}{R_i I_s} - \ln \frac{e_{ref}}{R_i I_s} \right)$$

Remember that subtracting logs is the same as dividing their arguments.

$$V_{load} = 26mV \ln \left(\frac{\dfrac{e_{in}}{R_i I_s}}{\dfrac{e_{ref}}{R_i I_s}} \right)$$

$$V_{load} = 26mV \ln \frac{e_{in}}{e_{ref}}$$

The reference level is now set with a single external voltage source. The circuit's dependence on device parameters and temperatures has been removed.

In either analog calculations or signal processing, once the signal has been properly conditioned, often it must be decoded. To convert these log "encoded" signals back into real-world terms, you must take the antilog. This is done with the antilog amp shown in Figure 10-25.

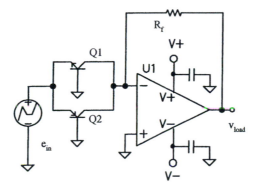

Figure 10-25 Simple antilog amp

Remember, for the log amp, one decade of change in the input signal causes 1V change at the output (assuming that you have applied the constant correction from ln to log). For the antilog amp, one voltage change at the input causes a decade of change at the output. So, in practical terms, you must restrict the size of the input signals to less than 1V. This makes sense since the input is taken across a transistor's base emitter junction. Simple models taught you that this voltage is **about** 0.6V.

When the input goes positive, current flows through Q2, to the virtual ground at the op amp's inverting input, and through R_f, producing an output voltage proportional to Q2's collector current. Negative input voltage causes similar results through Q1.

It was shown when looking at the log amp,

$$e^{\frac{V_e}{26mV}} = \frac{I_C}{I_s}$$

The emitter voltage is e_{in}. Making this substitution and rearranging a bit,

$$I_C = I_s \, e^{\frac{e_{in}}{26mV}}$$

$$V_{load} = -I_C R_f$$

$$V_{load} = -I_s R_f e^{\frac{e_{in}}{26mV}}$$

The output voltage is directly proportional to I_S. But you have already seen that I_S varies widely with transistor characteristics and temperature. Alone, the results from the circuit in Figure 10-25 are unpredictable. The solution is to include the transistors for the log encoding circuit (Figure 10-23) and the antilog decoding circuit (Figure 10-25) on the same wafer. I_S for both the encoding and the decoding are the same. In the log amp, I_S appears in the denominator. In the following antilog amp, it is in the numerator. Any error produced by I_S divides out.

10-4 Multipliers

Analog multiplication is too often overlooked as a direct, simple solution to complex signal processing problems. Monolithic integration has lowered the cost of multiplier ICs considerably. Frequency doubling and phase angle detection as well as real power computation can be done with multipliers. Multipliers can be configured to output DC whose value equals the true RMS of a signal. Voltage control of an amplifier's gain, an oscillator's frequency, and a filter's cut-off can all be obtained with multipliers. Multipliers can also be used to improve the data acquisition by taking the ratio of two signals. This divides out mutual error.

10-4.1 Multiplier Characteristics

A basic multiplier is shown in Figure 10-26. Two signal inputs (e_x and e_y) are provided. The output is the product of these two inputs divided by a reference voltage, e_{ref}.

$$V_{load} = \frac{e_x e_y}{e_{ref}} \qquad\qquad e_x \text{ and } e_y \leq e_{ref}$$

Normally, e_{ref} is internally provided (often 10V). As long as you insure that both input voltages are below the reference, the output of the multiplier does not saturate.

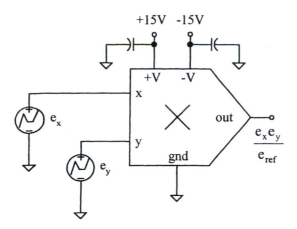

Figure 10-26 Basic multiplier schematic symbol

One way to implement multiplication relies on the property of logarithms. The sum of two logs is equal to the log of the products of the arguments. The difference of two logs is equal to the quotient of the two arguments.

$$\log(x \cdot y) = \log(x) + \log(y)$$

$$\log\left(\frac{a}{b}\right) = \log(a) - \log(b)$$

$$\log\left(\frac{x \cdot y}{z}\right) = \log(x) + \log(y) - \log(z)$$

By log encoding e_x, e_y, and e_{ref}, you can implement the basic multiplier block with a summer and an inverter. Then, to recover the output, take the antilog of the summer's output. This is illustrated in Figure 10-27. Remember, a log amp is not well behaved as its input goes through ground (log 0 is undefined). This technique of multiplication should be used only for signals that do not pass through ground.

The acceptable polarities of the input define the number of quadrants of operation of the multiplier. Look at Figure 10-28. If both inputs must be positive (as with the log implementation), the multiplier is said to be one-quadrant. A two-quadrant multiplier functions properly if one input is held positive and the other is allowed to swing both positive and negative. If both inputs may be either positive or negative, the IC is called a four-quadrant multiplier. Be sure in selecting a multiplier that you specify one that properly responds to the polarity inputs that you are using.

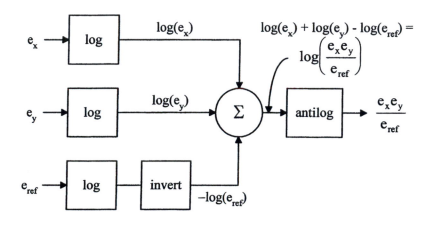

Figure 10-27 Log/antilog implementation of multiplication

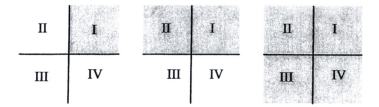

(a) one quadrant   (b) two quadrants   (c) four quadrants

Figure 10-28 Multiplier quadrants of operation

10-4.2 Multiplier Applications

The AD633 is a low cost, four quadrant multiplier made by Analog Devices. Both the x input and the y input are preceded by a difference amplifier. This is particularly handy when working with instrumentation and control systems since many signals are produced as small differences in potential.

The voltage controlled amplifier is one of the simplest applications of a multiplier. Its schematic is shown in Figure 10-29. The gain of the signal applied to the Y input(s) is controlled by the voltage applied to the X input. This allows remote automatic or digital control of an analog system's performance.

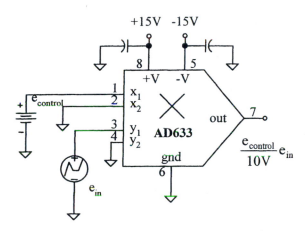

Figure 10-29 Voltage controlled amplifier

The average power dissipated by a load is measured by taking the average of the instantaneous product of the voltage across the load times the current through the load. Regardless of the characteristics of the load, the signal's wave shape, or how either of these changes, this relationship defines power. The circuit in Figure 10-30 accurately monitors average power.

The voltage across the load, R_l, is connected directly into the x_1-x_2 differential input of the multiplier IC. Current through the load is converted into a voltage by the 0.01Ω resistor. This resistor must be much smaller than the load resistor to assure that the measurement does not affect the load. Op amp U1 amplifies the small voltage dropped across the sense resistor. Select the gain of U1 so that at maximum peak load current, U1's output voltage is 10V (the maximum input into the AD633).

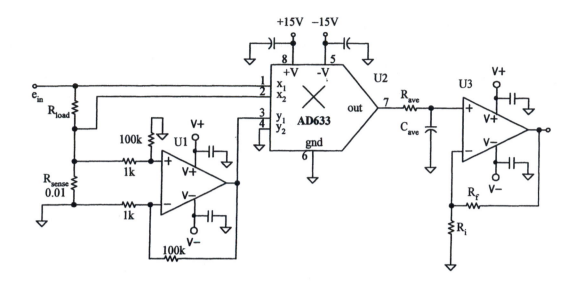

Figure 10-30 Average power measurement *(Courtesy of Analog Devices)*

U1 has been configured as a differential amplifier to measure the voltage **directly** across the sense resistor. It is tempting to just use a noninverting amplifier because the sense resistor is tied to ground, right? However, the return wire from the bottom of the sense resistor to system common may be several feet long. This could have 10mΩ to 100mΩ or more of resistance. That uncertainty in ground voltage may be as large as the voltage across the sense resistor. Configuring U1 as a difference amplifier allows you to remove those effects from your measurement of I_{load}.

The multiplier outputs a voltage equal to the product of its two input voltages, divided by 10V. But what you want is **average** power, so set the filter at U2's output to remove any AC variation, leaving only the DC (i.e., the average). The final amplifier, U3, completes the scaling. Remember, the multiplier also divides by 10V, so U3 multiplies by 10. This yields an output of 1V for each watt of power dissipated by the load. Depending on your system, you will need to alter the gain of this amplifier to assure that maximum power is a meaningful number, but that it does not saturate any of the ICs.

Example 10-4

Verify the performance of the real power circuit of Figure 10-30 by manual calculation and by simulation. Set $e_{in} = 9V_{RMS}$, 60Hz, $R_{load} = 8\Omega$, $R_{ave} = 100k\Omega$, and $C_{ave} = 1\mu F$.

Solution

$$V_{Rload} = 9V_{RMS} \frac{8\Omega}{8\Omega + 0.01\Omega} = 8.988V_{RMS}$$

$$V_{Rsense} = 9V_{RMS} \frac{0.01\Omega}{8\Omega + 0.01\Omega} = 0.01124V_{RMS}$$

$$V_{U1out} = 0.01124V_{RMS}\left(1 + \frac{100k\Omega}{1k\Omega}\right) = 1.135V_{RMS}$$

$$V_{U2ave} = \frac{e_{x\,RMS} \times e_{y\,RMS}}{10V} \cos\theta$$

where θ is the angle between the two signals. In this case, since the load is resistive, there is no phase shift.

$$V_{U2out} = \frac{8.988V_{RMS} \times 1.135V_{RMS}}{10V} = 1.020V_{DC}$$

$$V_{U3out} = 1.020V_{RMS}\left(1 + \frac{9k\Omega}{1k\Omega}\right) = 10.2V_{DC}$$

The power dissipated by the load is

$$P = \frac{\left(9V_{RMS}\right)^2}{8\Omega} = 10.13W$$

The *Electronics Workbench*[TM] simulation is shown in Figure 10-31.

For sine waves the effect of multiplication can be described mathematically as

$$\sin\theta \sin\phi = \tfrac{1}{2}\cos(\theta - \phi) - \tfrac{1}{2}\cos(\theta + \phi)$$

For
$$e_x = E_{x\,P}\sin(2\pi f_x t) \quad \text{and} \quad e_y = E_{y\,P}\sin(2\pi f_y t)$$

Applying this to the multiplier IC,

$$V_{load} = \frac{E_{x\,P}E_{y\,P}}{2E_{ref}}\cos 2\pi\left(f_x - f_y\right)t - \frac{E_{x\,P}E_{y\,P}}{2E_{ref}}\cos 2\pi\left(f_x + f_y\right)t$$

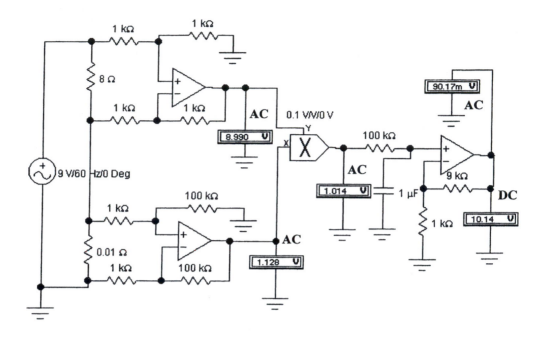

Figure 10-31 *Electronics Workbench™ simulation for Example 10-4*

You can take advantage of this "sum and difference frequencies" relationship to produce a circuit that doubles the frequency of a sine wave. Look at Figure 10-32. The first term is just DC $(f - f = 0)$. The RC coupler at the output blocks that. Be sure to set its cut off at least a factor of 10 below the lowest output frequency. The second term is a sinusoid shifted in phase from the input (-cos rather than sin) and at twice the frequency of the input $(f + f = 2f)$.

Amplitude modulation (as in radio transmission) is just a matter of allowing the audio signal to alter the amplitude of the high frequency carrier. That is precisely what a multiplier does. Apply the rf carrier to the y input, and the audio information to the x. As x (the audio) varies up and down, the gain alters accordingly. The multiplier outputs the carrier, but since the gain is varying at an audio rate, the amplitude of the output follows suit.

If you are familiar with communications electronics, you may recognize the sum and difference frequencies equation just developed as balanced modulation. Demodulation, to recover the audio signal, is difficult.

Standard amplitude modulation adds the carrier to the sum and difference signals.

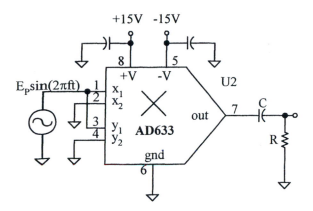

Figure 10-32 Frequency doubler

$$v_{load} = E \sin 2\pi f_y t + \frac{E_x E_y}{2E_{ref}} \cos 2\pi\left(f_y - f_x\right) - \frac{E_x E_y}{2E_{ref}} \cos 2\pi\left(f_y + f_x\right)$$

This produces an output that is easier to demodulate. Rectification eliminates the negative half. A low pass filter removes the carrier frequency variations, leaving only the envelop audio.

The carrier could be added to the multiplier output with a summer. The carrier would be connected to one input of the summer, and the balanced modulated signal (output of the multiplier) to the other. However, building carrier frequency circuits is more expensive and difficult than building circuits for DC and audio frequency. An alternative approach for adding the carrier is to add an offset to the audio signal **before** it is applied to the multiplier. This can be easily and cheaply done with the summer presented in Chapter 2.

10-4.3 Dividers

The first technique you saw for multiplication was to add two log encoded signals and then decode with an antilog amp. A similar technique can be used to divide two signals. Remember that

$$\log\left(\frac{y}{x}\right) = \log(y) - \log(x)$$

This technique for electronic division is shown in Figure 10-33.

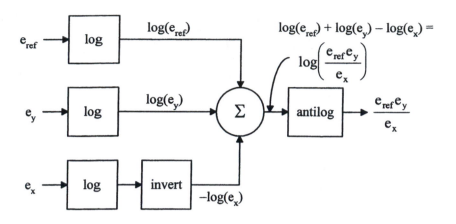

Figure 10-33 Log/antilog implementation of division

In reality, this is **exactly** the same block diagram as that for the multiplier. The role of e_{ref} (often $10V_{DC}$) and the input signal e_x have been exchanged. As with the log implemented multiplier, the signals in Figure 10-33 must be offset from ground, since $\log(0)$ is undefined, and the log amp outputs a transient step when its input passes through ground.

You saw that to build the antilog amp an antilog element (the transistors) was placed in the input loop. The complementary function, the log amp, can be built by placing the antilog circuit element in the **feedback** loop of the op amp.

The same is true for multipliers and dividers. Division, the complement of multiplication, can be accomplished by placing the multiplier circuit element in an op amp's **feedback** loop. Refer to Figure 10-34.

The multiplier IC, U2, has been placed in the feedback loop of the op amp, U1. The $10k\Omega$ resistor and the 22pF capacitor are to limit high frequency gain, increasing stability. Otherwise, they do not figure into the circuit's performance. The output voltage is

$$V_{load} = \frac{10V\ e_{in}}{e_x}$$

With properly applied negative feedback, there is no significant difference in potential between the op amp's noninverting and inverting inputs. So

$$e_{fb} = e_{in}$$

Assuming that the $10k\Omega$ resistor has no effect on the signal, the output of the multiplier is also e_{fb}, but that output is set by the products of its inputs.

$$e_{in} = e_{fb} = \frac{e_x e_y}{10V}$$

and the Y input of the multiplier is connected to the circuit's output.

$$e_{in} = \frac{V_{load} e_x}{10V}$$

Solving this equation for V_{load}, we have

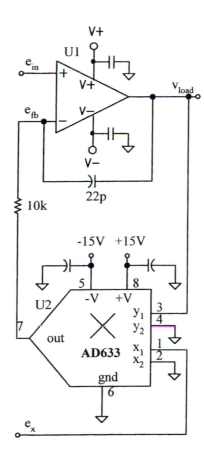

Figure 10-34 Divider circuit using the AD633 multiplier IC
(Courtesy of Analog Devices)

$$V_{load} = \frac{10V \; e_{in}}{e_x}$$

Division by zero is, of course, prohibited. Electronically, setting e_x to zero causes the output, V_{load}, to go into saturation. Since e_x cannot go to zero, it obviously cannot be allowed to pass through zero. Dividers are, at best, two-quadrant devices.

How small the denominator can become is a critical specification for dividers. Multipliers with 1% error usually will accept denominator signals of 0.1V or greater. Even so, offsets and drifts are typically 2 to 10 times worse for a divider with a small denominator signal than a multiplier made with the same IC.

The major advantage of using the inverse multiplication technique for division is low cost. If offset and drift errors, noise, bandwidth, or minimum denominator signal size are unacceptable, you can purchase a dedicated divider IC.

10-4.4 Operational Transconductance Amplifier

The operational transconductance amplifier (OTA) is an IC that outputs a current that is proportional to the difference in potential between its input pins.

$$i_{out} = g_m(e_{NI} - e_{INV})$$

The "gain" or output divided by input is

$$g_m = \frac{i_{out}}{e_{NI} - e_{INV}}$$

and must be in units of siemens (conductance).

The CA3280A is a dual programmable OTA. Its transconductance or gain (g_m) is directly proportional to the current driven into its amplifier bias control pin. At room temperature,

$$g_m = \frac{0.11}{V} I_{ABC}$$

See Figure 10-35.

This relationship holds for

$$I_{ABC} < 500\mu A$$

Combining these relationships, you obtain

$$i_{out} = \frac{0.11}{V} I_{ABC}(e_{NI} - e_{INV})$$

This holds true for $i_{out} < 400\mu A$ (typical)

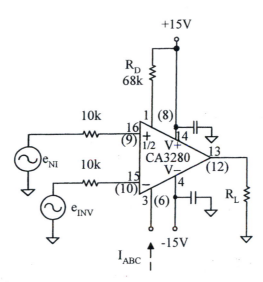

Figure 10-35 CA3280A operational transconductance amplifier *(Courtesy of Intersil Corporation, reprinted with their permission)*

Current through resistor R_D activates a linearizing network at the input. This assures a linear variation in output current as the input difference in potential varies up to 8V ($\pm$4V). Otherwise, a 100mV input swing causes a full scale, highly nonlinear variation in output current. Resistor R_L converts this output current into a voltage.

The series input 10kΩ resistors are required. Unlike an op amp, negative feedback is not used, and there may be a significant ($\pm$4V) difference between e_{NI} and e_{INV}. The inputs to the OTA IC (pins 15 and 16) are the bases of a transistor difference amplifier. This means that there can be only a tenth of a volt or so difference in potential between the input pins. The series 10kΩ resistors drop the remainder of the difference between e_{NI} and e_{INV}.

The amplifier bias current (I_{ABC}) directly affects the output. This feature provides the OTA's unique applications. There are several ways to set I_{ABC} as shown in Figure 10-36. The I_{ABC} pin is held two diode drops (1.2V) above the negative supply. So, if you do not plan to alter I_{ABC}, connect a resistor between its pin and ground. I_{ABC} is then determined by the negative supply and this resistor.

You can control I_{ABC} with an externally generated voltage, $e_{control}$ as shown in Figure 10-36(b). Changing $e_{control}$ changes I_{ABC}, which in turn changes g_m, the amplifier's gain.

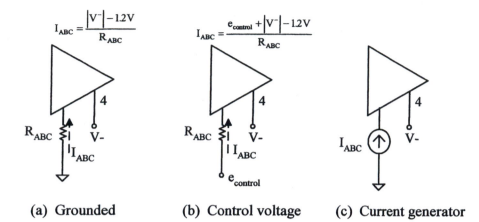

(a) Grounded   (b) Control voltage   (c) Current generator

Figure 10-36 I_{ABC} setting circuits

$$I_{ABC} = \frac{e_{control} + |V^-| - 1.2V}{R_{ABC}}$$

$$i_{out} = \frac{0.11}{V} I_{ABC} (e_{NI} - e_{INV})$$

$$i_{out} = \frac{0.11}{V} \left[\frac{e_{control} + |V^-| - 1.2V}{R_{ABC}} \right] (e_{NI} - e_{INV})$$

This shows that the output current is proportional to both the input voltage difference (e_{NI} - e_{INV}) and the control voltage. In other words, the gain of the amplifier can be remotely (electronically) altered by $e_{control}$. But with V^- and the 1.2V offsets, the control is not as simple as you might wish.

The third option for setting I_{ABC} is to set it directly with a current source. This current source could be made with a field effect transistor, a bipolar transistor, an op amp, an IC current source IC, or the other half of the CA3280.

Example 10-5

Calculate the output current for the circuit in Figure 10-37 for $e_{in} = 4V_{DC}$ and

   a. $e_{control} = +10V$
   b. $e_{control} = 0V$
   c. $e_{control} = -10V$

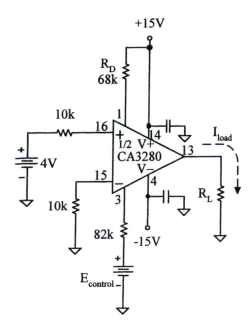

Figure 10-37 Schematic for Example 10-5

Solution

$$i_{out} = \frac{0.11}{V}\left[\frac{E_{control} + |V^-| - 1.2V}{R_{ABC}}\right]E_{in}$$

a.
$$i_{out} = \frac{0.11}{V}\left[\frac{10V + 15V - 1.2V}{82k\Omega}\right]4V = 128\mu A$$

b.
$$i_{out} = \frac{0.11}{V}\left[\frac{15V - 1.2V}{82k\Omega}\right]4V = 74\mu A$$

c.
$$i_{out} = \frac{0.11}{V}\left[\frac{-10V + 15V - 1.2V}{82k\Omega}\right]4V = 20.4\mu A$$

There is a linear relationship between $E_{control}$ and i_{out}. Changing the control voltage by 10V changes the output current by 54µA. But there is a 74µA offset which complicates any potential application.

The input to an OTA is voltage, and its output is current, up to about 400µA for the CA3280. I_{ABC} should also be on the order of 400µA. The 74µA offset problem you have just seen can be removed by using the second OTA in the IC package to convert an input control voltage into the I_{ABC} to control the first (main) OTA. Look at Figure 10-38.

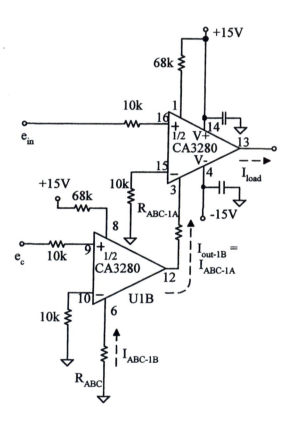

Figure 10-38 Setting I_{ABC} with a control voltage

For U1B,
$$I_{ABC-1B} = \frac{|V^-| - 1.2V}{R_{ABC}} \approx \frac{|V^-|}{R_{ABC}}$$

$$I_{out-1B} = \left(\frac{0.11}{V}\right) I_{ABC-1B} e_c$$

$$= \left(\frac{0.11}{V}\right) \frac{|V^-|}{R_{ABC}} e_c$$

For U1A, $I_{ABC-1A} = I_{out-1B}$

$$i_{load} = \left(\frac{0.11}{V}\right) I_{ABC-1A} e_{in}$$

$$i_{load} = \left(\frac{0.11}{V}\right)\left[\left(\frac{0.11}{V}\frac{|V^-|}{R_{ABC}}\right) e_c\right] e_{in}$$

The output current is linearly controlled by the product of two input voltages. The control voltage, e_C, must always be positive in order to drive I_{out-1B} **out** and into U1A's I_{ABC} pin. One way to view this relationship is as a two-quadrant multiplier. If you add a resistor from the output to ground, i_{load} is converted into v_{load} by Ohm's law.

$$v_{load} = \left(\frac{0.0121}{V^2}\frac{|V^-|}{R_{ABC}} R_{load}\right) e_c e_{in}$$

Remember from the section on multipliers that

$$v_{load} = \frac{e_x e_y}{e_{ref}}$$

So $$e_{ref} = \frac{R_{ABC}}{\dfrac{0.0121}{V^2} \times |V^-| \times R_{load}}$$

Example 10-6

Select the values of R_{ABC}, R_{ABC-1A}, and R_{load} to configure the OTA circuit of Figure 10-38 as a multiplier.

Solution

The maximum value for I_{ABC} is 400µA. So with ±15V supplies, set

$$I_{ABC-1B} \leq 400\mu A = \frac{15V - 1.2V}{R_{ABC}}$$

$$R_{ABC} \geq \frac{15V - 1.2V}{400\mu A} = 34.5k\Omega$$

Pick $\qquad$ R_{ABC} = 38kΩ. This sets I_{ABC-1B} = 363µA.

At a maximum input of e_C = 4V, the current out of U1B is

$$I_{out-1B} = \frac{0.11}{V} 363\mu A \times 4V = 160\mu A$$

Select R_{ABC-1A} so that at the maximum current out of U1B (and into I_{ABC-1A}) the voltage at the output of U1B is about +10V. Then for lower values of output current, the output of U1B drops toward the V^- + 1.2V that is held on the top of R_{ABC-1A} by U1A. This keeps U1B out of saturation.

$$R_{ABC-1A} = \frac{(-15V + 1.2V) - 10V}{160\mu A} = 149k\Omega$$

Pick $\qquad$ R_{ABC-1A} = 150kΩ.

Finally, pick R_{load} to convert the output current into an appropriate voltage level, establishing e_{ref}. You can set e_{ref} to whatever level is needed for your application. One common approach is to set it equal to the full scale input. So

$$e_{ref} = \frac{R_{ABC}}{\dfrac{0.0121}{V^2} \times \left|V^-\right| \times R_{load}}$$

$$R_{load} = \frac{R_{ABC}}{\dfrac{0.0121}{V^2} \times \left|V^-\right| \times e_{ref}} = \frac{38k\Omega}{\dfrac{0.0121}{V^2} \times 15V \times 4V} = 52.3k\Omega$$

Pick $\qquad$ R_{load} = 52kΩ.

Check at the half scale input.

$$e_{load} = \frac{e_x e_y}{e_{ref}} = \frac{2V \times 2V}{4V} = 1V$$

With $e_c = 2V$,

$$i_{out-1B} = I_{ABC-1A} = \frac{0.11}{V} \times 363\mu A \times 2V = 79.9\mu A$$

$$i_{out-1A} = \frac{0.11}{V} \times 79.9\mu A \times 2V = 17.6\mu A$$

$$e_{load} = 17.6\mu A \times 52k\Omega = 0.92V$$

In many circuits, the purpose of a resistor is to convert an input voltage into a current (that is the purpose of R_i in an inverting amp). The relationship of input voltage to current out of the resistor is set by the value of the resistor. That resistance, in turn, controls the gain, critical frequency, etc., of the circuit in which it is used.

The voltage-to-current conversion performed by the resistor is just like the input-to-output relationship of the OTA. It has an effective resistance. For the simple OTA,

$$i_{load} = \frac{0.11}{V} I_{ABC} e_{in}$$

$$R = \frac{e_{in}}{i_{load}} = \frac{1}{\frac{0.11}{V} I_{ABC}} = \frac{9.09V}{I_{ABC}}$$

Setting I_{ABC} with a control voltage, as in Figure 10-38, yields

$$i_{load} = \left(\frac{0.0121}{V^2} \frac{|V^-|}{R_{ABC}} \right) e_c e_{in}$$

$$\frac{i_{load}}{e_{in}} = \left(\frac{0.0121}{V^2} \frac{|V^-|}{R_{ABC}} \right) e_c = \frac{1}{R}$$

$$R = \frac{R_{ABC}}{\frac{0.0121}{V^2} \times |V^-| \times e_c}$$

One use of the electronically tunable resistor is in an active filter. Variation of the control voltage shifts the critical frequency of the filter, as illustrated in block form in Figure 10-39. This gives you remote (or microprocessor) control of the filter without having to use a switched capacitor filter that adds switching noise to the signal.

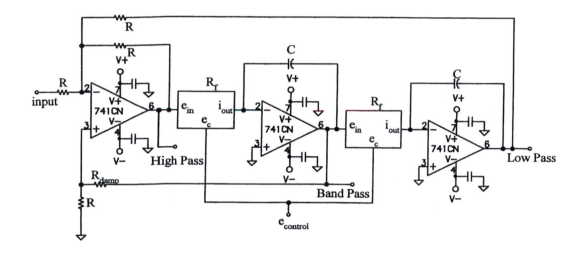

Figure 10-39 Electronically tunable active filter

Variation in I_{ABC} alters several characteristics. These include offset voltage, input bias current, power supply current demanded, input resistance, capacitance, output resistance, and slew rate. This means that changing I_{ABC} to adjust the gain, to achieve electronically controlled resistance, or to implement multiplication also significantly alters many of the other parameters of the amplifier. Input and output buffering with voltage followers helps, but you must also carefully review the effects that varying I_{ABC} has on slew rate.

Temperature also alters the performance of the CA3280A. The proportionality constant, 0.11/V, is valid only at room temperature. More properly, it should be

$$g_m = \frac{q\alpha I_{ABC}}{2kT}$$

where the terms are defined as they were for the log amp. Notice that g_m's (and therefore the entire OTA's) performance is inversely proportional to temperature (T).

Summary

Normally you work hard to produce a linear relationship between the input and the output of any circuit you build. However, under certain circumstances, a specific, well defined, nonlinear input-output (transfer) curve is needed.

Signal rectification and peak detection require a diode. Normally, the first 600mV of the signal is lost across the diode, since that is the voltage necessary to forward bias the diode. This is totally unacceptable for low level signals. The problem is solved by placing the diode within the negative feedback loop of an op amp. Assuming an open loop gain of 200k, the first 3μV of the input signal is multiplied by 200k (since the diode is off and the op amp is operating open loop). This produces the 600mV offset necessary to bias the diode on. Only 3μV, rather than 600mV, of the input signal has been lost to get the diode on. Care must be taken to insure that the op amp does not go into saturation during the off cycle. This same technique can be employed to build both a half wave and a full wave rectifier. Driving a capacitor from a voltage follower with diodes in its feedback loop allows you to capture and hold the peak of a signal.

Given a nonlinear transfer curve, you can build a general transfer function synthesizer. By using a resistor, diode, and bias network in the input loop of an inverting op amp amplifier, you can increase the gain with increasing input amplitude. Although this is a piece-wise approximation, for input signals larger than several volts, a close fit to a given nonlinear transfer curve can be produced. Placing the resistor, diode, and bias network in the feedback causes the amplifier's gain to decrease as the output goes up. This produces the oppositely shaped transfer function.

The root mean squared value of a signal is a measure of the power that can be delivered to a load. As such, it is the most valid measure for comparison among signals of widely different shapes. True RMS converter ICs produce a DC output that is numerically equal to the RMS of the input signal. Unlike the "rectify and filter" technique of producing DC, these ICs compute the actual root mean squared value of the input signal. Carefully review your application to select the most appropriate model. Usually only one or two external components are needed.

Log amps find applications in measurement and computational circuits. The basic log amp relies on the log relationship between a transistor's collector current and its base to emitter voltage. To compensate for other transistor parameters and for variations in temperature, two matched transistors and several op amps are needed. Antilog amps (exponentiation) can be built by placing the "antilog element" in the op amp's input loop.

Multiplication of two signals is also a nonlinear operation. If both signals are restricted to one polarity, the output falls into one quadrant. If one of the inputs can change sign, the output can be in either of two quadrants. If both inputs may change sign, the output may be in any of four quadrants. With a little imagination, you can use multipliers to simply solve problems that would otherwise require very complex circuits. Among these are frequency doubling, true RMS and real power calculations, voltage controlled amplification, AM modulation, division, squaring and square root extraction, and automatic gain control.

The operational transconductance amplifier (OTA) outputs a current proportional to its input voltage. The relationship between i_{load} and e_{in} (the transconductance) is programmable. Although there are some restrictions on voltage and current magnitudes, OTAs allow you to build programmable gain voltage amplifiers, voltage controlled resistance, voltage controlled active filters, and two-quadrant multipliers.

Problems

10-1 For the circuit in Figure 10-40, calculate Z_{in}, $V_{out\ P}$, $V_{out\ ave\ (DC\ value)}$.

Hint: $V_{DC} = \dfrac{V_P}{\pi}$.

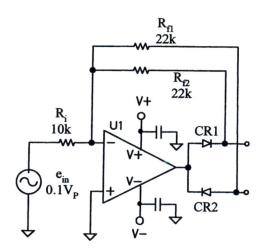

Figure 10-40 Schematic for Problem 10-1

10-2 Design a full wave ideal rectifier that outputs an average value numerically equal to the RMS of the input.

10-3 For the circuit in Figure 10-40, if a 100kΩ load is connected to the anode of CR2, and if R_{f2} is opened, draw the output. You must calculate **each** half cycle.

10-4 A peak detector must drive a 100kΩ load. Calculate the value of the capacitor needed to hold the droop of $10V_P$ to less than 1% during the first 10ms.

10-5 If the maximum current out of the op amp in the peak detector is 1mA, how rapidly can it charge a 0.1μF capacitor? Answer in V/μs.

10-6 Draw the transfer curve for the circuit in Figure 10-41. Accurately calculate and draw each slope and each break point.

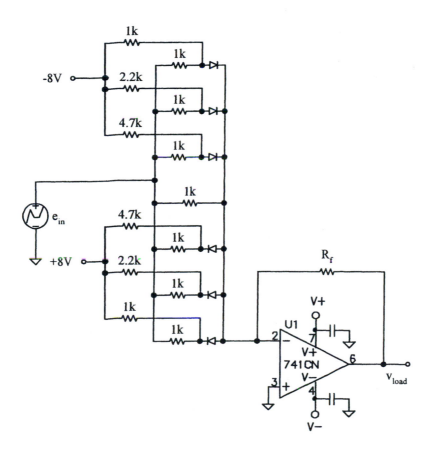

Figure 10-41 Transfer curve synthesizer for Problem 10-6

10-7   Draw the transfer curve for the circuit in Figure 10-42. Accurately calculate and
draw each slope and each break point.

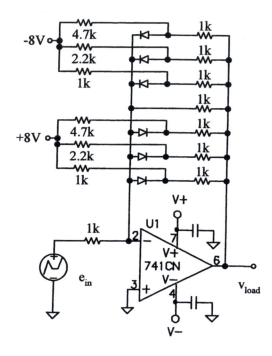

Figure 10-42 Transfer curve synthesizer for Problem 10-7

10-8   Design a transfer curve synthesizer whose output is

$$v_{load} = e^{(v_{in}-1)}$$

$$0.6V \leq v_{in} \leq 2.5V \qquad e = 2.71828$$

10-9   Design a transfer curve synthesizer whose output is

$$v_{load} = 20 \, \log_{10}\left(\frac{v_{in}}{0.775V}\right)$$

$$0.2V < v_{in} < 3.1V$$

10-10 Calculate the true RMS value of a

    a. $3V_p$ triangle wave.

    b. 0V to 5V symmetric square wave.

10-11 Select a true RMS converter IC that provides

    a. 0.3% accuracy at 300kHz.

    b. 0.5% accuracy at 100kHz.

    c. 0.1% accuracy at 100kHz.

10-12 **a.** Draw the schematic necessary to allow an AD736 to be used to provide a DC output equal to the RMS of a $3V_{RMS}$ signal.

    b. What is the resolution of this circuit?

10-13 **a.** Add to the circuit of Problem 10-12 the values of the capacitors needed for general purpose RMS conversion.

    b. What settling time do you expect?

10-14 Add to the schematic of Problems 10-12 and 10-13 the components needed to block any DC that the AC may have. Indicate appropriate values, assuming that the lowest frequency to be measured is 20Hz.

10-15 Modify the schematic of Problems 10-12 through 10-14 to accept a differential input.

10-16 **a.** The derivations in the section on log amps indicate that an antilog amp outputs a voltage proportional to 10 raised to the input voltage. Trace through the derivation to determine what must be done to produce

$$V_{out} = Ae^{-B \cdot v_{in}}$$

    where A and B are some constants and e = 2.71828.

    b. Determine the value or equations for A and B.

10-17 Design a circuit so that by throwing a switch, the output is either

$$V_{load} = dBm = 20 \; log_{10}\left(\frac{e_{in}}{0.775V}\right)$$

or
$$V_{load} = dBV = 20 \; log_{10}\left(\frac{e_{in}}{1V}\right)$$

10-18 Design a circuit that outputs a DC voltage equal to the power factor $(\cos\theta)$. Provide all component values, IC numbers, pin numbers, and other details needed to build it.

10-19 Design a first-quadrant divider using log amps to implement

$$V_{load} = 10V \frac{e_x}{e_y}$$

Draw the **full** schematic and specify all component values.

10-20 Design a circuit using OTAs. The circuit has a gain of 10 when $e_{control}$ is 1V and a gain of 20 when $e_{control}$ is 2V. Specify all component values.

10-21 Design an electronically tunable active filter whose f_o can be adjusted across the audio band. Specify all component values. Also indicate the relationship between $e_{freq\ control}$ and f_o.

10-22 The block diagram in Figure 10-43 is an automatic gain control circuit. The output amplitude is converted to DC by the AC to DC block. Any difference between this DC and the amplitude control DC causes the integrator to ramp up or down, changing the "gain" of the multiplier. Complete this schematic using op amps and OTAs. Use an ideal rectifier and a filter for the AC to DC converter. The frequency is between 300Hz and 3kHz. Peak amplitude is less than $4V_P$.

10-23 Draw the schematic of a true power converter using the OTA multiplier. *Hint*: You will have to offset e_y to keep it from going negative.

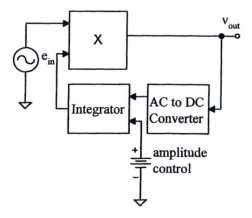

Figure 10-43 Automatic gain control block diagram

Ideal Rectifier Lab Exercise

A. Single Diode Rectifier

    1. Build the circuit in Figure 10-44.

    2. Measure the voltage across the resistor with a multimeter. Explain the results.

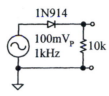

Figure 10-44 Simple diode rectifier

B. Simple Ideal Rectifier

    1. For the circuit in Figure 10-45, connect one channel of your oscilloscope to the output of the op amp, and the other across the load. Measure the load voltage with your digital multimeter.

    2. Explain the differences between these measurements and the ideal results.

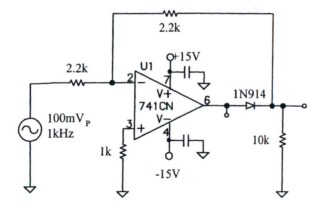

Figure 10-45 Simple ideal rectifier

C. Practical Ideal Rectifier

1. Build the circuit in Figure 10-46.

2. Connect one channel of your oscilloscope to the output of the op amp, and the other across the 10kΩ load. Also measure the voltage across the load with your digital multimeter. Record the two waveforms and the DC voltage.

3. Record the theoretical DC value and the % error.

4. Repeat steps C2 and C3 for the negative output. Be sure to move the 10kΩ load when you move your instruments.

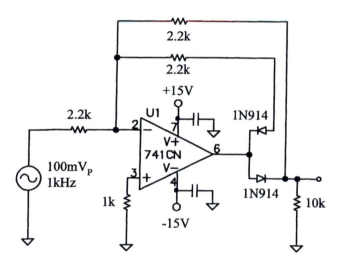

Figure 10-46 Practical ideal rectifier

D. RMS Converter

1. Design, build, and test a circuit (based on Figure 10-46) that outputs a smooth DC value equal to the RMS of the sine wave input. Be sure to set the gain properly to compensate for the difference between DC and RMS. Also include appropriate filtering to provide a smooth DC. The lowest frequency of interest is 100Hz.

2. Tabulate the output DC voltage and % error with a $100mV_{RMS}$ input sine wave at the following frequencies: 100Hz, 1kHz, 30kHz, 100kHz, and 300kHz. Plot the circuit's frequency response.

3. Explore ways to improve the circuit's frequency response.

Operational Transconductance Amplifier Lab Exercise

A. Introduction

For the circuit in Figure 10-47:

1. Calculate the value of R_{ABC} to set:
   a. $I_{ABC} = 20\mu A$.
   b. $I_{ABC} = 200\mu A$.

2. Calculate the output current with $E_{in} = 4V_{DC}$, and:
   a. $I_{ABC} = 20\mu A$.
   b. $I_{ABC} = 200\mu A$.

3. Build the circuit in Figure 10-47 using the larger R_{ABC} calculated in step A1 (to produce the smaller I_{ABC}).

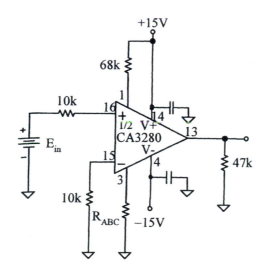

Figure 10-47 Schematic for the initial OTA set-up

B. Characteristic Measurements

1. Set $E_{in} = -4V_{DC}$.

2. Measure the output DC voltage on the digital voltmeter. Also monitor the output on the oscilloscope for any noise. Complete the first part of Table 10-4 (for $I_{ABC} = 20\mu A$).

3. Repeat step B1 for the smaller R_{ABC} completing the columns for $I_{ABC} = 200\mu A$.

4. Plot e_{in} versus $I_{out\ theory}$ for $I_{ABC} = 20\mu A$. On the same graph plot e_{in} versus $I_{out\ measured}$.

5. Repeat step B3 for $I_{ABC} = 200\mu A$.

6. Discuss the performance of the basic OTA.

Table 10-4 Characteristic measurements for the CA3280A

| e_{in} V | V_{out} V | $I_{ABC} = 20\mu A$ | | | V_{out} V | $I_{ABC} = 200\mu A$ | | |
		I_{out} μA	I_{theory} μA	error %		I_{out} μA	I_{theory} μA	error %
-4.0								
-3.5								
-3.0								
-2.5								
-2.0								
-1.5								
-1.0								
-0.5								
0								
0.5								
1.0								
1.5								
2.0								
2.5								
3.0								
3.5								
4.0								

C. Voltage Controlled Amplifier

1. Build the voltage controlled amplifier shown in Figure 10-48.

2. Set $e_{in} = 100mV_{RMS}$ and $E_{control} = 4V_{DC}$.

3. Adjust R_L to give a gain of 16.

4. Adjust $E_{control}$ to 1V, 2V, 3V, and 4V. At each control voltage setting, record v_{load}, the actual gain, the theoretical gain, and the % error of the gain.

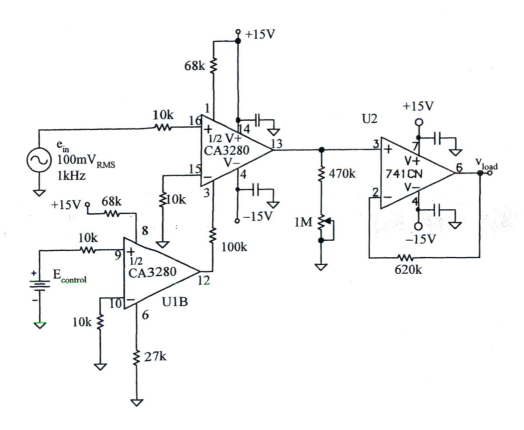

Figure 10-48 Voltage controlled amplifier for the OTA lab exercise

Answers to Odd-Numbered Problems

Chapter 1

1-1 The monolithic IC has all of the components diffused into the silicon wafer, where as the hybrid has discrete components placed on a ceramic wafer and then interconnected.

1-3 Monolithic ICs cost less than hybrid ICs because monolithic ICs can be mass produced in a large batch. Hybrid ICs must be made individually.

1-5 Comparators, speech synthesizers, cellular phones

1-7 a. $9.9\mu V$ **b.** $5\mu V$ **c.** $10\mu V$

1-9 The LM741C will not work. The required bandwidth is 1.8MHz, which is too large.

1-11 a. $\pm 10V$ **b.** $\pm 12V$ **c.** $\pm 18V$

Chapter 2

2-1 a. The output is a square wave whose levels are $\pm 10V$. The output is high when the input is above ground. The output is low when the input is below ground.

  b. The output varies from part a by going high when the input is above 3V.

  c. The output is high whenever the input is above $-4V$.

  d. The output is always at $-10V$.

2-3 The output is $+10V$ when e_{NI} is above e_{INV}. The output is $-10V$ when e_{NI} is below e_{INV}.

2-5 a. 2.24V **b.** $-0.286V$

2-7

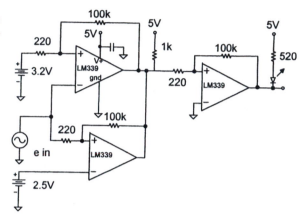

2-9 a. $1.5k\Omega$ **b.** 65ns

2-11 A_{OL} goes down; v_{out} goes down; e_f goes down; e_d goes up; v_{out} goes back up

2-13 Use a noninverting amplifier. R_f is a $10k\Omega$ potentiometer. $R_i = 50k\Omega$.

2-15 a. $5V_{PP} - 2V_{DC}$ **b.** sine, +1V positive peak, lower 20% clipped at $-7V$.

2-17 3.53V

2-19 $A_{common\ mode} = 18.88E\text{-}3$, $v_{out} = 94.4mV$

2-21 $e_{out} = \dfrac{R1+R2}{R1}e_2 - \dfrac{R3}{R1}e_1$

2-23 V_A=2.55V, V_B=2.10V, V_C=0.191V, V_D=0.191V, V_E=0.491V, V_F=2.46V, V_G=2.77V, V_H=2.46V, V_I=3.47V

Chapter 3

3-1 Omitted

3-3 Omitted

3-5 Omitted

3-7

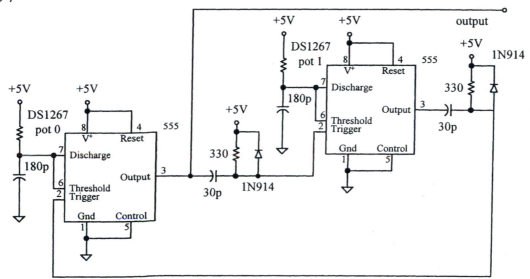

a. R0=30.58kΩ, R1=71.36kΩ,
   code=0 10110110 01001110
b. R0=86.65kΩ, R1=15.29kΩ,
   code=0 00100111 11011101

3-9 Omitted

3-11 Place a voltage follower op amp
amplifier between the voltage divider
and the REFAB input to the MAX512.
Also add a capacitor directly to the
REFAB input. Monitor that point for
oscillations. Consider changing the op
amp or the size of the capacitor.

3-13

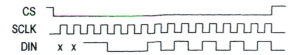

Chapter 4

4-1 $v_P = -7.5V$, $v_{min} = -7.1V$,
$v_{ripple\ PP} = 0.4V_{PP}$, $V_{DC} = -7.3V_{DC}$

4-3 $v_P = -16.4V$, $v_{min} = -11V$,
$v_{ripple\ PP} = 5.4V_{PP}$, $V_{DC} = -13.7V_{DC}$

4-5 $V_{load} = 5.3V$, $V_{out\ op\ amp} = 6V$,
$I_{out\ op\ amp} = 1.5mA$, $V_P = 7.5V$, $V_{min} = 7.1V$,
$V_{ripple\ PP} = 0.4V$, $V_{DC\ in} = 7.3V$, $P_Q = 95mW$

4-7 $I_{max} = 250mA_{DC}$

4-9 $R1 = 100\Omega$, $R2 = 1150\Omega$, $V_P = 24.1V$,
$V_{min\ needed} = 17V$, $C = 1000\mu F$,
$V_{min\ actual} = 19V$, $V_{DC} = 21.6V$,
$\Theta_{JA\ max} = 1.7°C/W$

4-11 $R1 = 120\Omega$, $R2 = 100\Omega$, $R3 = 330\Omega$,
$R4 = 120\Omega$, $R5 = 1.4k\Omega$, $R6 = 1k\Omega$,
$R7 = 1k\Omega$, $R8 = 2.4k\Omega$, $R9 = 120\Omega$

570

4-11 (cont.)

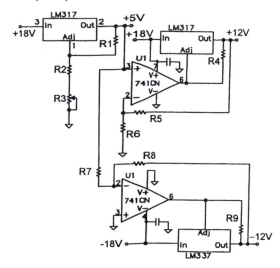

4-13 $v_{sec}=6.3V_{RMS}$, $C_{filter}=2200\mu F$, $R_i=1k\Omega$,
$R_f=5.5k\Omega$, $0.35\geq D\geq 0.0625$; for
$V_{in}=7.5V$, $L=45\mu H$, $I_{ind\,ave}=1.2A$,
$I_{ind\,max}=1.3A$, $R_C\leq 1.4k\Omega$, $C_{out}>80\mu F$,
$ESR<0.05\Omega$, $C_C>16\mu F$, $P_{IC}=140mW$,
no heat sink needed

4-15 omittted

4-17 The pwm controller for the forward
converter outputs a single wave, whose
pulse width is proportional to the
desired output voltage. The controller
for the half bridge must output two
signals, of opposite phase. There must
also be a dead time so that both
transistors are off before either is
turned on.

Chapter 5

5-1 omitted

5-3 The inverting input is at ground. With
the signal source at 0V, there is no
difference in potential across R_i. $I_{Ri}=0$.
All of I_B^- must flow through R_f.

5-5 $800\mu V$

5-7 $V_{Rcomp}=-727\mu V$, $V_{Rf}=-727\mu V$,
$I_{Ri}=72.7nA\uparrow$, $I_{Rf}=12.7nA\leftarrow$,
$V_{Rf}=728\mu V$, $V_{load}=0V$

5-9 Set $R_f=10k\Omega$, $R_i=1k\Omega$, $R_{comp}=909\Omega$.
$V_{Rcomp}=-72.7\mu V$, $V_{Ri}=-72.2\mu V$,
$I_{Ri}=72.2nA\uparrow$, $I_{Rf}=12.7nA\rightarrow$,
$V_{Rf}=127\mu V$, $V_{load}=200\mu V$

5-11 $R_f=1.999V/I_{B\,max}$ Use Fig 5-11.

5-13 GBW=1MHz, $A_{OL\,20Hz}=50k$,
$A_{OL\,15kHz}=67$, $A_{OL\,650kHz}=1.5$,
$f_H=12.5kHz$, $f_{1\%}=1.8kHz$, $f_{10\%}=6.3$

5-15 Build two amplifiers, each with a 741,
each with a gain of 7.07.
GBW=283kHz

5-17 slew rate=$40V/\mu s$

5-19 $A_o=60$, GBW=4.2MHz, SR=$0.27V/\mu s$

5-21 omitted

5-23 Use Fig. 5-22. $R_i=10k\Omega$, $R_f=200k\Omega$,
$C_f=39pF$

5-25 omitted

5-27 omitted

5-29 $V_{ce\,max}>112V$

5-31 $V_A=2.000V$, $V_B=2.000V$, $I_{R6}=0A$,
$V_C=2.000V$, $I_{R2}=1.9900\mu A\leftarrow$,
$V_D=9.9502mV$, $V_E=4.9502mV$,
$I_{R4}=72.7978\mu A\downarrow$, $I_{R3}=72.7978\mu A\downarrow$,
$V_F=2.4073V$

Chapter 6

PCB layout omitted.

Chapter 7

7-1 $v_{load}=5.75V_P$ on the positive half cycle.
Everything below +2V is clipped off.

7-3 Fig. 7-5, R_f=300kΩ, R_i=20kΩ, R_A=R_B=100kΩ, C_i>0.27µF

7-5 Fig. 7-11, R_f=300kΩ, R_i=20kΩ, R_A=R_B=100kΩ, C_i>0.1µF, C_f>0.27µF

Chapter 8

8-1 t_{high}=55.2µs, t_{low}=22.8µs, T=78µs, f=12.8kHz

8-3 Fig. 8-3, R_A=R_B=10kΩ, C=241µF, C=7.3nF

8-5 omitted

8-7 12V/ms

8-9 **a.** ±4.7V **b.** ±9V

8-11 **a.** f=1.39kHz, **b.** rate=12.2V/ms, **c.** $v_{triangle}$=±2.2V, **d.** v_{square}=±10V

8-13 The output is a positive going ramp that starts at 0V, at every line voltage zero crossing, and rises to 4.5V.

8-15

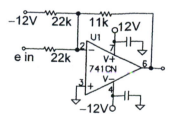

8-17 Fig.8-21, C_t=0.33µF, R_{t2}=2.9kΩ, R_{mult}=23.5kΩ

8-19 discussion

8-21 V_{DD}=5V, R=10kΩ, f_{center}=100kHz→C=0.22nF, f_{center}=1kHz→C=22nF, pick 1% ripple, $T_{longest}$=50ns, τ=2.5s, C=100µF, R=25kΩ

8-23 Fig. 8-33, 1MHz crystal oscillator, 5 7490 dividers & a divide by 2 → 5Hz clock, 16 bit feedback divider, C_t=3.3nF, 0.33µF, 33µF

8-25 3Hz→10rev/min

8-27 0.112°/address→3217 addresses, 2^{12} provides 4096 addresses

8-29 omitted

Chapter 9

9-1 **a.** −30.46dB, −18.06dB **b.** 0.07943, 7.94

9-3 omitted

9-5 omitted

9-7 omitted

9-9
$$\frac{Z_2 Z_4}{Z_1 Z_2 + Z_1 Z_3 + Z_2 Z_3 - Z_1 Z_4}$$

9-11 f_o=6.58kHz, A_o=1.27, α=1.73, Bessel, f_{-3dB}=5.17kHz

9-13 **a.** f_{o1}=3.29kHz, f_{o2}=7.23kHz, **b.** $α_1$=0.9, $α_2$=0.2, **c.** 3dBChebyshev, **d.** f_{-3dB}=7.8kHz, **e.** −24dB/oct, **f.** A_o=5.91=15.4dB

9-15 f_o=6.58kHz, A_o=1.27, α=1.73, Bessel, f_{-3dB}=8.38kHz

9-17 GBW=20kHz, slew rate=0.18V/µs

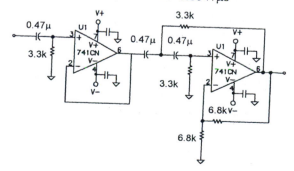

9-19 $\dfrac{A_o \dfrac{1}{RC} s}{s^2 + 3\dfrac{1}{RC}s + \dfrac{1-A_o}{R^2C^2}}$

9-21 Cascade two single op amp band pass filters, as shown in Fig 9-21. Pick $C=0.22\mu F$, $R3=37k\Omega$, $R1=5.9k\Omega$, $R2=90\Omega$, GBW>0.8MHz

9-23 Use Fig. 9-24. $R=1.5k\Omega$, $C=0.022\mu F$, $R_A=29k\Omega$, $R_B=10k\Omega$, $f_{-3dB\ low\ pass}=7kHz$, $f_{-3dB\ high\ pass}=3.6kHz$

9-25 $f_o=60Hz$, $A_{o\ LP}=-1$, $A_{o\ HP}=-1$, $\alpha=0.0958$, $Q=10.4$, $A_{BP}=10.4$

9-27 Use Fig.9-30. $R3=110k\Omega$, $R1=55k\Omega$, $R2=290\Omega$, $R4=R5=R6=1k\Omega$

9-29. Use Fig. 9-32. $R=10k\Omega$, $R_f=265k\Omega$ $C=10nF$, $R_g=10k\Omega$, $R_A=100k\Omega$, $R_B=3.3k\Omega$ (See prob 9-25).

9-31 Low pass of Fig. 9-36 sets f_h. $R=1.2k\Omega$, $C=0.033\mu F$. Connect its output into the input of Fig. 9-37. It sets f_l. $R=4.8k\Omega$, $C=0.033\mu F$.

9-33 Use a clock that sweeps from 10Hz to 100kHz. Set $f_{o\ antialiasing}=5kHz$ with $C=0.01\mu F$, $R=3.3k\Omega$ **a**. Use Fig. 9-45(b). **b**. Use Fig.9-46. $C_{in}=22\mu F$.

Lower R in bias voltage divider = $3k\Omega$. Upper R in bias voltage divider =$100k\Omega$. C in bias voltage divider = $100\mu F$.

9-35 $f_o=800Hz$, $Q=4.7$, $\alpha=0.21$, $A_{o\ LP}=-1$, $A_{o\ BP}=-4.7$, $A_{o\ HP}=-1$, $A_{P\ LP\ \&\ HP}=4.7$

Chapter 10

10-1 $Z_{in}=10k\Omega$, $V_{out\ P}=\pm220mV_P$, $V_{out\ DC}=\pm70mV_{DC}$

10-3.On the positive input half-cycle, the output is the negative half cycle of a sine wave, $220mV_P$. On the negative input half cycle, the output is also the negative half cycle of a sine wave, $76mV_P$.

10-5 $10V/\mu s$

10-7 Transfer curve is decreasing slope Fourth quadrant is the mirror image of the second quadrant. $S1=-1$, $S2=-0.5$, $S3=-0.33$, $S4=-0.25$, $V_{B2}=V_{load}=-1.7V$, $V_{B3}=V_{load}=-3.6V$, $V_{B4}=V_{load}=-8V$

10-9

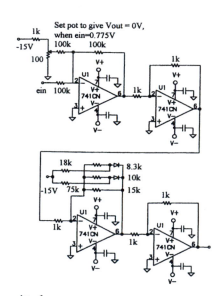

10-11 omitted

10-13 a. General purpose, 20Hz $C_{av}=33\mu F$, $C_f=10\mu F$ **b**. 360ms

10-15 See Fig. 10-22. Add the voltage follower to drive C_C, instead of grounding C_C.

10-17 Use Fig. 10-24. Set $R_i=10k\Omega$ for all three amps. For the difference amp, set $R_f=33k\Omega$. Set $e_{ref}=0.775V$ or 1V.

10-19 omitted

10-21 Use Fig. 9-36. R_f=2.2kΩ, C_f=3.3nF, f_o=2.19kHz $E_{control}$ log encoding $E_{control}$ before it goes into the filter gives finer control across the full audio range.

10-23 omitted

Index